ARISTOTLE'S *PHYSICS* ALPHA

Aristotle's *Physics* Alpha

Symposium Aristotelicum

edited by

KATERINA IERODIAKONOU,

PAUL KALLIGAS,

and

VASSILIS KARASMANIS

OXFORD

UNIVERSITY PRESS

OXFORD
UNIVERSITY PRESS

Great Clarendon Street, Oxford, OX2 6DP,
United Kingdom

Oxford University Press is a department of the University of Oxford.
It furthers the University's objective of excellence in research, scholarship,
and education by publishing worldwide. Oxford is a registered trade mark of
Oxford University Press in the UK and in certain other countries

First Edition published in 2019
Impression: 1

Published in the United States of America by Oxford University Press
198 Madison Avenue, New York, NY 10016, United States of America

British Library Cataloguing in Publication Data
Data available

Library of Congress Control Number: 2019933438

ISBN 978–0–19–883099–3

Printed and bound by
CPI Group (UK) Ltd, Croydon, CR0 4YY

PREFACE

The *20th Symposium Aristotelicum*, dedicated to the first book of Aristotle's *Physics*, was held in Delphi from 26 July to 2 August 2014. The meetings took place at the European Cultural Centre of Delphi (ECCD), which functions under the supervision of the Greek Ministry of Culture and the auspices of the Council of Europe. The ECCD is a well-known conference centre in Greece. Its premises lie on the western side of the modern town of Delphi, at walking distance from the archaeological site and the museum, and is situated in a landscape of rare natural beauty, overlooking the famous olive grove of Amphissa and the gulf of Itea. We would like to thank the director of the ECCD, Prof. Athanasios Markopoulos, and the staff of the ECCD for their help and hospitality during the works of the Symposium.

The organization of the Symposium would not have been possible without the generous financial support of the Stavros Niarchos Foundation which covered most of the expenses. We would also like to express our gratitude to Peter Momtchiloff from Oxford University Press, as well as to the anonymous referees of the Press for their valuable suggestions. Finally, we would like to thank Anna Ntinti for her help in transforming the various papers into a homogeneous text and for producing the bibliography and the indices.

This volume is dedicated to the memory of Michael Frede who died in the area of Delphi in the summer of 2007, during another international conference on ancient philosophy. Michael Frede was particularly fond of Delphi (and the ECCD) and he had often expressed his wish to have one of the Symposia organized there. Unfortunately, this was bound to happen only after his death.

Katerina Ierodiakonou, Paul Kalligas,
and Vassilis Karasmanis

CONTENTS

20TH SYMPOSIUM ARISTOTELICUM, DELPHI, 26 JULY–2 AUGUST 2014 LIST OF PARTICIPANTS

Contributors

Andreas Anagnostopoulos, Ludwig-Maximilians-Universität München
Gábor Betegh, Christ's College, Cambridge
Sarah Broadie, University of St Andrews
Michel Crubellier, Université de Lille
Sylvain Delcomminette, Université libre de Bruxelles
Mariska Leunissen, University of North Carolina at Chapel Hill
Hendrik Lorenz, Princeton University
Stephen Menn, McGill University/Humboldt-Universität Berlin
Benjamin Morison, Princeton University
Carlo Natali, Università Ca' Foscari Venezia
Diana Quarantotto, Sapienza—Università di Roma

Other Participants

Francesco Ademollo, Università degli Studi di Firenze
István Bodnár, Eötvös Loránd University/Central European University
Marcelo Boeri, Pontifica Universidad Católica de Chile
Tim Clarke, University of California, Berkeley
Alan Code, Stanford University
John M. Cooper, Princeton University
Frans de Haas, Universiteit Leiden
Panos Dimas, University of Oslo
Jean-Baptiste Gourinat, Centre Léon Robin, CNRS/Université Paris Sorbonne (Paris IV)
Katerina Ierodiakonou, National and Kapodistrian University of Athens/ Université de Genève
Thomas Johansen, University of Oslo
Paul Kalligas, National and Kapodistrian University of Athens
Vassilis Karasmanis, National Technical University of Athens
Sean Kelsey, University of Notre Dame
André Laks, Université Paris Sorbonne (Paris IV)/Universidad Panamericana, Mexico, DF
Oliver Primavesi, Ludwig-Maximilians-Universität München
Christof Rapp, Ludwig-Maximilians-Universität München
Marwan Rashed, Université Paris Sorbonne (Paris IV)
Gabriela Rossi, Universidad Adolfo Ibáñez
David Sedley, Christ's College, Cambridge
Stasinos Stavrianeas, University of Patras
Christian Wildberg, Princeton University/University of Pittsburgh

SYMPOSIUM ARISTOTELICUM

1. (1957) *Aristotle and Plato in the Mid-Fourth Century*, edited by I. Düring and G. E. L. Owen, Göteborg: Almqvist & Wiksell 1960.
2. (1960) *Aristote et les problèmes* de méthode, publié par S. Mansion, Paris/Louvain: Nauwelaerts 1961.
3. (1963) *Aristotle on Dialectic: The Topics*, edited by G. E. L. Owen, Oxford: Clarendon Press 1968.
4. (1966) *Naturphilosophie bei Aristoteles und Theophrast*, herausgegeben von I. Düring, Heidelberg: Lothar Stiehm Verlag 1969.
5. (1969) *Untersuchungen zur Eudemischen Ethik*, herausgegeben von P. Moraux und D. Harlfinger, Berlin: Walter De Gruyter 1971.
6. (1972) *Études sur la Métaphysique d'Aristote*, publié par P. Aubenque, Paris: Vrin 1979.
7. (1975) *Aristotle on Mind and the Senses*, edited by G. E. R. Lloyd and G. E. L. Owen, Cambridge: Cambridge University Press 1978.
8. (1978) *Aristotle on Science: The Posterior Analytics*, edited by E. Berti, Padova: Antenore 1981.
9. (1981) *Zweifelhaftes im Corpus Aristotelicum: Studien zu einigen Dubia*, herausgegeben von P. Moraux und J. Wiesner, Berlin: Walter de Gruyter 1983.
10. (1984) *Mathematics and Metaphysics in Aristotle = Mathematik und Metaphysik bei Aristoteles*, herausgegeben von A. Graeser, Bern/Stuttgart: P. Haupt 1987.
11. (1987) *Aristoteles Politik*, edited by G. Patzig, Göttingen: Vandenhoeck & Ruprecht 1990.
12. (1990) *Aristotle's Rhetoric: Philosophical Essays*, edited by D. Furley and A. Nehamas, Princeton: Princeton University Press 1994.
13. (1993) On Aristotle's De interpretatione [not published].
14. (1996) *Aristotle's Metaphysics Lambda*, edited by M. Frede and D. Charles, Oxford: Oxford University Press 2000.
15. (1999) *Aristotle's On Generation and Corruption Book I*, edited by F. de Haas and J. Mansfeld, Oxford: Oxford University Press 2004.
16. (2002) *Aristotle's Metaphysics Beta*, edited by A. Laks and M. Crubellier, Oxford: Oxford University Press 2009.
17. (2005) *Aristotle's Nicomachean Ethics Book VII*, edited by C. Natali, Oxford: Oxford University Press 2009.
18. (2008) *Aristotle's Metaphysics Alpha*, edited by C. Steel, Oxford: Oxford University Press 2012.
19. (2011) *Aristotle's De Motu Animalium*, edited by Ch. Rapp and O. Primavesi, Oxford: Oxford University Press 2019.
20. (2014) *Aristotle's Physics Alpha*, edited by K. Ierodiakonou, P. Kalligas, and V. Karasmanis, Oxford: Oxford University Press 2019.

Introduction: Some General Remarks on Aristotle's Physics *I*

CARLO NATALI

I.

The first book of the *Physics* seems to be a well-constructed λόγος. *Physics* I starts by establishing a general thesis: τῆς περὶ φύσεως ἐπιστήμης πειρατέον διορίσασθαι πρῶτον τὰ περὶ τὰς ἀρχάς, 'it belongs to the science of nature to try to distinguish firstly what relates to principles' (184a15–16), and continues by listing different opinions by other philosophers about the *number* of principles. The discussion of their positions leads naturally to the question of how many principles there are in reality, and what they are. The book ends saying that the question has been answered: ὅτι... μὲν οὖν εἰσὶν ἀρχαί, καὶ τίνες, καὶ πόσαι τὸν ἀριθμόν, διωρίσθω ἡμῖν οὕτως, 'now, that there are principles, what they are and how many in number, may be taken as settled in this way' (192b2–4). So the discussion seems to start from a problem and to arrive at a solution.

This is not to say that the book is in itself complete and unconnected with the rest of the *Physics* and with the physical treatises in general. It says many things that will be useful for the entire study of the natural world; however, at the end of book I Aristotle thinks he has accomplished an important first step. The results of his research are too general, to be sure, and need further

I would like to thank Sarah Broadie, Christopher Mirus, Benjamin Morison, Marwan Rashed, and an anonymous referee for many useful suggestions on the content and form of this introduction.

specification, but he has freed the ἐπιστήμη of the natural world from confusions and errors deriving from his predecessors' ingenuities.

To begin, let us examine the general structure of the book and its connection to *Physics* II. In its present form, *Physics* I is divided into nine chapters, and, with one exception, this division has been followed by the contributions to the present volume. However, the reader should not attribute too much weight to this element. In particular, there is no need to ask questions about the aim and unity of individual chapters.

The fact is that the current division of the *Physics* into chapters was established by Michael Isengrin, the publisher of the third Basel edition of the complete works of Aristotle in Greek (1550). He also added short apocryphal titles to each chapter. As he says, the division was made on the basis of medieval translations, where available, and of suggestions by the Swiss bibliophile and scientist Conrad Gessner (1516–65). Isengrin's aim was eminently practical: he wanted to give to the reader, first, some resting point in the long process of reading the complex text; and second, a way to find a specific passage easily and to memorize it when found.[1] The resulting division is often sensible, but occasionally it interrupts the flow of argumentation. In particular, its practical aim has the effect of dividing the book into units that are too small, for reasons of convenience: in a shorter section it becomes easier to find the passage one is looking for.

In fact, as the following essays will clearly show, some of Isengrin's chapters are closely connected and should not to be divided in the present way, whereas others are mixed bags, composed of different sections. Some examples: according to Crubellier, chapter 2 is composed of (1) a general introduction to the dialectical discussion of Aristotle's predecessors that continues until chapter 5 (184b15–25), and (2) the beginning of a discussion of the Eleatic position (184b25–186a3). According to Quarantotto, the criticism in chapter 2 and the solution of the Eleatic puzzles in chapter 3 form the two sections of a diptych, until line 186b35. After that, in the same chapter 3 there is a short criticism of some post-Eleatic philosophers who gave in to two Eleatic arguments (187a1–11). Betegh remarks that the division between chapters 3 and 4 cuts across a μέν–δέ connection, and suggests that the discussion of the physicists in chapter 4 is

[1] On the reverse of the frontispiece there is an address from the editor, entitled *Isingrinus candido lectori*, where we read: 'Praeterea totum autorem in capita distinximus et Latinis translationibus, ubi ea extabant, accomodovimus: ubi vero desiderabantur, distinctiones ipsi addidimus, et brevibus argumentis illustravimus, adjuti Gesneri opera, qui id in suo libro magna ex parte fecerat, ut lectores has capitum quasi stationes in quibus respirent habeant et quae velint facilius discreta locis inveniant, inventaque melius ac tenacius mandent memoriae'.

connected to the criticism of the Eleatics in chapter 3.[2] Delcomminette notes that chapter 5 starts with a δή (188a19) that creates a strong connection to the end of I 4, and with a sentence that draws a conclusion from the discussion of I 4 and introduces I 5. In general, it is not easy to find clearly separate chapters in the lines 184b15–189a10. After that the distinctions become clearer.[3]

Even if the division in chapters is not to be completely trusted, nonetheless book I is not a single extremely long argument. It is possible to identify at least four main stages in its search for the first principles of physical change:

1. Chapter 1 is an introduction both to the following chapters and to the rest of the *Physics*. It establishes a fundamental methodological principle, as we will see later, that is followed in the rest of the book, to pass from statements that are true but not clear, because too general and indeterminate, to statements that are true and clear because more precise (ἀκριβής) and detailed.

2. Chapters 2–5 are mostly characterized by *verba dicendi* related to other people: λέγουσι (185b7, 187a12, 189a3), λέγοντες (188a19, 189b2) φασί (187b1) etc. This applies also to chapter 6, οἴεται (189a16), but in a lesser measure. Because of this, some commentators have considered this section, I 2–6, as a mere doxographical discussion. But things are not so simple, for in fact there are many positive statements of Aristotle's own position in these pages, and it might be safer to put the distinction in the following way: up through chapter 6 Aristotle proposes his theories in a continuous discussion with earlier philosophers, whereas in chapter 7 he speaks only in his own voice. Aristotle's solution to the puzzle about principles of change is prepared for not only by the criticism of his predecessors, but also by the formulation of a puzzle that progressively emerges in this section. More on this later.

3. A perceptible change in tone occurs at the beginning of chapter 7. It starts emphatically with Ὧδ' οὖν ἡμεῖς λέγωμεν, 'Let us then give an account' (189b30).[4] The phrase is usually understood as opposing Aristotle's opinion,

[2] Cutting the connection between μέν and δέ was a standard practice for those who invented the chapter divisions for Aristotelian treatises. For instance, in *Physics* II it can be found at 3–4, 5–6, 8–9; III, 1–2, 5–6; IV 1–2, 2–3, 3–4, 6–7, 7–8, 8–9, 9–10, 11–12, 13–14, etc. In *Metaphysics* A the construction is present at every beginning of a chapter, except at the beginning of A 8 and 10. The same phenomenon can be seen in Z, Θ, Λ (second part), M. This transitional device is less frequent in other books of the *Metaphysics*. I don't know if this particularity of style reveals anything about the composition of the books and their chronology.

[3] Chapter I 6 starts with Ἐχόμενον δ' ἂν εἴη λέγειν, 'After that we must say' (189a11), ending with ἀπορίαν ἔχει πολλήν, 'it is a question of considerable difficulty' (189b29); chapter I 7 starts with Ὧδ' οὖν ἡμεῖς λέγωμεν, 'Let us now give our own account' (189b30), ending with ἐκ τούτων θεωρείσθωσαν, 'can be seen from these points' (191a22). Each chapter has a clear starting point and conclusion.

[4] Cf. Horstschäfer 1998, 248.

as expressed in chapter 7, to the discussion of other philosophers' opinions examined before; but it could also be interpreted as responding to the phrase at the end of chapter 6: ἀπορίαν ἔχει πολλήν, 'it is a question of considerable difficulty' (189b29). Nonetheless, it differentiates this section of the book from the preceding one. This chapter is the most important part of the book, because it is where Aristotle solves the puzzle about change.

4. The following chapters, I 8–9, are the conclusion of the book and are clearly divided from what precedes them by a transition formula, λέγωμεν μετὰ ταῦτα, 'we should discuss next' (191a24). They contain an endoxical confirmation of the theory established in I 7, by showing up to what point his predecessors were right and where they went wrong. This is a strategic move described in *Nicomachean Ethics* VII:

> We must not only state the truth but also the cause of the falsehood: for this contributes to conviction, for whenever the explanation for why something appears true, while not actually being true, seems to be reasonable, this produces more conviction for the truth. (1154a22–5)

In the present case Aristotle discusses first (I 8) the most ancient philosophers and next (I 9) the Platonists, whose presence was still visible in the debate of Aristotle's time, as Broadie rightly says. The division of the section into two chapters can be justified by the fact that two different groups of philosophers are discussed, but Isengrin's choice is not absolutely necessary. By solving the puzzles they fell into, and by giving his predecessors their due, Aristotle shows the superiority of his position and his fairness in considering theirs.

A word on the connection to the second book of the *Physics*. The commentators of the twentieth century almost completely agreed in judging the first book of the *Physics* as an independent early treatise, and some of them estimated it to be more connected to Aristotle's ontology in general than to his physical science specifically.[5] This opinion is no longer dominant, but the problem still exists of how to connect book I to book II, since it seems that book II makes no reference to book I.[6]

Contrary to that limitative judgement, there in fact seem to be some conceptual connections between parts of book I and book II, as Menn and Lorenz

[5] See Ross 1936, 5; Mansion 1946, 54; Düring 1966, 220; Berti 1977, 292; Horstschäfer 1998, 482. Aubenque says: 'la recherche sur les principes, qui occupe tout le livre I de la *Physique,* est une recherche ontologique et non physique' (1962, 423). *Contra,* Couloubaritsis, who thinks that 'la continuité entre les l. I et II ne fait aucun doute' (1980, 213). A middle position in Boeri (1993, 17–18) and Pellegrin, who thinks that book I could have been written as an independent treatise, but 'il forme avec les livres suivants un ensemble cohérent' (2000, 40).

[6] Ross writes: 'The beginning of book ii presents the appearance of being the beginning of a separate work' (1936, 5).

say in their chapters. Book I discusses change in general, making no distinction between natural and artificial changes until line 190b17, where the principles of change for natural entities become the centre of the enquiry. There is no indication in book I of what constitutes a natural entity, but the question is amply discussed in II 1, and so this chapter studies in depth a notion that was simply assumed in I 7. In my impression book II is both continuous to and independent from the first book.

This is not as strange as it appears. It is a normal practice in Aristotelian treatises to conclude an inquiry under one point of view, and afterwards to continue it under another point of view. In my opinion this is what happens in the passage between the two first books of the *Physics*. This can be seen from an attentive reading of the text.

People have already tried this way. Diels estimated that, with the exception of book VII, 'alle Bücher mit einander verknüpft sind'. As proof of the connection between I and II he quoted the fact that the ms. E (Paris. gr. 1853) at the end of book I, 192b4, adds a phrase identical to the beginning of book II: *λέγωμεν τῶν γὰρ ὄντων τὰ μέν κτλ.* He thought that the *γάρ* strictly connects the end of book I to the beginning of book II.[7] This opinion has been refuted by Ross, who considers the *γάρ* 'a later attempt to produce at any rate an external connection'.[8] It is possible that the phrase in E is a *reclamans,* as those we find in the ms. A^b (Laurentianus Med. 87,12) of the *Metaphysics.*[9] A similar case in the ms. E is at the end of IV, 224a16, where there are written the first words of the following book (see Ross, in apparatus).[10]

In my opinion they were looking at the wrong point of the text: at the end of I 9 Aristotle says: *πάλιν δ' ἄλλην ἀρχὴν ἀρξάμενοι λέγωμεν*, 'let us discuss again, making a fresh start' (192b4). The phrase has scarcely been noted by commentators, who have nothing to say about it, but this is a very frequent transition formula in the Aristotelian treatises and has a precise function.[11] The general idea conveyed by such a phrase is to continue the discussion of the same subject as before, but from a new standpoint.[12]

Aristotle in II 1 continues the enquiry of book I, but he does so from a new point of view. He starts the new enquiry with the study of a subject that was

[7] Diels 1882, 41.

[8] Ross 1936, 5.

[9] Cf. Jaeger 1957, ix–x: they could be the mark of an ancient version in papyrus rolls, and were meant to facilitate the reader to find the successive roll.

[10] I owe this last reference to M. Rashed.

[11] Normally it is found at the beginning of a chapter and not at the end of the preceding chapter, as here, but its function is always the same.

[12] Interesting other examples in *EE* 1218b31 and 1222b15. In general see Bonitz 1870, 111b,35–45.

simply assumed in I 7. So, the phrase in 192b4 at the same time separates the discussion of II 1 from, and connects it to, what precedes.

The transition from the analysis of *ἀρχαί* in book I to the analysis of *αἰτίαι*, which begins in chapter II 3, is neither easy nor smooth:

- at the end of I 9, 192b4, there is the indication of a new start, and we find in II 1 an analysis of what 'nature' is;
- at the beginning of II 2, 193b22, Aristotle says *μετὰ τοῦτο*, 'after this', a phrase by which he usually indicates the passage to another point of the discussion, and tackles the question of the difference between the study of nature and the study of mathematical realities;[13]
- finally, at II 3, 194b16, there is again a new fresh start, *Διωρισμένων δὲ τούτων*, 'These distinctions having been drawn', which finally introduces the long discussion about *αἰτίαι* of II 3–8.

In this section we find at least three changes of perspective (192b4, 193b22, 194b16) before engaging on the main enquiry of book II, causation.

In conclusion, in my opinion book I is not an independent treatise nor is it dedicated to general ontology but, as the authors of the chapters in this volume say, it is about the principles of changing things in general and it constitutes an introduction to a general study of physical science.

2.

Since *Physics* I is a well-constructed *logos*, we can ask to what kind of public it is addressed. In some of his works Aristotle gives some hints about this, but not here.[14] However, we can infer some characteristics of the intended hearers of the present *logos* from some of its aspects.

In the first chapter Aristotle does not argue in favor of the interest of doing physical science. This is an indication of the fact that the public to whom the speech is directed is not the common wealthy Athenian citizen, but a more restricted group of hearers already interested in natural philosophy, who don't need to hear an exhortation to the study of nature. They are supposed to know the main theses of the Presocratics, as Quarantotto says, and to have some knowledge of

[13] This formula, or the equivalent *μετὰ ταῦτα*, is used in the *Topics*, at the beginning of books IV and VIII, and in other works which are, for the most part, descriptive, such as the *Historia animalium* and the *Athenaion politeia*. It appears, with the same meaning in *Metaphysics*, *Physics*, *De caelo*, *Rhetoric*, etc. See also Bonitz 1870, 458a21–4.

[14] Cf. *NE* I 3, 1095a2.

the *Categories* and of dialectical arguments such as the distinction of the many senses of a term and the distinction *per se/per accidens*.

The first passage of chapter I 2 also indicates a cultivated audience; in fact here Aristotle recapitulates a traditional doxographical move, well known to the Athenian cultivated public, namely the dispute about how many ὄντα there are, and applies it in an original way to the question of the ἀρχαί.[15] He says explicitly that he is doing so, at 184b22–5: his division is modelled on the divisions of early ontology. Moreover, the criticism of Academic theories in I 9, as usual in Aristotelian treatises, is very concentrated. It presupposes, to be fully understood, some knowledge both of Plato's dialogues and of the Academic discussions starting from them. It looks as if the same thing is happening as when in a contemporary discussion someone quotes, e.g., the case of Cambridge change[16] and doesn't feel the need to explain what it is, assuming that the hearers are expert enough to understand what he is speaking about.

In addition, the image of the path, ὁδός, is connected to the question of the public. In chapter I 1 Aristotle makes it clear that we are at the beginning of a search, of an ὁδός which is 'according to nature', as Aristotle himself says, proceeding from a question to its answer.[17] The metaphor of the way, ὁδός, immediately reminds the modern reader of Parmenides' narrative, but it is not strictly connected with it. It was also used by Aristotle and the Peripatetics independently from any Eleatic connection.[18]

However, it is not an ὁδός they make together: at the beginning of the λόγοι the public does not yet know where they will arrive, but Aristotle knows. He is not in the process of discovering the principles when writing one chapter after another; he already has the solution in mind before beginning to write. He re-enacts for the public the steps of a search he has already completed, in order to convince them of the correctness of his results. The passage from what is clearer to us to what is clearer by nature is not only a heuristic procedure but also a didactic device that makes his lectures more effective.

[15] The discussion by philosophers about how many ὄντα there are was a stable laughingstock for orators and comedians. See the parodies in comic authors as Antiphanes, (fr. 122 Kock), and orators as Isocrates (*Hel.* 3, *Antid.* 268); see also Xenophon (*Memorabilia* I 14).

[16] Cf. in this volume, Morison.

[17] The metaphor of the road is present, but with different meanings, in different cultural contexts. In Greek philosophers ὁδός, both here and in Parmenides, indicates an intellectual search for the truth; on the contrary, *dao* in ancient Chinese thought is concerned with the truth, but always in connection to a set of rules on how to organize a coherent and dignified life, cf. Lloyd and Sivin 2003.

[18] *EN* 1095a30–b1. See also the testimony by Aristoxenus, *Elementa harmonices* II 1 (39 Da Rios): 'It will be good perhaps to consider before all what type of treatise this is (τὸν τρόπον τῆς πραγματείας); so that, with such a foreknowledge, as if it was of a road on which we must travel (ὥσπερ ὁδὸν ᾗ βαδιστέον), will enable us to recognise each stage as we reach it and so travel in a more leisurely way (ῥᾷδιον πορευώμεθα).'

We have the impression that Aristotle wants to lead his public by hand, step by step, to the right philosophical position. He takes care, especially in I 7, to explain himself in a smooth and accessible way, and at some point stops to indicate to the hearers the progression of his argument and at what point he has arrived (189b16, 191a15–19, 23, 35).

So, the book seems addressed to hearers interested in physical science, with some preparation in dialectic, but who need some argument and some convincing to accept Aristotle's position. They also need to be told more about the positions of the ἀρχαῖοι (193a23), because it seems that they are not as familiar with the history of the problem as they are with Platonic positions. They are, in my opinion, the πεπαιδευμένοι about which the beginning of *Parts of Animals* has so many things to say.[19] Let us pass now to questions connected to the content of the book.

3.

The chapters in this volume are very rich and it would be impossible to present in a short introduction to every interesting position they contain. I will limit myself to a few points, noting that occasionally many authors agree on them.

The importance and function of the first section of the book have been well illustrated by Menn. Aristotle presents his account of ἀρχαί in competition with the accounts of earlier physicists and of the Academy, but with a different approach. Aristotle does not offer a cosmogonic narrative, but describes principles in a more abstract sense, as universal functions and types. This is also Betegh's, Delcomminette's and Broadie's opinion, when they discuss Aristotle's polemics against other philosophers. It is well known that, when Aristotle says here, 'we must proceed from universals to particulars for the whole is more knowable to sensation and the universal is a kind of whole, for the universal encompasses many things as parts' (184a23–6), 'universal' is apparently used in a nonstandard way, as 'what is more knowable by us', instead as 'what is more knowable by nature' (see *APo* 71b29–72a5). Instead of interpreting universals as confused data of experience and particulars as their composing elements or ἀρχαί, as usual, Menn, following a suggestion by Alexander of Aphrodisias, proposes another reading. According to this reading, Aristotle means to say that we start both from the cognition of things that are derived from principles, and from an obscure and extremely general cognition of the principles themselves. This interpretative choice gives a better connection between this section and the

[19] *PA* I 1, 629a1–16.

following chapter, where Aristotle lists in a very abstract and general way the theories of his predecessors, the ancient physicists and the Academics. So, Aristotle's argument turns out to be less peculiar than it has been judged by most of the critics, and the entire first section appears well connected to the rest of the book.

If this is true, the comparison of people who start the search to little children who learn to speak properly may also not be as innocent as it appears. Aristotle says: 'little children at first call all men fathers and all women mothers, and afterwards they distinguish each of these' (184b12–14); now, at the end of the first book of *Metaphysics* he speaks about his predecessors as speaking like 'someone who lisps, being youthful and in the beginning' (993a15–16). The metaphor is the same, and indicates that according to Aristotle his predecessors had insights, but could not reach a coherent view on the question of principles.[20] They were misled as children thrust onto the wrong road by their inexperience (191a24–7). Aristotle seems to think that philosophy had an infantile age, characterized by confused language and insufficient theorical distinctions; after this the following generation, the Platonists, worked towards the right solution, but in an insufficient way (191b35–6).[21] Now, in *Physics* I, Aristotle arrives at a true definition of the number and nature of the principles, albeit still in a general formulation.

Let us pass now to the second section. There Aristotle first lists the possibilities relative to the number of principles—one or many, limited in number or not—and so on, and connects the different possibilities to the names of ancient *φυσιολόγοι* who have inquired about principles. The inquiry starts in a systematic way, but then develops in a quasi-historical narrative.

A typical feature of Aristotelian treatises is that Aristotle's own position never appears at the beginning of the discussion, but is preceded by many preparatory analyses and arguments. The move we find in *Physics* I 2 depends on the status of the discipline in Aristotle's time. In some fields, such as ethics and logic, he is the first to write a systematic treatise, but in physical science Aristotle has many predecessors, and because of this he feels the need to situate his theory within the framework of the common search for physical principles.[22] In this way he manages to show that his solution supersedes everything proposed before him, but, at the same time, to recognize and build on what is good in the theories of others.

The discussion starts from the Eleatics, whose opinions are described and criticized up until the end of chapter 3. Crubellier and Quarantotto analyse Aristotle's arguments in detail, and show how the discussion of Parmenides and Melissus is useful for determining what the principles are. They think that

[20] I follow the text and interpretation given by Cooper 2012, 341–9. Cf. also Horstschäfer 1998, 33.

[21] But this is controversial, and according to Broadie (this volume) Aristotle considers Plato to be engaged in a completely different project.

[22] Cf. Wieland 1993, 127 and 139.

Aristotle considers the Eleatic position to be the result of serious reflection on some ἀπορίαι concerning plurality and change, and not just the fruit of a purely eristic position; this is why the Eleatics deserve such a long treatment. The dialectical aspect of chapters 2–3 is evident: Aristotle refutes most of the Eleatic premises and arguments, and explicitly presents his discussion as dialectical (185a19–20). However, the discussion is not a pure refutation in the Socratic style: it is rather a λύσις, the solution of a puzzle, as both Crubellier and Quarantotto say. In carrying it out, particularly in chapter 3, he touches on problems and makes statements very similar to those found in the discussions of οὐσία in *Met.* Z. Even though Aristotle provides a lengthy justification for discussing the Eleatics here, the modern reader of these pages cannot help wondering why Aristotle did not discuss some of the Eleatic theses in the *Metaphysics* as well, since, as both Crubellier and Quarantotto show in their chapters, the solution of the Eleatic puzzles involves many basic concepts of Aristotelian ontology and dialectics, such as the distinction between per se and accidental predicates, essence and accident, substance and attribute, touching closely the problems of first philosophy.[23] There is common ground between Aristotle and the Eleatics, namely the admission of the validity of the principle of non-contradiction (PNC) and the idea that nothing can come from absolute nothingness. The main differences between them are the assumption by Aristotle of the existence of natural change and of the multivocity of 'being'. The interpreters observe that Aristotle starts from his own thesis of the plurivocity of the term 'being', and that on the basis of this he arrives at the thesis of the plurality of being itself. Both Crubellier and Quarantotto aptly note that Parmenides' position not only denies change, but also denies the very concept of a principle, since in an absolutely unitary conception of being there is no room for the distinction between a principle and what depends on it. Quarantotto in particular underlines the metaphor of a *path* (ὁδός) used by Aristotle more than once in the book. The inquiry itself is a path: we start from what is better known to us and arrive at what is more knowable per se, making it more knowable to us as well.

At the end of both chapters 2 and 3 (at lines 185b25–186a3 and at lines 187a1–11), Aristotle refers briefly to others who, reacting to Eleatic positions, either violated the PNC by admitting that the one was many (186a2), or admitted that there is absolute not-being (187a2).[24]

[23] Lorenz (this volume) notes that in the second part of I 7 Aristotle hints at a number of philosophically important ideas, fully developed only in the *Metaphysics*.

[24] Aristotle names only Lycophron in the first group; according to Crubellier they were not immobilist monists but people influenced by Eleatic arguments. He offers no name for the second group of people influenced by Eleatic arguments. Quarantotto does the same in her chapter.

After this, in chapters 4–5, Aristotle discusses the positions of those who consider being as a unity comprising parts. So, the analysis of the Eleatics in chapters 2–3 is connected to the discussion of Empedocles and Anaxagoras in chapter 4 as the first arm of a possible dichotomy, as already discussed.

As Betegh says in his chapter, chapter 4 does not continue the discussion of the previous philosophers about principles, following the division established at the beginning of chapter 2. This division was only a systematic grid of different possible options and not the blueprint of the following discussion. Chapter 4 examines the different ways in which the physicists looked for ways to produce multiplicity out of the One. The chapter continues Aristotle's review of the ways in which different philosophers have tried to solve the monistic challenge, which began with his earlier discussion of those people who denied the PNC or admitted that something can come out of nothingness.

The Ionian philosophers, but also Anaxagoras, Empedocles, and Democritus, accepted that 'what is' is one, but tried to have a more constructive approach to the problem of change, proposing different ways to introduce distinction and plurality into the one. In doing so, they discovered two important notions: the idea of an ὑποκείμενον, a substrate underlying empirical changes, and the role of the opposites. In a dim way they discovered the main principles of Aristotle's position, but with many shortcomings. The discussion in chapter 4 follows with great liberty the approach by Plato in the *Sophist*, and in some passages requires an audience cognisant of the *Timaeus* and other dialogues, as we have already seen.

In chapter 5, discussed by Delcomminette, Aristotle takes a momentary break from the question of the number of principles to tackle the question of the relationship among principles. On this point there is an agreement among all the preceding philosophers, from the Ionian philosophers to Plato: they all saw the principles as contraries. But such a *consensus omnium* is not by itself sufficient to establish what the principles are; it needs a confirmation. Aristotle provides this confirmation first with an εὔλογον, an appropriate justification, and then with an argument of his own. In doing so, he passes from a doxographical record to his own version of the thesis attributed to all philosophers, revealing a completely new and fully Aristotelian way of conceptualizing change. He already points to his own position, according to which not some particular pair of contraries but *contrariety itself* is a principle. The argument appears to be an induction, starting from empirical evidence and arriving at a universal conclusion, analogically valid for every kind of change (189a1); the idea of analogy reappears at the end of I 7 (191a7–8). In the final section of the chapter Aristotle recalls his distinction between the prior *per se* and the prior with respect to us, and distinguishes contraries according this criterion. Contraries understood as couples of existing things are closer to perception but posterior per se to contraries as functions, which are instead better known according to reason. Both at the beginning of chapter 4 and at the end of the present chapter, Plato is praised for having already carried out the first steps of this progression (187a17–18; 189a8).

At the end of this section, chapter I 6 eliminates some exceedingly absurd positions and concentrates on the more plausible theories that pose a limited number of principles. It provides arguments for the thesis that there are at least two contrary principles, and perhaps also a third, that is non-contrary and underlying. But earlier, at the end of I 5, it was said that the principles are contrary (189a9–10). This creates an ἀπορία, a puzzle, which will be solved in the following section, as Anagnostopoulos clearly shows.

The third section, corresponding to chapter I 7, provides Aristotle's way out of the Eleatic riddles and of the puzzle that arose in chapters 5 and 6. It has been divided in two sections, the first, 189b30–190b17, dedicated to every kind of change, and the second, 190b17–191a22, dedicated especially to natural change. Here, as Morison and Lorenz aptly say, Aristotle gives us an account of the principles of the *things* that change and not only of the change itself.

The first section is very general in its formulations, and there is no trace of the normal Aristotelian vocabulary of substance, accidents, and predicates. It starts from the way in which we speak about change in common language, but it aims at describing how things behave in the process of changing. Aristotle describes the subject and the terms that delimit a single change. Morison interprets the item from which something comes to be as a complex reality composed by a substrate and the lack of a quality: the final term of the change will be the same substrate with the relevant quality, and in this way at the end of change something will be different but also something will be the same. The two extremes are qualified as contraries, and in this way Aristotle connects his discussion to the main theme of chapter I 5, contraries as principles (190a17–21, b13–17). This analysis is proposed twice in the chapter: first in the case of a substance acquiring a predicate and then in the case of the generation of a substance (190a31–b10 and 191a3–22).

In the second part of I 7, the analysis of change is connected to the question of how many principles there are. This passage focuses on the concept of form, rather than on the lack of form, because the lack of form cannot be understood as a constituent of a natural substance. The section applies the model described in the first part to the specific case of natural things. Morison and Lorenz agree in considering the opposites, privation and form, as sheer attributes or qualifications, and not as denotations of an ὑποκείμενον under two different descriptions, e.g. as 'unmusical man' and 'musical man'. In this they depart from the recent commentary tradition.[25]

[25] E.g. see Charlton 1970, 70–9; Quevedo 1998, 205.

There is some ambiguity in this section, and Aristotle's solutions are far from being simple and straightforward. Lorenz shows this clearly. The principles are in one way three and in another only two; in a way they are contraries and in a way they are not: the substrate often is described as a complex reality composed by the thing that remains in change and the privation, but sometimes the term *ὑποκείμενον*, 'substrate' is used for the matter alone;[26] sometimes the privation is considered a bona fide principle of change and sometime it is not, especially when the change is explained only by the presence and absence of form (191a5–7). This ambiguity is due to the ontological situation of privation, which is understood by the two commentators as an attribute (*συμβεβηκός*, 190b27) and in some sense a positive characterization of the substrate, but is also as a form of not-being.[27] Others, including Leunissen in the present volume, follow tradition in preferring the usual translation 'accident'. This situation makes it impossible to consider privation as a constituent of a natural substance and difficult to consider it as a full-fledged principle of change. This is a problem the concept of potentiality doesn't seem to have to deal with.

Let us stop for a moment to consider this element. In the metaphysical treatises the element opposed to form is a *δύναμις*; here in *Physics* I Aristotle prefers the different concept of *στέρησις*.[28] The concept of privation, *στέρησις*, is central in this book, and the commentators usually use the term *στέρησις* also when commenting on the second section (I 2–6), particularly in the analysis of the first of the two contraries which determine the change, the one which describes the substrate as not-being an X, or the 'not'-contrary (henceforth NC). But the term *στέρησις* appears only in line 190b28, in the context of the discussion of natural change.[29] In this last section of I 7, and only there, Aristotle introduces his usual terms such as *μορφή*, *εἶδος*, and *ὕλη* (190b20, 25, 28).[30] This is due, in my opinion, to his tendency to speak in the most general terms possible, and to the fact that 'privation' is already a specification of NC.

[26] Cf. Chapter 5; Chapter 8.

[27] They follow here Clarke 2015. The opposite position appears in Quevedo 1998, 200–1, whose interpretation is based on examples as the incultivated man: 'La unidad del hombre y de lo inculto es . . . una unidad numerica *per accidens*'. Against Suarez (*Disp. Metaph.* LIV, 5. 7) she insists that privation is only a principle *per accidens.*

[28] Aristotle distinguishes many kinds (*τρόποι*) of *στέρησις*, absolute (*ἁπλῶς*), primary (*πρώτη*), according to the language, i.e. to the use of a privative (*ἀπό τοῦ ἀ*)—and a full study of this concept is still a desideratum. The only exception, as far as I know, is Quevedo 1998 who has collected all the relevant passages and has proposed a first analysis of this elusive notion.

[29] See, e.g., Wieland 1993, 161 e 165; Quevedo 1998, 206. There is an agreement among most commentators on the fact that this use of the idea of *στέρησις* is different from the conception of privation in *Cat.* 10 and *Met.* Δ 22. *Contra*, Irwin 1996, 110.

[30] In fact in his discussions of privation Aristotle insists on the idea that *στέρησις* often can be constructed as the absence of an attribute one would expect to be naturally present, or of an attribute that is naturally possessed, or *per se* (*Cat.* 10, 12a29–31; *Met.* Δ 22, 1022b22–31; Θ 2, 1046b12–13, I 4, 1055b5). There are other kinds of privation, but this is the most important.

In the second section of the book and in the first part of I 7, Aristotle describes the NC in terms of contrariety, that is to say, as a determinate negation and not a mere contradictory negation of a form. It differs from negation, as Delcomminette says, for two reasons: it belongs to a determinate genus and to a determinate recipient. There is, in the contemporary critical literature, a tendency to interpret the concept of privation in terms of potentiality, as the example of the seed could suggest. Now, it is true that both concepts, privation and potentiality, converge in being relative to a form which the substrate may or may not assume, in a dynamic conception of the realization of an entity. In fact NC applies to the substrate before the change, and form, in turn, after the change. Both privation and potentiality are different from the complete negation of the form and indicate a lack starting from which the change can take place. However, one could object that Aristotle himself clearly distinguishes the two concepts: *εἷς μὲν δὴ τρόπος οὗτος, ἄλλος δ' ὅτι ἐνδέχεται ταὐτὰ λέγειν κατὰ τὴν δύναμιν καὶ τὴν ἐνέργειαν*, 'this is one way of handling the matter, another possible way is to express the same things in terms of potency and actuality' (191b28–9). Besides, potentiality is strictly connected with the material substrate, whereas privation is completely distinct from the substrate, *ὑποκείμενον*, which in turn is identified with matter at 190b25. NC is to be carefully distinguished, in my opinion, on one side from undetermined negation and on the other side from the concept of potentiality.

There have been two main ways of interpreting *στέρησις*. The first one is to consider NC, and its specification as privation, only as a way of looking at things. In this interpretation privation cannot be individuated outside the analysis of a single change, and to which it is relative. In other words, only from the point of view of the achieved result is it possible to tell what the privation was. From this point of view the translation of *συμβεβηκός* as 'accidental' makes sense.[31] The second interpretation, preferred by Lorenz and Morison, is to consider NC, and consequently privation, as a part of the being of a subject. In this interpretation the translation of *συμβεβηκός* as 'attribute' is fully justified.

In my opinion both options have some justifications, because each brings to the light a different feature of the substrate which makes the change, i.e. (1) the coming from a kind of not-being which disappears, and (2) the fact that the substrate is the sort of the thing which is capable of receiving some form. Aristotle's examples go in both directions. The example of the seed, as already said, is wholly consistent with a positive consideration of NC, because it is in the nature of an acorn to become an oak tree, if nothing opposes its realization. But in the case of non-natural changes there is no such natural connection between, say, a lump of bronze or some construction material and the final

[31] E.g. Wieland 1993, 166–7.

realization, be it a statue of Hermes or the house of Socrates, and the connection seems to be accidental. To be sure, the form to be realized requires some particular kind of matter and not another, but the lump could also become a shield, and the construction material could also become a temple, a porch, or some other thing. There is in the nature of bronze or building material no tendency to assume the form present at the end of the process. In this way one could maintain that the privation has an accidental connection to the form.[32]

In the final section, as I already said, there is a dialectical confirmation of the truth of Aristotle's theory. The examination of other philosophers' mistakes proceeds in an historical way, first discussing the older philosophers and next the Platonists. The metaphor of the way reappears, in the form of discovering the error as a form of being sidetracked from the proper road. The most ancient philosophers were impeded by their inexperience with dialectic, as Leunissen says, and by their incapacity for making the appropriate distinctions. They took not-being in a simple way, and did not distinguish the different possibilities of coming from what is and what is not. Leunissen describes Aristotle's argument in detail, showing how he finds in these thinkers a conceptual confusion which prevented them from getting the phenomena right.

In the last chapter Aristotle discusses a Platonist theory of natural coming-to-be. In Broadie's interpretation, this theory is something Aristotle considers possibly actual in its conceptual articulation. It would be a possible extension to the realm of physical change of the Platonic metaphysical theory of the opposition between the One and the Great-and-Small, an extension not made by Plato himself. Broadie describes at length the Platonic background of the criticized theory, especially in the *Philebus* and *Timaeus.*

In fact, this position is dangerous to Aristotle, because in a way the Platonists too understood the principles of change as a triad, considering the Great and the Small as two things. But it was the wrong way to understand the triad, Aristotle thinks, and he considers it as a confused deviation from his schema, because there is no place for the notion of privation in it. His discussion uses many Platonic metaphors, in a polemical way. According to Broadie, Aristotle's polemic would hardly scratch the skin of convinced Platonists, because he only shows how their set and number of principles are not the same as his. But the criticism gains impact if there really was an attempt, in the Platonic school, to extend their theory of principles to the corporeal realm of living things.

After this Broadie extends her discussion to two cases in which Aristotle's triad of principles apparently does not apply. The first is the movement of the heavenly spheres, in which it seems difficult to individuate a privation that completely disappears, and the second is the causality of Aristotle's prime mover.

[32] On this point, see also Bostock 1982, 190.

In her view, in this case efficient and final causality cannot be pulled apart; she describes very well the final causality of prime mover, and adds some reasons for admitting also that it exercises efficient causality. Now, it is completely true that in an Aristotelian perspective final causes and efficient causes are not exclusively opposed, but cooperate to determine a single movement, together with material and final causes. I don't think anybody supposes that the heavenly bodies' circular movement is only produced by the final causality of the prime mover. The question is rather whether the moving cause should be placed in the entity which acts as final cause, as Broadie proposes, or in the soul of the living object which is moved by it, as the traditional interpretation maintains.[33]

4.

It should be clear by now that Aristotle's position is very different from everything that has been said by his predecessors, and that he is very proud of his achievements (191a15–20).

As was already said, Aristotle 'decosmologized' the question of principles. We don't have a cosmogony in *Physics* I, because Aristotle's world is eternal. We also do not find an identification of the first principles of natural change in terms of some particular stuff or entity, such as water or *nous*. In positing as principles two opposites and a substrate, Aristotle passes from the consideration of the principles as *things*, material or immaterial, to the consideration of the principles as *functions* that can be assumed by different realities, or universal *types* of things to be differently instantiated in different episodes of change, with the specifications that one of the two opposites is the privation of the other, and that they never exist alone, but only as qualifications of the underlying matter. In the same way, in his practical philosophy, Aristotle poses as *horos* of the right mean that it is a virtue: not an external good as health or money, but the right functioning of the practical intellect.

It is interesting to look at the examples Aristotle employs in his analysis. He provides us with examples of complex events—much more complex than those of classical physics such as falling apples, billiard balls moving on the green pool, swinging pendulums, etc. Rather, Aristotle has chosen the learning of a τέχνη by a human being (188a31–b6; 189b6–190a13). This approach is in accordance with Aristotle's tendency to find the most fundamental distinctions, on which his theories must be based, in complex phenomena and not in simple cases, like those often examined in contemporary philosophy, as, for instance, the

[33] For a defence of the traditional interpretation, see also Natali 1997; Gourinat 2004; Lagnerini 2015.

classical example of 'raising an arm' in Wittgensteinian discussions. In simple cases, Aristotle thinks, the relevant distinctions tend to be imperceptible because the examples do not leave enough room for a clear separation of the constitutive elements. According to Aristotle, to find the fundamental elements in the theoretical description of natural events one must look at examples on a more elevated level. As he himself says in the final chapter of the *Meteorologica* IV 12, 390a14–20, it is not easy to determine the form and function of a reality at the elemental level and to state accurately the form and function of the four elementary bodies, because their function is not determinate. On the contrary, we can do it with realities and events at a higher level, as in the case of natural and artificial entities like a living being or a house, and in the case of complex processes like learning to be musical.

Already in chapter I 5 it is said that all physical transformations cannot be referred to a random antecedent. For every kind of being there is a fixed sphere of possible movements that cannot be broken. What is red can become blue, a square form can become round, but nothing can pass from 'red' to 'square'. In chapter I 7, the conditions of the regularity of events according to Aristotle are spelled out. In contrast to the main stream of modern philosophy of nature, Aristotle considers becoming not only as a succession of events regularly connected under a covering law and individuated by their position in the causal series. He considers natural movements as the succession of changes in an enduring subject. Such a subject works as a principle of selection of possible events, according to its powers. The existence of a fixed scale of possible movements of a being implies that the event is neither totally nor exclusively determined by the conditions that provoked it causally. From Aristotle's perspective, the regularity of transformations comes from the nature and powers of the changing substratum considered as the complex of matter plus privation.[34] The function of the moving cause is only hinted at here, and will be developed in the following book.

This implies that in Aristotelian physics not all descriptions of a process are equivalent: 'Socrates learns to play' and 'the white-haired moves one of its extremities on the keyboard' are not the same. Only the first statement can give a precise idea of the process in act, regarding all the essential qualities and capacities that qualify the beings that are not accidentally implied in the process of this becoming. The notion of essence here is not context-dependent, but absolute: an essential description of any item involved in the process can be found and is different from the many possible accidental descriptions of the same change. If an event

[34] A similar position has been revived in the last decades by some contemporary non-empiricist and neo-Aristotelian approaches to the notion of causality. They go under the general label of 'causal dispositionalism', for which see Mumford 2012, who says: 'the dispositionalism developed here is one intended to be strongly anti-Humean in its acceptance of a thoroughgoing realism about dispositions'.

can be identified, it can be identified as an actualization of a particular capacity of a specific substrate. This is because understanding what kind of thing 'playing' is depends on understanding what kind of thing a human being is.

5.

In conclusion, we can say that the first book of the *Physics* is a well-constructed *logos* but it is not an easy read. Like a well-marked mountain path, it is long and complex, contains some challenging passages, and seems at times to wander in ways that are difficult to understand. But when we arrive from the ascending section of chapters I 2–6 to the wide plateau of chapter I 7, the progression becomes a little easier and the point of arrival now seems worth the fatigue. The present book, like a good guide, can indicate to the interested reader all the difficulties and the beauties of the ὁδός of discovery in *Physics* I, and will be useful to bring along for people initiating their journey into Aristotle's physical theorizations.

I

The Path to the Principles

Physics I 1

STEPHEN MENN

Physics I is an introduction to the *Physics*—perhaps just to *Physics* I–IV, perhaps to *Physics* I–VIII, perhaps to something even larger—and *Physics* I 1 is an introduction to *Physics* I. It is supposed to motivate, and to give methodological guidance for, the investigation of the ἀρχαί, and more specifically of the ἀρχαί of natural things. Unfortunately the chapter is more difficult, perhaps merely through extreme compression, than the book it is supposed to be introducing, and has left a legacy of puzzlement among ancient and medieval as well as modern commentators.[1] The puzzlement has centred on Aristotle's saying here that universals (τὰ καθόλου) are more knowable to us while particulars (τὰ καθ' ἕκαστα) are more knowable by nature, which is at least verbally the opposite of

I would like to thank all the participants in the Symposium for oral discussion, and Andreas Anagnostopoulos and Michel Crubellier and Sean Kelsey and Gabriela Rossi and (outside the Symposium) Lucas Angioni and a publisher's reader for written comments. István Bodnár and Andrea Falcon and Henry Mendell gave helpful references, and Falcon also showed me his unpublished paper on this chapter (Falcon 2018), and talked through the issues with me. Crubellier's and Kelsey's comments were especially helpful.

[1] On this chapter, and *Physics* I generally, the best ancient commentaries are those of Simplicius (CAG IX) and Alexander (lost, but Simplicius generally reports its contents). There is also the commentary of John Philoponus (CAG XVI), and the brief paraphrase of Themistius (CAG V, 2), which tends to dodge controversial issues of interpretation. (There was also a commentary by Porphyry, which Simplicius often reports.) The most important medieval commentaries are Averroes' *Long Commentary*, lost in Arabic but extant in Hebrew and Latin translations (the Latin is available in volume 4 of the Juntine edition, Averroes 1550) and Thomas Aquinas 1954. Both of these texts are easily found online. Much of the Arabic commentary literature is summarized in Lettinck 1994. Of recent literature, the most useful source on book I in general is Horstschäfer 1998. On chapter 1 in particular, see Mansion 1979, Bolton 1991, Angioni 2001, Cerami 2009, and Falcon 2018.

what he says elsewhere. The chapter's ambiguities and tensions may ultimately be such that it will not give us clear guidance for how to read the rest of the book. But working through the difficulties, both difficulties internal to *Physics* I 1 and difficulties about how its programme connects with what Aristotle actually does in *Physics* I and subsequent books, will at least alert us to interpretive issues that we should keep in mind in reading *Physics* I. I think that two popular proposals for resolving the difficulties of I 1—namely, that the root of the difficulty lies in Aristotle's using the word καθόλου here in a nonstandard way, and that he is here just describing the methodology of physics in general, not book I's investigation of form, matter, and privation—are both mistaken. Even if no alternative solution wins full conviction, there will be at least a negative benefit in not accepting a too easy solution. And I will make some positive suggestions, although I will probably not convince everyone.

I will divide *Physics* I 1 into three sections, 184a10–16, 184a16–23, and 184a23–b14, translating and commenting on each of them in turn. The first section is relatively uncontroversial. In the *Physics*, as in the *Metaphysics* and *Nicomachean Ethics*, Aristotle begins with a general γνώμη about (theoretical or practical) knowledge, and then makes an argumentative transition to the discipline he is here concerned with. Thus:

> *General* γνώμη: Since knowledge and scientific understanding (τὸ εἰδέναι καὶ τὸ ἐπίστασθαι) arise, in all disciplines which have ἀρχαί or causes or στοιχεῖα, from recognizing (γνωρίζειν) these (sc. ἀρχαί or causes or στοιχεῖα),
>
> *Supporting enthymeme*: For we think that we know (γιγνώσκειν) each thing when we recognize (γνωρίζειν) [its] first causes and first ἀρχαί, as far as the στοιχεῖα,
>
> *Application to the present case*: It is clear that within the science of nature too we must try to determine first [the part or the questions] concerning the ἀρχαί. (184a10–16)

There are some puzzles here, which gave rise to differences of interpretation reported already in Simplicius' commentary. It seems to be agreed that τὸ εἰδέναι καὶ τὸ ἐπίστασθαι restricts knowledge to ἐπιστήμη in the sense of the *Posterior Analytics* (whether we think, as Alexander seems to, that ἐπίστασθαι is one species of εἰδέναι, or, as Simplicius proposes, that εἰδέναι in the *strict* sense is being identified with ἐπίστασθαι)—whereas γνωρίζειν must be knowledge in a broader sense, or there would be an infinite regress.[2] But then the apparently restrictive clause 'in all disciplines which have ἀρχαί or causes or στοιχεῖα'

[2] For Alexander and Simplicius on εἰδέναι and ἐπίστασθαι see Simplicius *In Physics* 12,14–13,13. I am provisionally translating γνωρίζειν as 'recognize': I'll discuss another possible interpretation below. Burnyeat 2012 has some discussion of Alexander and Simplicius on how the verbs are related in this sentence, as well as much on the relations between εἰδέναι and ἐπίστασθαι and γιγνώσκειν more generally, and between the different nouns that may or may not correspond to these verbs.

seems odd: surely *all* sciences in the sense of the *Posterior Analytics* have ἀρχαί or causes or στοιχεῖα? (Or, more precisely, surely the objects known by these sciences have ἀρχαί or causes or στοιχεῖα.)[3] So perhaps the clause is not really restrictive: perhaps those disciplines which have ἀρχαί or causes or στοιχεῖα are the disciplines in which ἐπιστήμη in the sense of the *Posterior Analytics* comes about (and perhaps only these are really *disciplines* (μέθοδοι)), and this is how such ἐπιστήμη comes about.[4] Or, as both Alexander and Simplicius suggest in different ways, perhaps ἀρχαί and causes and στοιχεῖα are not synonymous, and while all sciences have at least one of these, not all sciences have all three. Simplicius takes Aristotle to be dividing ἀρχαί into causes-in-the-strict-sense and στοιχεῖα, and, surprisingly, construes ὧν εἰσὶν ἀρχαὶ ἢ αἴτια ἢ στοιχεῖα as 'of which the ἀρχαί are either causes or στοιχεῖα' or perhaps 'of which there are ἀρχαί, either causes or στοιχεῖα'.[5] But the basic thought remains unaltered by these disputes. First Aristotle says that all scientific knowledge comes from recognizing the ἀρχαί of things; then he applies this to the study of nature, infers that here too we will have scientific knowledge only once we have determined the ἀρχαί of natural things, and draws the practical conclusion that this is what we should devote ourselves to first.

There is no reason to expect that these ἀρχαί of natural things will be the ἀρχαί in the strict sense, the first of all things, the ἀρχαί and first causes of all things universally, which *Metaphysics* A 1–2 says will be the objects of wisdom. (Or, if Speusippus is right that there are no first causes of all things universally, since natural things and unchanging things have different first causes, then the objects of wisdom will be the ἀρχαί of unchanging things, not the ἀρχαί of natural things. Only if there is nothing prior to natural things should we expect that the ἀρχαί of natural things will be the ἀρχαί in the strict sense.) So what are these ἀρχαί of natural things?

The most obvious answer is that they are the form, matter, and privation discussed in *Physics* I. Indeed, the last lines of *Physics* I, 'let us thus have determined that there are ἀρχαί, and what they are, and how many in number; now let us make a new start and say' (I 9, 192b2–4), seem to close off the investigation of the ἀρχαί of natural things announced in I 1, and to make a transition to something else in book II. This interpretation is also presupposed by Codex E, which gives book I the title περὶ ἀρχῶν and book II the contrasting title περὶ

[3] At 184a11, ὧν εἰσὶν ἀρχαὶ ἢ αἴτια ἢ στοιχεῖα, it seems that grammatically the antecedent of ὧν can only be the preceding τὰς μεθόδους; but we might want to say that there is a slight anacoluthon, and that Aristotle really means to refer back to the things we know, which are the objects of these disciplines.

[4] See Simplicius 13,14–21 and 14,9–28.

[5] See Simplicius 13,21–33 and 10,8–24 and 11,16–12,5; on Simplicius on causes vs. στοιχεῖα, see discussion below.

αἰτίων.[6] On this interpretation, book II would not be adding further ἀρχαί, but telling the physicist how to explain natural things by connecting them back to the ἀρχαί (ἀνάγειν εἰς τὰς ἀρχάς, II 3, 194b22–3), namely the ἀρχαί already discussed in book I, form and matter and privation.[7] You might complain that this is too restrictive a list of ἀρχαί: why shouldn't the efficient and final causes also count as ἀρχαί, especially since Aristotle commonly refers to the efficient cause as ἀρχὴ κινήσεως? But we might reply, drawing on *Physics* II 7, that the efficient and final causes do not give further natural ἀρχαί beyond matter and form, since your efficient cause is not strictly your father but your father's soul, and this was already a formal ἀρχή (one and the same ἀρχή can be a formal and final cause of your father and also an efficient cause of you); and the first unmoved mover, which is not a formal ἀρχή, is not a nature or a natural thing, and so it is not a *natural* ἀρχή.[8]

Despite these indications in *Physics* I–II, Aristotle clearly sometimes thinks of the investigation of physical ἀρχαί as something broader than *Physics* I. *De Caelo* I 7 refers back, for a proof that there is no infinite body, to 'our λόγοι . . . ἐν τοῖς περὶ τὰς ἀρχάς: for there too already it was determined about the infinite, in what way it exists and in what way it does not exist' (274a19–24). So on this conception *Physics* III is part of a broader Περὶ τὰς ἀρχάς extending beyond *Physics* I, and presumably including at least *Physics* I–IV. The first sentence of the *Meteorology* also presupposes something like this when it summarizes the *Physics* as περὶ τῶν πρώτων αἰτίων τῆς φύσεως (I–IV?) καὶ περὶ πάσης κινήσεως φυσικῆς (V–VIII?), before going on to the *De Caelo*; and Adrastus in *On the Order of Aristotle's Treatises* says that some people entitle the first five books of the *Physics* Περὶ ἀρχῶν (Simplicius *In Physica* 4,11–15).[9] And this conception

[6] Codex E is available at the website of the Bibliothèque Nationale: <http://gallica.bnf.fr/ark:/12148/btv1b84192492>. The subtitle περὶ ἀρχῶν (in a later hand) is not easy to read, and there may be something else there too; περὶ αἰτίων at the beginning of book II is clear. Ross is apparently wrong in reporting that Codex F gives Ἀριστοτέλους φυσικῆς ἀκροάσεως ἢ περὶ ἀρχῶν *A*. This page is available at: <http://teca.bmlonline.it/ImageViewer/servlet/ImageViewer?idr=TECA0001111841#page/1/mode/1up> and it seems to say only Ἀριστοτέλους φυσικῆς ἀκροάσεως *A*, although there is water damage and an illegible scholium. I thank Pantelis Golitsis and Henry Mendell for advice, and Pantelis for looking at Codex F *in situ*.

[7] Warning: book II says almost nothing about privation, mentioning it only at *Physics* II 1, 193b19–21.

[8] For the first thought see *Physics* II 7, 198a24–7, and for the second 198a27–31 and 198a35-b4.

[9] Presumably it is the first *four* books which are plausibly called Περὶ ἀρχῶν. But Adrastus is making the mistake of the Peripatetic tradition beginning with Andronicus, until Porphyry corrected it, of putting the break after *Physics* V rather than after *Physics* IV. I think Ross's diagnosis of the mistake in his *editio maior* (Ross 1936, 1–4) is right: they knew from early Peripatetic references that the broader *Physics* contained a treatise called περὶ ἀρχῶν or φυσικά or περὶ φύσεως followed by a treatise περὶ κινήσεως, but they also had references (at least that of Damas, cited by Simplicius *In Physica* 924,12–14) to a three-book περὶ κινήσεως; so they assumed that the περὶ κινήσεως must be the last three books of the *Physics*, and that the rest was the περὶ ἀρχῶν or φυσικά proper. But in reality the three-book περὶ κινήσεως must have been *Physics* V, VI, and VIII. (Codex E, very strangely, guesses that *Physics* VII, VI, and VIII in that

of the task can be supported by texts of *Physics* III–IV which make clear that one motive for examining infinity and place and so on is that these things are at least plausibly thought to be ἀρχαί. (Thus 'all those who seem to have touched on this kind of philosophy in a way worth mentioning have given an account of the infinite, and all of them posit it as an ἀρχή of the things that are', III 4, 203a1–4; Aristotle argues against these philosophers that the infinite cannot be actual and an οὐσία, and therefore cannot be an ἀρχή, III 5, 204a20–1. In IV 1 he gives a plausible argument that place is 'prior to all things', i.e. that is an ἀρχή, since 'that without which none of the other things exists, but it [can exist] without the others, must be first', but it is plausible that anything that exists is in some place, and 'the place does not perish when the things in it are destroyed', 208b27–209a2; Aristotle suggests that this is why Hesiod made Chaos first.) The difference between *Physics* I and *Physics* III–IV cannot be that *Physics* I is about the real ἀρχαί of natural things while *Physics* III–IV are about things plausibly but wrongly thought to be ἀρχαί of natural things, since *Physics* I also considers proposed ἀρχαί that Aristotle will reject. So perhaps the difference is that *Physics* I is about things which are, or are thought to be, ἀρχαί in a stricter sense, while *Physics* III–IV are about ἀρχαί in a broader sense, namely things which are (or are thought to be) necessary conditions of motion (so III 1, 200b12–25), and are therefore things without which nothing else would come-to-be out of the ἀρχαί in the strict sense. We might compare Lucretius book I, where only atoms are ἀρχαί in the strict sense, in Lucretius' Latin the *primordia rerum*, but void and infinity are ἀρχαί in a broader sense, since if there were no void the atoms could not move, and if there were not an *infinity* of atoms, with infinite space and time to move in, the worlds would not arise from them.[10]

order are the three books of the περὶ κινήσεως.) Porphyry drew the correct division I–IV/V–VIII (Simplicius *In Physica* 802,7–13), and was followed not only by Philoponus (*In Physica* 2,13–3,10) but also by Simplicius himself in his earlier *De Caelo* commentary (226,19–23), but Simplicius goes back to the older tradition in his *Physics* commentary: see his discussions, and his references to earlier writers, 4,11–16 and 801,13–802,13 and 923,7–924,16.

[10] Here I have spoken of *Physics* I–IV as a unified treatise on the ἀρχαί of natural things. There are some clear indications for this, but there are also problems about the internal unity of *Physics* I–IV. I have written about this in an unpublished paper, '*Physics* II and its Relation to *Physics* I–IV'; let me here briefly summarize the problem. *Physics* I seems to look forward to *Physics* II, in that: (i) the end of *Physics* I 9 as we have it can only be a transition to *Physics* II; (ii) *Physics* I 7, 191a19–20 'whether the form or the ὑποκείμενον is οὐσία is not yet clear' refers ahead to *Physics* II 1, which resolves the issue; and (iii) while *Physics* I *says* that we are investigating the ἀρχαί of natural things, its argument is not in fact restricted to natural as opposed to artificial things (it never defines what things are natural, and never uses any premise about natural things beyond that they change or come-to-be), and thus seems to be looking forward to *Physics* II, which defines natural things and narrows the investigation to them (see discussion below). So too the beginning of *Physics* III, which lays out the program for III–IV, clearly refers back to the II 1 definition of nature (200b12–13), and attaches itself to the μέθοδος περὶ φύσεως (200b13). On the other hand, *Physics* II never refers back to *Physics* I, and there are several apparent discrepancies if *Physics* I has preceded it; and *Physics* II seems, like *Parts of Animals* I 1 to which it is often a longer parallel, to be

One way Aristotle sometimes distinguishes the *ἀρχαί* of *Physics* I from *ἀρχαί* in a broader sense is to say that the *ἀρχαί* of *Physics* I, and only these, are *στοιχεῖα*. Many Greek philosophers, starting with Democritus, compare their *ἀρχαί* with the *στοιχεῖα*, the letters of the alphabet (or phonemes in speech), or even describe their *ἀρχαί* metaphorically as *στοιχεῖα*.[11] But this way of describing the *ἀρχαί* entails a way of thinking about how they are related to other things: namely, that the *ἀρχαί* are *present in* or *constituents of* the things they are *ἀρχαί* of, as the *στοιχεῖα* are constituents of the *συλλαβαί*. And so Aristotle insists that some *ἀρχαί* are *στοιχεῖα* and others are not. The crucial difference between what Aristotle says about *ἀρχαί* in *Metaphysics* Δ 1 and what he says about *στοιχεῖα* in Δ 3 is that 'of [*ἀρχαί*] some are *ἐνυπάρχουσαι* (present in, or constituents of, the thing they are *ἀρχαί* of) and others are external' (Δ 1, 1013a19–20), while 'it is common to all that the *στοιχεῖον* of each thing is the first *ἐνυπάρχον* of each thing' (Δ 3, 1014b14–15). Aristotle is consistent in drawing this distinction between *ἀρχαί* in general and *στοιχεῖα* in particular, but he is not consistent in how he applies it. Sometimes he applies it in such a way that the matter is a *στοιχεῖον* and form is not; and some ancient commentators assume that that is what Aristotle meant in *Physics* I 1, 184a11 and a13–14 in speaking both of *ἀρχαί* and of *στοιχεῖα*. But sometimes he applies the distinction in such a way that both matter and form are 'constituents' of a thing, and are therefore its *στοιχεῖα*, while notably its efficient cause is not; and some commentators assume that that is what he means in *Physics* I 1.

In *Metaphysics* Z 17 Aristotle uses the *ἀρχή*/*στοιχεῖον* distinction to solve the problem of the unity of a definition. A definition is a *λόγος*, and it is plausible to think of giving a thing's *λόγος* as spelling it out into its *στοιχεῖα*. But there must be some cause why these *στοιχεῖα* are not just many things, but constitute a single 'syllable': if the *λόγος* leaves this out, then it is incomplete, and if this cause of unity must be added to the *λόγος* as a further *στοιχεῖον*, there will be an infinite regress. Aristotle solves the problem by saying that a

introducing a physics of form and final causality, which fits badly with *Physics* III–VIII. My conclusion has been that *Physics* II was originally an independent monobiblion, perhaps a sequel to *On Philosophy*; Aristotle subsequently decided to incorporate this monobiblion into a longer treatise on the *ἀρχαί* of natural things, *Physics* I–IV, and wrote the longer proemium *Physics* I for that occasion. I also suggest that the original proemium to *Physics* II when it was a monobiblion is still preserved (since no one had the heart to throw it out), namely as *Metaphysics* α 3. (The suggestion about α 3 had also been made, independently and almost certainly before me, in an unpublished paper by Daniel de Montmollin.)

[11] Democritus may have said *γράμματα* rather than *στοιχεῖα*, but in any case he used the letters-comparison. See Aristotle, *On Generation and Corruption* I 2, 315b6–15, where Democritus said either that a tragedy and a comedy come to be from the same letters, or that *τραγῳδία* and *τρυγῳδία* (an archaic word for comedy) come to be from the same letters except one: for the emendation from *κωμῳδία* to *τρυγῳδία* see West 1969. For Plato on *στοιχεῖα* or *γράμματα* see Menn 1998. The sixth and ninth *aporiai* of *Metaphysics* B (B 3, 998a20-b14, B 4, 999b24–1000a4) seem to compare and contrast a Democritean with a Platonic application of the letters-analogy.

thing's material constituents are its *στοιχεῖα*, and that the 'syllable' is not simply its *στοιχεῖα* but something over and above them, and the *οὐσία* of the syllable (which its definition must express) is whatever causes these *στοιχεῖα* to be this syllable. So the thing's *οὐσία* is the cause of unity to its many *στοιχεῖα*, and if this cause were a further *στοιχεῖον*, or were composed of further *στοιχεῖα*, there would be a regress to a further cause of unity. So the *οὐσία*, i.e. the form, is a cause which 'is not a *στοιχεῖον* but an *ἀρχή*: a *στοιχεῖον* is that into which [something] is divided, *ἐνυπάρχον* as matter, like *α* and *β* in the syllable' (1041b31–3).[12] Eudemus and Alexander followed this lead in interpreting the *ἀρχή*/cause/*στοιχεῖον* distinction from *Physics* I 1, saying that *στοιχεῖον* here means the matter, that *ἀρχή* most properly means the source of motion, and that 'cause' most properly means the formal or final cause.[13] So on this interpretation not all the *ἀρχαί* discussed in *Physics* I would be *στοιχεῖα*, but only the matter.

However, in *Metaphysics* Λ 2–5 Aristotle argues that corruptible compound substances are not 'out of' eternal substances, i.e. do not have eternal substances as constituents, either as their matter or as their form,[14] but that they do have other causes that lead up to eternal substances, namely the heavenly bodies and their movers. He expresses this by saying that, while the constituent *στοιχεῖα* of different kinds of things are not the same, except analogically the same, 'as if someone were to say that there are three *ἀρχαί*, form and privation and matter' (Λ 4, 1070b18–19), and while the *στοιχεῖα* of non-eternal things are non-eternal, nonetheless all things can share eternal non-constituent *ἀρχαί*.

> Since not only the constituents (*ἐνυπάρχοντα*) are causes, but also some external things, like the mover, it is clear that *ἀρχή* and *στοιχεῖον* are different, and both are causes, and *ἀρχή* is divided into these (sc. into constituent and non-constituent), and the mover or what stops motion is not a *στοιχεῖον*, although it is an *ἀρχή*. So there are by analogy three *στοιχεῖα*, but four *ἀρχαί* and causes. (Λ 4, 1070b22–6)[15]

[12] On these issues see Menn 2001.

[13] For Eudemus and Alexander see Simplicius *In Physica* 10,8–24. Thomas Aquinas *ad locum* follows Alexander in identifying the *ἀρχή* with the efficient cause, the cause with the final and formal causes, the *στοιχεῖον* with the material cause. Averroes *ad locum* follows Alexander in identifying the *ἀρχή* with the efficient cause, and the 'cause' with the final cause, but agrees with Simplicius against Alexander in identifying the *στοιχεῖον* with the material *and formal* causes. (I don't mean that Simplicius was Averroes' source—as far as we know Simplicius' commentary wasn't translated into Arabic.)

[14] This picks up the tenth *aporia* of B, B 4, 1000a5–1001a3, cited at Λ 10, 1075b13–14: if both corruptable and eternal things are composed out of eternal *ἀρχαί*, why are some of their compounds corruptible and others not?

[15] Reading [*στοιχεῖον οὐκ ἔστιν*] *ἀρχή τις οὖσα* (fallen out by *saut du même au même* from *ἱστὰν* to *ἐστιν*?), or Jaeger's conjecture *ἀρχή τις οὖσα* [*οὐκ ἔστι στοιχεῖον*], for the *ἀρχή τις οὖσα* printed and obelized by Jaeger, and transmitted (according to the apparatus of Alexandru 2014) by virtually all manuscripts other than A^b (E, J, M, C, and the whole δ family), and apparently presupposed by both Arabic versions. Ross's *ἀρχή τις καὶ οὐσία*, transmitted apparently only in A^b (and, according to Alexandru, presupposed by Bessarion's translation) is pretty clearly wrong, a scribal attempt to fix the damage.

These 'four causes', in context, are not the standard four, but matter and form and privation and the external moving cause. The external moving cause can, like the matter and form, be specifically different for different species (my father and Bucephalus'), or numerically different for different individuals of the same species (my father and yours). But in some cases, where the moving cause is not conspecific with its effect, there can be a numerically single eternal cause for a whole species and across species, and even across categories. In this way 'the sun and the oblique circuit' (Λ 5, 1071a15–16) and 'that which, as first of all things, moves all things' (Λ 4, 1070b34–5) are causes of the generation of all sublunar substances, and even of accidents. Aristotle's point in Λ is that earlier philosophers have gone wrong in looking for the first ἀρχαί as material or formal constituents of things, since we will not reach a numerically single eternal ἀρχή as a shared constituent of the many corruptible things. Aristotle proposes instead to look for a numerically single eternal ἀρχή as a non-constituent moving cause, of the heavens and thus of sublunar things. So here he treats both form and matter as constituents (ἐνυπάρχοντα) and thus as στοιχεῖα, contrasting them with the moving cause, while in Z 17 he treats only the matter as a constituent and a στοιχεῖον, and contrasts it with the form. And Simplicius, against Eudemus and Alexander, follows this application of the ἀρχή/στοιχεῖον distinction in *Physics* I 1: the matter and form are στοιχεῖα, and they do not lead up beyond natural things to the first ἀρχαί of all things, whereas the efficient and final cause are not στοιχεῖα and do lead up, stage by stage, to first ἀρχαί beyond the natural domain.[16] On this interpretation, the ἀρχαί discussed in *Physics* I would all be στοιχεῖα, and Aristotle would go on in subsequent books to discuss ἀρχαί that are not στοιχεῖα. *Physics* I 1 would not be explicitly drawing this distinction, and the end of I 9 (192b2–3) speaks as if we have now discussed all the ἀρχαί, but the opening of *Physics* I 1, in mentioning both ἀρχαί and στοιχεῖα, would be deliberately allowing the distinction to be drawn should it be needed at some later stage in the argument. And certainly on any interpretation the

[16] Simplicius *In Physica* 10,8–24 and 11,16–23. Simplicius adds, following Proclus, that matter and form are not properly causes but συναίτια (11,29–32)—it seems bizarrely un-Aristotelian to say that form is not a true cause, but Simplicius, like Proclus, distinguishes the 'formal cause', an immanent form, from the 'paradigmatic cause', a separate form, and thinks that only the latter leads up to the first ἀρχαί. (For Proclus on αἴτια and συναίτια see notably his *In Timaeum* I,2,1–3,23 and *In Parmenidem* 888,15–35; followed also by Philoponus, *In Physica* 5,7–16, a clear statement.) When Aristotle says that to know a thing scientifically we must know τὰ αἴτια τὰ πρῶτα καὶ τὰς ἀρχὰς τὰς πρώτας καὶ μέχρι τῶν στοιχείων, Simplicius suggests that this means not that we must know the first causes and principles all the way *up* to the στοιχεῖα, but that we must know *both* the *first* or highest principles, which are the first causes, and all the way *down* to the last or lowest principles, the στοιχεῖα, which are not properly causes at all (11,32–12,5); both Averroes and Thomas follow this fairly bizarre suggestion. (I take it, instead, that both καί's are epexegetic, paraphrasable as: 'to know the first causes that explain a given thing, we must trace it back to the first beginnings, I mean all the way back to the στοιχεῖα'. Aristotle is critical elsewhere of describing the first ἀρχαί of a thing as its στοιχεῖα, with the possible implication that they are its constituents, but he is not being critical of that description here.)

transition to the first unmoved mover as the first ἀρχή, which Aristotle makes in *Physics* VIII, will depend on recognizing non-constituent ἀρχαί.

Both Aristotle's narrower account of ἀρχαί in *Physics* I and his broader account in *Physics* I–IV are setting themselves up in competition with the accounts περὶ ἀρχῶν of earlier physicists (so first with the division of views on the ἀρχαί, I 2, 184b15–25). These philosophers, like Aristotle, had started by trying to determine about the ἀρχαί before giving their overall account of nature; and Aristotle cites their views in developing his own both in I and in III–IV (and less systematically in II on nature and chance). At the same time, it is not obvious that they mean the same thing by ἀρχή that Aristotle does. Setting aside the question whether they used ἀρχή as a technical term, what Aristotle describes as their ἀρχαί are the things that they posited, at the beginning of their accounts, as existing 'in the beginning', before the ordered cosmos came to be out of them. So they are 'beginnings' of a cosmogonic narrative. Aristotle gives no such narrative, both because he thinks it would be non-scientific and because he thinks the world is eternal; as I have put it elsewhere, he is 'denarrativizing' physics. Still, in the *De Caelo* and *Generation and Corruption* and *Meteorology* and into the biology, he is following roughly the sequence that would have been followed in these narratives, as they would recount the formation of the cosmos, the rotation of the heavenly vortex, the separation of light from heavy bodies, then astronomical and meteorological phenomena and the formation of particular things including plant and animal species within this world-system. But the *Physics*, unlike these more specialized treatises, does not follow even a quasi-narrative order. Further, the *Physics* not only denarrativizes, but also decosmologizes. As Simplicius puts it, where Plato and some earlier philosophers 'raised the questions that are treated here as if they were questions about the cosmos and its parts, and did so in writings on the cosmos… Aristotle… teaches, as if there were no cosmos, about natural body itself in its own right' (*In Physica* 7,27–34), before going on to consider the cosmos in the *De Caelo*. So *Physics* I will deal with ἀρχαί of natural things, not in the sense of things temporally prior to the cosmos which contains all natural things, but with ἀρχαί that are 'first' in some more abstract (causal?) sense; and, rather than numerically *single* things presupposed by one great original process of coming-to-be, they will be universal *types* of things presupposed by the many small individual comings-to-be.

Despite these differences, Aristotle and earlier physicists share a methodological problem in their accounts of the ἀρχαί: how can we establish these ἀρχαί, if there are no prior ἀρχαί to derive them from? Narrative physicists, when they do not simply lay down their ἀρχαί without argument, argue that the phenomena of the world, under some broad description, require this particular kind of ἀρχαί, or that, if we posited different ἀρχαί, the world would not arise. Thus Lucretius argues that the fact of motion requires the existence of void, the fact that things do not irreversibly decay requires incorruptible and therefore indivisible ἀρχαί, and so on; the *Timaeus* argues that because the world is sensible it must have come-to-be and so must have a maker, that because the world is beautiful its maker must be good and must use an eternal model, and that because earth, water, air, and fire can be transformed into each other they must share a common neutral substratum. So we cannot just immediately 'determine first concerning the ἀρχαί', by stipulation or direct observation, and then go on to discuss the things that arise from the ἀρχαί. Rather, as Aristotle says, 'we must *try* (πειρατέον) to determine first concerning the ἀρχαί' and in this trying we must proceed methodically, not immediately leaping to answer the question about the ἀρχαί, but beginning by describing the phenomena and making inferences about what ἀρχαί they require. So there will be two phases of argument: 'it should not escape us that there is a difference between arguments from the ἀρχαί and arguments to the ἀρχαί. Plato too rightly used to pose this question and to ask whether the path (sc. that we are on at any given moment) is from the ἀρχαί or to the ἀρχαί, as in a race-course from the judges to the goal[-post] or back' (*NE* I 4, 1095a30–b1). In *Physics* I, the argument to Aristotle's favoured ἀρχαί will start from the fact that things come-to-be, and Aristotle will develop his positive view in discussion with the views of the Presocratics and Plato, not only on what the ἀρχαί are and on how coming-to-be happens, but on how we can use the fact of coming-to-be to make inferences to the ἀρχαί.[17]

Aristotle starts by saying:

> The path goes naturally (πέφυκε . . . ἡ ὁδός)[18] from the things that are more knowable and clearer to us to the things that are clearer and more knowable by nature: for it is not the same things that are knowable to us and [knowable] without qualification (ἁπλῶς). For this reason we must make progress (προάγειν)[19] in this way from the things that are less clear by nature but clearer to us to the things that are clearer and more knowable by nature. At first it is rather confused things that are evident and clear to us,[20] but afterwards, [starting] from these, the στοιχεῖα and ἀρχαί become knowable [to us when we] divide these (sc. confused) things.[21] (184a16–23)

[17] E.g. can we infer from the fact of coming-to-be that there is a single eternally persisting ὑποκείμενον? That the contraries between which a change takes place must already exist before the change? That there must be a non-being from which coming-to-be begins?

[18] In πέφυκε . . . ἡ ὁδός, I suppose that ὁδός is effectively the verbal noun of ἰέναι, so the sense is 'we naturally go', 'the direction we naturally go in is . . . '.

[19] On Aristotle on προάγειν, see Menn 2012, 220, n38, with primary and secondary references.

[20] This English version keeps a syntactic ambiguity about the role of μᾶλλον: (i) the confused things, rather (or more) than the unconfused ones, are evident and clear to us; or (ii) 'rather confused' (or more confused) things are evident and clear to us. In support of the latter construal, Andrea Falcon points to *Historia Animalium* I 15: *Μάλιστα δ' ἔχει διωρισμένα πρὸς τοὺς κατὰ φύσιν τόπους τὰ ἄνω καὶ κάτω ἄνθρωπος τῶν ἄλλων ζῴων· τά τε γὰρ ἄνω καὶ κάτω πρὸς τὰ τοῦ παντὸς ἄνω καὶ κάτω τέτακται. Τὸν αὐτὸν τρόπον καὶ τὰ πρόσθια καὶ τὰ ὀπίσθια καὶ τὰ δεξιὰ καὶ τὰ ἀριστερὰ κατὰ φύσιν ἔχει. Τῶν δ' ἄλλων ζῴων τὰ μὲν οὐκ ἔχει, τὰ δ' ἔχει μὲν συγκεχυμένα δ' ἔχει μᾶλλον* (494a26–32).

[21] Someone at the Symposium suggested the possibility of taking διαιροῦσι as a finite verb, so that ὕστερον δ' ἐκ τούτων γίγνεται γνώριμα τὰ στοιχεῖα καὶ αἱ ἀρχαὶ διαιροῦσι ταῦτα would mean 'starting

This text sounds very familiar to us, and probably it sounded familiar already to many of its first hearers and readers—it stands in very close relations to things Aristotle says in several other places. These relations include both similarities and contrasts: I will try to bring out the main points without too much repetition.

In speaking here of a 'path' (ὁδός), which must in context be a path to the ἀρχαί, Aristotle stays close to *NE* I 4, 1095a30–b1. He also stays close to that passage, and to several others, in distinguishing two senses of 'knowable' (γνώριμον). These passages contrast with other passages where Aristotle says, *without* distinguishing, that in arguing or defining we should start from the more knowable in order to prove or clarify the less knowable, or that the ἀρχαί or στοιχεῖα must be more knowable than what arises from the ἀρχαί or is composed of the στοιχεῖα, since we can know the στοιχεῖα without knowing the composites but cannot know the composites without knowing their στοιχεῖα. Thus he argues in the *Protrepticus*:

> Prior things are always more knowable than posterior things...And prior things are more *causes* than posterior things: for when [the former] are removed (ἀναιρούμενα), the things that have their existence (οὐσία) from them are also removed, lengths when numbers [are removed], surfaces when lengths [are removed], when the letters (στοιχεῖα) [are removed] the so-called syllables...[22] So likewise [in the knowledge] about nature: for the knowledge of the causes and στοιχεῖα must be much prior to that of posterior things. For [the latter] are not among the highest things, nor do the first things arise out of them, rather from [the former] and through them the other things manifestly come-to-be and are constituted. For whether fire or air or numbers or some other natures are causes and firsts of the other things, it is impossible to know (γιγνώσκειν) any of the others if we do not know these: for how would someone know (γνωρίζειν) either a phrase (λόγος) if he did not know syllables, or understand these if he did not know (εἰδέναι) any of the στοιχεῖα? (Part of B33 and all of B35–6, Düring)

Here, for the idea that we cannot know syllables without knowing their στοιχεῖα, Aristotle seems to be following *Theaetetus* 206a1–b11.[23] From the fact that the

from these, the στοιχεῖα become knowable, and the ἀρχαί divide these'. This is grammatically possible, but I have not been able to do anything with it.

[22] Here the manuscript (of Iamblichus' *Protrepticus*) has the obviously impossible στοιχεῖα δὲ τῶν ὀνομαζομένων συλλαβῶν. I read, with Düring's apparatus to B33 (in Düring 1961, 60), στοιχείων δ' αἱ ὀνομαζόμεναι συλλαβαί, following the suggestion of a marginal scholium in Iamblichus (but the scholium is proposing this as the underlying syntax, not as an emendation). The alternative would be to write συλλαβαὶ δὲ τῶν ὀνομαζομένων στοιχείων.

[23] Aristotle does not explicitly make the point here that we can know the στοιχεῖα without knowing the composites, but he does notably at *Rhetoric* II 24, 1401a28–31, scorning 'Euthydemus' argument, e.g. he knows that there is a trireme in the Piraeus, because he knows each of these (sc. trireme and Piraeus), and the person who knows the στοιχεῖα knows the ἔπος, for the ἔπος is the same (as its constituent στοιχεῖα)'. (Aristotle also cites 'the argument (λόγος) of Euthydemus' at *SE* 20, 177b12–13. The source may be *Euthydemus* 277a1–b1, arguing that you learn what you already know, since when you learn some exercise it is just a lot of letters and you already know all the letters. But in both the *Rhetoric* and the *SE* the 'λόγος of Euthydemus' seems to involve an argument about knowing triremes in the Piraeus, which is not in the *Euthydemus*.)

existence of one knowledge entails the existence of the other knowledge but not vice versa, we infer that the knowledge of the στοιχεῖα or of the prior things is prior to the knowledge of the composites or of the posterior things; and, if the prior things are the cause of knowability to the posterior things, they must be more knowable than the posterior things. But in the *Protrepticus* Aristotle is trying to convince a potential patron that wisdom or the knowledge of the ἀρχαί is easy, without taking any controversial stand on what the ἀρχαί are. By contrast, in the *Physics* and other texts, where he is trying to actually acquire (for himself or for his students) a knowledge of the ἀρχαί, he needs more sophisticated methodological reflection. So in *Metaphysics* A 2, while he says that 'the first things and the causes are most knowable, for the other things are known on account of these and out of these, not these things through what is subordinate to them' (982b2–4)—and so far this is very close to *Protrepticus* B35–6—he now adds that these first things and causes are also 'most difficult for humans to know' (982a24), so that what is 'most knowable' and what is easiest for *us* to know come apart.

Thus in *NE* I 4, after citing from Plato the metaphor of the path from the judges to the goal and back again, he says:

> We must begin from what is knowable, and this is [said] in two ways: for some things are knowable to us, and others [knowable] without qualification (ἁπλῶς). So presumably *we* must begin from what is knowable to *us*. For this reason, someone who is going to listen in an appropriate way [to lectures] about what is noble and just, and generally about political things, should have been brought up in good habits. For the ἀρχή (i.e. the starting point *for us*) is the ὅτι, and if this is sufficiently manifest he won't lack the διότι any further: and this sort of person possesses the ἀρχαί or will grasp them easily enough. (1095b2–8)

This passage is more methodologically sophisticated than it might at first appear. The talk of the ὅτι and the διότι shows that Aristotle is relying on something like the view developed in *Posterior Analytics* I 13 and II 1ff., that when we investigate the διότι, *why* S is P, in the hope of discovering a cause and an argument in proper demonstrative order that S is P, we have to begin from first knowing (or first grasping in *some* way) the ὅτι, *that* S is P.[24] Aristotle suggests, here in *NE* I 4 and in *Posterior Analytics* I 13, that beginning from the ὅτι, which is more knowable to us, we can infer to its cause, which is more knowable in the sense that the cause is the source of *scientific* knowability to the things which we know scientifically by deducing them from the cause. Being brought up in good habits should give someone at least broadly right beliefs about what things

[24] Presumably this is in turn Aristotle's development of the *Meno* suggestion that we can start with true δόξα *that* p and be guided by it to knowledge, which we will reach by reasoning out the cause *why* p.

are good and noble, without knowing *why* these things are good or noble, and—in the context in *NE* I, and the *Eudemian* parallel—this gives us the starting point we need for reasoning to the good-itself (for humans if not without qualification), which will be most good and will explain the derivative and lesser goodness of the things generally acknowledged to be good.[25]

Goodness also comes up, in a somewhat surprising way, in the most famous parallel to *Physics* I 1, 184a16–23, namely *Metaphysics* Z 3, 1029a33–4, b3–12:

> It is agreed that some of the sensible things are substances, so we should begin investigating with them. For it is helpful to proceed to what is more knowable. For learning happens in this way for everyone, [going] through the things that are less knowable by nature to the things that are more knowable. And this is the task—as in action, [starting] from the things which are good for each person, to make the things that are good without qualification (ὅλως) good for each person, so too from the things that are more knowable to that person, [to make] the things that are knowable by nature knowable to that person. The things that are knowable and first for each person are often scarcely knowable, and have little or nothing of being; but nonetheless from things which are poorly knowable (γνωστά rather than γνώριμα, but the meaning seems to be the same), but knowable to oneself, one must try to know the things that are knowable without qualification (ὅλως γνωστά), passing (μεταβαίνειν), as has been said, through these same things (sc. the things more knowable to oneself).

Aristotle is here announcing a search for substances beyond the sensible world, which have 'more being' than sensible substances: correct method requires that we begin with agreed-on (ὁμολογούμεναι) substances, just as correct method in looking for the highest good requires that we begin with agreed-on goods, and then look for causes of their being or goodness respectively.[26] But here he is speaking not of a *highest* good (human or transcendent) but of ordinary goods such as wealth which are 'good without qualification (ἁπλῶς)' (so the parallel *NE* V 1, 1129b1–6) in that they would benefit a virtuous and healthy human being in normal circumstances, but might be harmful to a particular person at a particular time. It is not right for us to go about amassing as much of these goods as possible simply on the ground that they are 'good without qualification', just as it is not right for a sick person to eat without restraint the things that are 'good without qualification' and would be good for a healthy person.

[25] On a good-itself which is a cause to the other things of their being good, see *NE* I 4, 1095a26–8. It may not be clear in the *NE* whether Aristotle himself accepts such a good-itself, but he clearly does in *EE* I 8 (compare 1217b2–5 with 1218b10–14), although it is not an Idea of the Good (1218b7–10), and is introduced as a (final) cause only of goods achievable for humans.

[26] On starting from agreed-on substances in investigating non-sensible substances, compare *Metaphysics* H 1 1042a22–5; on starting from agreed-on goods in trying to 'show the good-itself', compare *EE* I 8, 1218a15–24.

'People pray for and pursue these things,[27] but they should not: rather, they should pray that the things that are good without qualification should also be good for them, but they should choose the things that are good for them' (*NE* V 1, 1129b4–6). What is good for a sick person might be the kind of diet that a sick person would be able to digest, but it might also be, e.g. surgery or cautery, which is not good without qualification and would not be good for a healthy person, but is good for a sick person just because it helps to restore him to health. Likewise, what is good for a not-yet-fully-virtuous person might be simply the things he can enjoy without damage to his virtue, but perhaps also things that can help move him towards virtue, so that what is good without qualification will become good for him too.

To draw the cognitive analogy that Aristotle intends in *Metaphysics* Z 3: when he says that 'the things that are knowable and first for each person are often scarcely knowable, and have little or nothing of being', the implication is that we are initially in a disordered cognitive condition, like the initially disordered medical or moral condition in the comparison cases. We would likely harm ourselves, and in any case get no benefit, if we immediately tried to know what is most knowable by nature, as if the sick person tried to eat the healthy person's diet. We must, instead, gradually strengthen our cognitive condition until the things that are more knowable by nature become knowable to us, like the people of *Republic* VII 516a–b who must gradually strengthen their sight until they can look at the sun. Studying some of the things that are knowable to us is supposed to help us in this, perhaps by instilling intellectual discipline (patience in following out long chains of reasoning without reliance on imagination or fear of contradicting common opinions), but mainly by allowing us to infer from these things to their less immediately accessible causes. Aristotle makes a similar point in *Topics* VI 4 in the context of definition: in definition, as in demonstration, we must start from what is prior and more knowable (141a26–30), but sometimes we must start not from what is more knowable without qualification but from what is more knowable to us. Thus when we define a point as the limit of a line, a line as the limit of a plane, and the plane as the limit of a solid, we are defining the prior through the posterior by nature, and we do so because the solid is more manifest to sensation (141b3–22); 'and perhaps (ἴσως) too what is knowable without qualification is not what is knowable to everyone, but what is knowable to those whose thought is well disposed, just as what is healthy without qualification is what is healthy to those whose body is well disposed'

[27] Taking ταῦτα in 1129b4 to refer back to the things that are good without qualification in b3. Alternatively, with Barnes in the Revised Oxford Translation (Barnes 1984), we could read ταὐτά, 'people pray for and pursue *the same* things', when they should pursue one set of things in their practice, while praying for something better.

(142a9–11). What marks our cognitive condition as disordered is the fact that the things most knowable to us 'have little or nothing of being' (*Metaphysics* Z 3, 1029b9–10), or that 'as bats' eyes are to the light of day, so is our souls' νοῦς to the things that are by nature most manifest of all' (*Metaphysics* α 1, 993b9–11, clearly drawing on the *Republic*'s Sun simile). Or, if talk of degrees of reality and luminosity seems unhelpful, we can say instead that we are compelled to start by inferring from effects to causes, rather than in the scientific order from causes to effects. Coming to have scientific knowledge in some domain is not simply acquiring new truths or new causal explanations, but reordering *what* we learn through *what*, bringing our cognitive order into alignment with the order in reality, so that we will know effects through knowing their causes, which are also the causes of their scientific knowability.[28]

Aristotle is in one way or another describing such a process of reordering our cognitions and reaching a healthy cognitive condition when he says, in *Physics* I 1, that we come to scientific understanding by γνωρίζειν the ἀρχαί or causes or στοιχεῖα, which are γνώριμα by nature but not to us, and when he says, not here but in the parallel passages, that these things will become γνώριμα to us as well. However, there is an issue about how γνωρίζειν and being γνώριμον are related. I translated γνωρίζειν above as 'recognize'. Perhaps this is the basic notion, and γνώριμον is defined in terms of γνωρίζειν: for X to be γνώριμον to person S is for S to be able to γνωρίζειν X, or to be able to do so well or easily, and so, when we become (better) able to recognize the ἀρχαί, they become (more) γνώριμα to us. But it is also possible, as David Sedley suggested at the Symposium, that γνωρίζειν is defined in terms of γνώριμον: for S to γνωρίζειν X is for S to make X γνώριμον for S.[29] If so, when Aristotle says that we come to scientific understanding by γνωρίζειν the ἀρχαί, he would mean not that we master a science through *recognizing* the ἀρχαί, but through making them *familiar* to ourselves, and more familiar than their effects, i.e. through *reordering* our cognitions. 'Familiar' cannot ultimately be the translation

[28] This point has been a theme of Robin Smith's work. Smith emphasizes in particular the foundations in Aristotle's syllogistic: if we collect all the truths in a given domain, and then try to work their inferential relationships, we will discover some A- and E-predications that are ἄμεσα, 'immediate' or 'unmiddled' in the sense that there are no middle terms, distinct from the two extremes, that would allow us to validly infer the proposition from any other two true propositions. Mastering the science involves, not just working forward to discover new truths, but working backward to discover what are the unmiddled true propositions within the domain, and what inferential relations (chains of syllogisms) allow us to derive the other true propositions from them. This reflects something about the historical development of Greek geometry and allied sciences, but also turns on technical peculiarities of Aristotle's syllogistic, which lie behind the arguments of the *Posterior Analytics* against circles or infinite regresses of deductions. See Smith 1993, 2009a, and 2009b.

[29] Γνωρίζειν can in fact be a ternary relation, with an indirect object, as at *Prometheus Bound* 486–7, where for S to γνωρίζειν X to T is to make T familiar with X or able to recognize X or able to use X as a sign.

of γνώριμον (what could it mean to say that something is familiar ἁπλῶς rather than to some person?) so for something to be γνώριμον would still be for it to be knowable or recognizable, but that knowing or recognizing would be called not γνωρίζειν but γιγνώσκειν, and γνωρίζειν would be a derivative term, 'making something γιγνωσκ-able'. But *Physics* I gives us no more evidence, and I have not been persuaded that this distinction between γιγνώσκειν and γνωρίζειν will work consistently in Aristotle.[30]

The real trouble, however, emerges when we juxtapose *Physics* I 1 with the other parallels, and especially *Posterior Analytics* I 2. Aristotle starts there from a characterization of scientific knowing (ἐπίστασθαι) as knowing a thing through its causes, as in *Physics* I 1, 184a10–14, and then infers a characterization of when a deduction is scientific or demonstrative: to be a demonstration, a deduction must be 'from things that are true and primary and immediate [or unmiddled] and more knowable (γνωριμώτερα) and prior and causes of the conclusion' (*Post. An.* I 2, 71b21–2). He now glosses 'more knowable and prior and causes of the conclusion':

> They must be causes and more knowable and prior: causes, because we know [a thing] scientifically whenever we know its cause (τότε ἐπιστάμεθα ὅταν τὴν αἰτίαν εἰδῶμεν, cp. *Physics* I 1, 184a12–14 τότε γὰρ οἰόμεθα γιγνώσκειν ἕκαστον, ὅταν τὰ αἴτια γνωρίσωμεν τὰ πρῶτα καὶ τὰς ἀρχὰς τὰς πρώτας καὶ μέχρι τῶν στοιχείων); and prior, if they are indeed causes; and known before, not only in the sense of acquaintance (ξυνιέναι, sc. with each of the terms) but also through knowing (εἰδέναι) that they are [the case]. But 'prior' and 'more knowable' are [said] in two ways: for 'prior by nature' and 'prior for us' are not the same, nor 'more knowable' and 'more knowable to us'. I call prior and more knowable to us the things that are closer to sensation, prior and more knowable without qualification the things that are more remote [from sensation]. The things that are most universal are most remote, the particulars closest; and these are opposite to each other. (71b29–72a5)

This is very close to *Physics* I 1, except that *Posterior Analytics* I 2 says that the more universal things are more knowable by nature and less knowable to us, whereas *Physics* I 1 says that the more universal things are more knowable to us and less knowable by nature. And *Metaphysics* A 2, which has many echoes with *Posterior Analytics* I 2, seems to side with the *Posterior Analytics* against the *Physics* here: the *Metaphysics* chapter says that the desirable characteristic of 'knowing (ἐπίστασθαι) all things' will belong 'to the person who most of all has universal

[30] Thus Bonitz in the *Index Aristotelicus* (Bonitz 1870) under γνωρίζειν, notes that 'alicubi γνωρίζειν videri potest idem significare ac γνώριμον ποιεῖν, δηλοῦν', but concludes that the distinction cannot be maintained, that 'omnino et significatione et usus ambitu γνωρίζειν et γινώσκειν vix possunt inter se distingui, ac pro synonymis vel in eodem sententiae contextu vel in iisdem formulis γνωρίζειν, γιγνώσκειν, γνῶσιν λαμβάνειν, μανθάνειν, εἰδέναι, ἐπίστασθαι leguntur'. After trying to impose the distinction on *Posterior Analytics* I 1, I 2, and II 19, and *Metaphysics* A 1, I am inclined to agree with Bonitz.

knowledge (ἐπιστήμη)...for this person knows, in a way, all the subject things (i.e. if we know that every B is A, we will, in a way, know of each of the B things that it is A); but the things that are most universal will also be just about (σχεδόν) the most difficult for human beings to know, for they are the most remote from the senses' (982a21–5). *Metaphysics* A 2 also connects universal knowledge with knowledge of causes, and the *most* universal knowledge with knowledge of the *first* causes, presumably on the ground that the highest causes are causes of the most things, so that in knowing them we will know the most universal effects.[31] *Physics* I 1, however, says that in looking for the ἀρχαί and causes and στοιχεῖα we *start* from the universals which are more knowable to us and proceed to the particulars which are more knowable by nature. This is the notorious problem of *Physics* I 1, to which I will turn, with the remaining section of the chapter (184a23–b14), after a brief comment on the last sentence of the section already quoted (184a16–23).[32]

To repeat that last sentence: 'at first it is rather confused things which are evident and clear to us, but afterwards, starting from these, the στοιχεῖα and ἀρχαί become knowable [to us when we] divide these (sc. confused) things' (184a21–3). Up to a point, this is what we would expect Aristotle to say. Initially

[31] Throughout this passage Aristotle is trying to show that 'from all the things which have been mentioned [i.e. all the distinguishing characteristics of wisdom and the wise person] the desired name [sc. 'wisdom'] will belong to the same science, for it must contemplate the first ἀρχαί and causes' (982b7–10); the argument that this science will be most universal must be intended to support this conclusion. When Aristotle says 'the first things and the causes are most knowable, for the other things are known on account of these and out of these, not these things through the ὑποκείμενα' (982b2–4, cited above), ὑποκείμενον seems to mean what is 'subject' to a cause, but at 982a23, where the person with universal knowledge 'in some way knows all the ὑποκείμενα', it seems to mean the subject of a universal. Aristotle seems not to perceive a tension, but that doesn't mean that he's confusing the universal–particular relation with the cause–effect relation: rather, if e.g. the sun is the cause of growth, then whatever is the subject of the universal 'growing thing' will be subject to the sun as cause, and the only way to have scientific knowledge of the ὑποκείμενα of the predicate will be to know what causes the ὑποκείμενα to have that predicate.

[32] Simplicius handles the problem by distinguishing two kinds of universals, one more knowable by nature and less knowable to us, which is a cause to the individuals, the other easily known to us by abstraction and posterior *in re* to the individuals. (On this see Menn 2010, and, on the neo-Platonic discussions lying behind Simplicius, Helmig 2012.) Philoponus prefers to speak of the more-knowable-to-us universal as a vague particular, e.g. the thing approaching which is recognized as an animal before it is recognized as a human, and recognized as a human before it is recognized as Socrates (*In Physica* 10,28–11,23, with defence through 13,14). Thomas resolves the conflict between *Physics* I 1 and *Posterior Analytics* I 2 by saying that the *Analytics* means by 'singularia' (καθ' ἕκαστα) sensible individuals, which are more knowable to us than universals because sensation is prior for us to intellection, whereas *Physics* I 1 means by καθ' ἕκαστα the *infimae species*, and by 'universals' the higher universals, so that in *Physics* I 1 he would be drawing a distinction within the realm of intelligible things rather than a distinction between intelligibles and sensibles. For Thomas, all the things that are called less-knowable-in-themselves involve some sort of potentiality, and are therefore of themselves only potentially intelligible; for this reason, they need the activity of the agent intellect to make them actually intelligible.

it is not the ἀρχαί which are clear to us, but other things, which are not clear by nature; starting by examining these things, we get clearer about the ἀρχαί. What may be more surprising about this sentence is the suggestion that we get clearer about the ἀρχαί by *dividing* the initially given things (rather than, say, by inferring from these things to their causes). And this is connected with another peculiarity, Aristotle's referring to the initially given things as 'confused' (συγκεχυμένα). This does not seem to be simply a synonym for 'unclear' (ἀσαφῆ). Rather, things are confused if they are unclear because they are mixed with each other; and, if this is the source of the difficulty, we will get clearer about them if we separate or divide them from each other. Things might be mixed with each other in a straightforwardly bodily way, like the ingredients of Anaxagoras' primordial mixture, where 'all things being together, none of them was evident, on account of their smallness: for everything contained both air and aether, both infinite' (B1). But Plato also says in *Republic* VII that sensation initially perceives contrary attributes mixed together in the same thing, and calls on reason to distinguish them: 'sight too sees the large and the small, not separated but as something confused (οὐ κεχωρισμένον ἀλλὰ συγκεχυμένον τι) ... In order to get clear about this (διὰ τὴν τούτου σαφήνειαν), thought (νόησις) was compelled to see large and small not confused (συγκεχυμένα) but delimited from each other (διωρισμένα), contrariwise to sight' (524c3–8). Without agreeing with either Anaxagoras or Plato about what the ἀρχαί are, or about how they are related to the things given to sensation, Aristotle has a general methodological agreement with them. He also agrees with them, more specifically, that the fact that sensible things *come to be* is a sign that allows us to recognize the compresence of contraries in them (although for Aristotle the compresence is just something's being F actually and not-F potentially), and thus allows us to infer that sensible things are not themselves primitives but arise from ἀρχαί, and allows us to say something about what these ἀρχαί are like.

The Anaxagorean and Platonic model of things initially given to us mixed together, and needing to be separated, also helps to explain why Aristotle speaks here not just of ἀρχαί but of στοιχεῖα *and* ἀρχαί: as we have seen, a στοιχεῖον is an ἀρχή thought of as a *constituent* in a thing alongside other στοιχεῖα, and we recognize the στοιχεῖα by dividing the syllable into them.[33] Aristotle presumably does not mean that the ἀρχαί *must* be found as ingredients of the initially given things, but he takes such cases as models for how we go about searching for the ἀρχαί. Something else that we can take from Plato and Anaxagoras: we do not *just* start with a confused cognition of the mixture and come to a clear cognition of its elements. Rather, we also start with a confused cognition *of the elements themselves* (so, in *Republic* VII, of large and small and similar pairs of

[33] On these themes in Plato see Menn 1998.

contraries), as they are given to us in the mixture (and are initially less knowable to us than the mixture), and come to a clearer cognition of these same elements. Inspired by this model—or so I will try to suggest in the next section—Aristotle wants to show in *Physics* I how we can gradually come to a clearer knowledge of the ἀρχαί of natural things.

Aristotle says:

> For this reason, we must proceed from universals (ἐκ τῶν καθόλου) to particulars (ἐπὶ τὰ καθ' ἕκαστα):[34] for the whole is more knowable to sensation, and the universal is a kind of whole (ὅλον τι), for the universal encompasses (περιλαμβάνει) many things as parts. And the same thing, in a way, also happens with names (or nouns, ὀνόματα) in relation towards their definition (λόγος): for [a name] like 'circle' signifies something as a whole and indeterminately (ὅλον τι καὶ ἀδιορίστως σημαίνει; or, reading ὅλον τι καὶ ἀδιόριστον σημαίνει, 'signifies something whole and indeterminate'), whereas its definition (ὁρισμός) divides into particulars. And little children at first call (προσαγορεύει; in some manuscripts ὑπολαμβάνει, so 'think' rather than 'call') all men fathers and all women mothers, and afterwards they distinguish (διορίζει) each of these (viz. men from fathers and women from mothers).[35] (184a23–b14)

[34] Or, in codex E (and Simplicius *In Physica* 208,29–32), εἰς τὰ καθ' ἕκαστα.

[35] ὑπολαμβάνει is in codex I, in the Arabic (which has 'the child at first supposes of each man it sees that he is its father, and of each woman that she is its mother'), and in codex E *ante correctionem*. προσαγορεύει is in the other manuscripts that Ross reports for *Physics* I (F, J, E *post correctionem*) and is presupposed by Simplicius *In Physica* 17,16 (καλοῦσι) and Themistius *In Physica* 2,18; but Philoponus *In Physica* 11,21 (νομίζει, and context 11,19–23) seems to presuppose ὑπολαμβάνει. A passage of Calcidius *In Timaeum* 208–9 (in Waszink 1962, 225–6), citing Aristotle, *may* be referring to *Physics* I 1, 184b12–14, but, because it cites things that are not in our passage (notably, Aristotle calling adults who think that doing injustice is beneficial *senes pueros*), it is often taken to be reporting an exoteric work which Aristotle would presumably also be drawing on in *Physics* I 1. (Ross in his *Aristotelis Fragmenta Selecta* (Ross 1955) prints some of the text as *Protrepticus* Fr. 17.) Either way, Calcidius has *Aristoteles . . . dicens initio pueros etiamnum lactentis aetatis omnes quidem viros putare patres, omnes autem feminas matres*, apparently reflecting ὑπολαμβάνει—and these children are probably too young to be *calling* anyone anything. (But Calcidius may well not be looking directly at any work of Aristotle's, but rather taking this from Porphyry's commentary on the *Physics*, or from a parallel in Porphyry's commentary on the *Timaeus*. Porphyry may also be the source of Philoponus' paraphrase with νομίζει.) There was lively discussion at the Symposium about προσαγορεύει or ὑπολαμβάνει, about the age of the children that Aristotle is thinking of, and about whether the children are addressing all women as 'mama' (or thinking that the women are *their* mother) or are saying or thinking that the women are *someone*'s mother (not having the linguistic or conceptual resources to distinguish 'a woman' from 'a mother'). Some participants urged that no child, or at any rate no child old enough to talk, could confuse other women with its own mother, and therefore that, at least if προσαγορεύει is right, Aristotle must mean merely that they say or think of each woman that she is *a* mother, *someone*'s mother. But in fact, if Aristotle wrote προσαγορεύει, he could be describing the phenomenon of 'overextension', much discussed in the recent child-language literature since Clark 1973. This literature gives many examples of children who start by calling all women or all women who help take care of them (and sometimes even their fathers) 'mama', and then gradually narrow down until this means only their mother. 'Overextension' seems typical of children aged 1 or 2 years; children this age can certainly distinguish their mother from other women, but may not yet have a vocabulary with both a broader and a narrower term. (A child may learn to make distinctions in various ways, e.g. by saying 'mama' with an accent on the penult for its mother and on the ultima for its father.) There is also a phenomenon, discussed by Piaget 1928, 74–107 (esp. 106–7), where a child

This passage picks up on the sentence we have just discussed, where the things initially given to us are wholes and must be divided for us to understand clearly the things we are most interested in understanding, the *ἀρχαί* and *στοιχεῖα* of the initially given things. He did not mean 'dividing' (or the wholes which must be given for us to divide) literally, or at any rate he did not mean wholes and dividing in the most straightforward sense. But, as he now clarifies, we must start from universals, and the universal is a kind of whole. That the universal is a kind of whole emerges also from the *Metaphysics* Δ chapter on wholes, Δ 26. There the sense of 'whole' as 'what encompasses (*περιέχει*) the things encompassed in such a way that they are one' applies in one way to wholes that encompass things 'in such a way that the one is [composed] out of them' (the more usual sense of 'whole'), but also in another way to wholes that encompass things 'in such a way that each of them is one thing', i.e. so that some one thing is predicated of all of them (1023b27–9). For, as Aristotle explains, 'the universal, and what is said universally (*ὅλως*) as being some whole (*ὅλον τι ὄν*), is universal as encompassing many things through being predicated of each, and all of them are one in the sense that each of them [is that one thing], e.g. human and horse and god [are some one thing] because they are all animals' (1023b29–32). And if the universal is a kind of whole, it is reasonable to speak of dividing it.

The case of people proceeding from the name to the definition of 'circle', and the case of the babies learning to distinguish their parents from the rest of adult humanity and restrain their enthusiasm in saying 'mama!' and 'papa!', are meant to be analogies from ordinary experience for what we (adults who know the definition of 'circle' as well as we know our parents) must do in searching for the *ἀρχαί*, or in searching for a clearer knowledge of the *ἀρχαί* than we have to begin with. It is striking how often Aristotle uses forms of *διορίζω* or *ὁρίζω*, three times in this passage. We start by perceiving something analogous to a whole or a syllable *ἀδιορίστως*, without articulations, and then come to a clearer knowledge by doing something analogous to discerning the individual letters or phonemes within the syllable, as we learn what the first syllable of the

may have difficulty in applying relative terms, which Piaget diagnoses as a difficulty grasping how things would seem from someone else's point of view: a child may say that all boys are 'brothers' (of someone), perhaps also that no adults are brothers. But this happens at a much later stage (Piaget's informants here are typically 7–9 years old), probably not what Aristotle's diminutive *παιδία*, 'little children', would pick out: 1-to-2-year-olds are not using relative terms at all. Aristotle in *Generation of Animals* I 2 says that 'we call an animal male if it generates in another, female if it generates in itself; which is why in the universe too, they think of (*νομίζουσιν*—perhaps "worship") the nature of the earth as a female and mother, but address (*προσαγορεύουσιν*) the heaven and the sun or anything else like this as generators and fathers' (716a13–17). This might just mean that heaven and earth are popularly represented as the father and mother of the other gods, or of vegetation, but it looks as if the worshippers are addressing heaven as 'Father', i.e. conceiving of it as having generated themselves as well as other things. There are, of course, many examples of this.

spoken name 'Theaetetus' is by breaking it down into the phonemes θ and ε (so that we can spell the syllable by writing each of these down separately), distinguishing θ from τ and so on. The ὁρισμός of 'circle' is thought of as adding this kind of articulation to our originally imprecise cognition of a circle, whether the ὁρισμός is a genus–species definition spelling the species-form out into its genera and differentiae, or whether it spells out the circle into parts in a more straightforward sense, the circumference and the centre and perhaps the radii, as in Euclid's definition of a circle as 'a plane figure surrounded by a single [curved] line, to which all the straight lines falling from a single point inside the figure are equal to each other' (I def. 15).[36] (Without intending to build anything on it, I note that Aristotle is to all appearances taking the example of circle from Plato, *Seventh Letter* 342–3, where we are seeking 'what is knowable (γνωστόν) and truly being', for which Plato's example is circle. There we start from the name (ὄνομα) and proceed to the λόγος—Plato gives it as 'what has an equal distance from the extremes to the middle in every direction'—and also consider the 'image', a concrete more or less circular object drawn in the sand or produced on a lathe. All these things are ἀσαφῆ and unstable—including the λόγος, because it is made out of nouns and verbs (or predicates, ῥήματα)—and so they impel us towards something further which will be knowable in the strongest sense.) When the child διορίζει, it is extensionally distinguishing its mother from other women whom it had incautiously addressed as 'mama!'; perhaps it is also distinguishing 'mama' from 'woman', or distinguishing, within the intension of 'mama', between the generic constituents 'human, adult, female' and the further differentiae of its mother.

Serious issues remain about what methodological lesson Aristotle is trying to support by speaking about wholes and parts, universals and particulars, names and definitions, and the example of the children. One fairly basic divergence is between two possible readings of the passage, which for lack of better terms I'll call Reading A and Reading B. According to Reading A, all of Aristotle's expressions and analogies are trying to get at the same thing, namely: we start with a crude cognition of things which are causally derived from some ἀρχαί, and which we cannot know scientifically until we come to a cognition of those ἀρχαί. These initially given things can be described at least metaphorically as if they were wholes and the ἀρχαί were their constituent στοιχεῖα; and what we do in coming to a cognition of the ἀρχαί can be described at least metaphorically as an articulation or demarcation, a διαιρεῖν or διορίζειν, of these constituents from each other within the whole. According to Reading B, by contrast, we do not simply start from a crude cognition of things which are derived from ἀρχαί.

[36] With Heiberg deleting the MSS's phrase 'which is called the circumference' after 'single [curved] line' and the MSS's phrase 'towards the circumference of the circle' after 'falling'.

Rather, we start from a cognitive state in which things that are derived from ἀρχαί are more knowable to us than their ἀρχαί, but we start not simply from a cognition of the derived things, but also from an obscure cognition of the ἀρχαί themselves. (So where Reading A takes ἐκ τούτων at 184a22–3—'[starting] from these (ἐκ τούτων), the στοιχεῖα and ἀρχαί become knowable'—to mean 'from the confused things', Reading B takes it to mean 'from the confused things' being better known to us', 'from our being in a cognitive condition where the confused things are better known to us'.) As we have seen, our passage speaks repeatedly of [δι]ορίζειν, where we start by perceiving something ἀδιορίστως and then come to a clearer knowledge. But already in the first sentence of the chapter Aristotle had said that we must first try to διορίσασθαι τὰ περὶ τὰς ἀρχάς (184a15–16), and this seems to imply that the ἀρχαί themselves, not just the things which arise from them, are given indeterminately and must be further determined. Because the derived things can be described at least metaphorically as if they were wholes composed of the ἀρχαί, the ἀρχαί can be described at least metaphorically as being present within the derived things, and thus as being cognized obscurely and collectively when we cognize the whole of which they are parts. When we initially know the ἀρχαί, we know them only under some vague and thus universal description—we might, for instance, start by knowing them only under the description ἀρχαί, or ἀρχαί *of natural things*—and we gradually come to know them under more precise descriptions, and these more precise descriptions will apply only to them, rather than applying universally both to them and to other things. Or we might come to know some one ἀρχή under a more precise description which will apply only to that ἀρχή rather than applying universally to all the ἀρχαί, and this might be described as articulating or demarcating the ἀρχαί from each other.

While I prefer Reading B, both readings have to face some difficulties. The main difficulty with Reading A is that while it can more or less explain why Aristotle would describe the things initially given as *wholes*, it seems unable to explain why he would describe them as *universals*. And the description of the things initially given as universals does not seem to be just one more metaphorical description (or description applying literally to some cases and metaphorically extended to others), like describing these things as wholes or comparing them to names (or comparing our initial cognition of things to knowing them merely by name). Rather, Aristotle's talk about wholes seems to be building up to a point about universals, and it is a major drawback of any interpretation of the chapter if it cannot explain this. Aristotle says at 184a23–6, 'For this reason (διὸ), we must proceed from universals to particulars: for the whole is more knowable to sensation, and the universal is a kind of whole, for the universal encompasses many things as parts'. The διὸ points back to the previous sentence, 184a21–3, where wholes (or 'confused things') are initially better known to us, and we make progress by dividing them into their στοιχεῖα. 'For the whole is more knowable to sensation' picks up this premise, and 'the universal is a kind of whole' adds a new premise, which Aristotle combines to yield the new conclusion 'we must proceed from universals to particulars'. 'For the whole is more knowable to sensation' recapitulates the argument of 184a21–3 that confused things are more knowable to us in our initial condition ('more knowable to sensation' need not mean that in all cases we have sense-perception of the whole, but that it is more knowable to us than other things while we are in the condition in which sense-perception is our standard of clear knowledge), but this is part of a chain of arguments whose conclusion is that we must proceed from universals to particulars.[37] (Or, more cautiously, that universals are *one* kind of whole, and so are *among* the things from which we must proceed.) Aristotle could very easily have given us an argument that we must proceed from wholes to parts, and stopped there. Instead he embeds this in an argument that we must proceed from universals to particulars. Notoriously there is at least apparent conflict between what Aristotle says here about the order of proceeding and what he says in *Posterior Analytics* I 2. If 'universals' here were just an alternative way of saying 'wholes', then why bother, why not just stop with the argument about wholes? Reading A, it seems to me, will have great difficulty in explaining this. Reading B, by contrast, can say that in coming to know the ἀρχαί we will proceed from universals, that is, from knowledge of the ἀρχαί under vague and universal descriptions, to knowledge of the ἀρχαί under more precise descriptions, which apply only to the ἀρχαί and not to other things, or which apply only to one ἀρχή and not to others. Here the analogy with other kinds of wholes may be helpful. We can start with a general apprehension of a wagon, which is a whole encompassing things 'in such a way that the one is [composed] out of them' (*Metaphysics* Δ 26, 1023b28–9), and then distinguish its hundred constituents from each other and from the whole; or we can start with a general apprehension of animal, or with a vague apprehension of elephant as an animal, where animal is a whole encompassing things 'in such a way that each of them is [some] one thing [namely animal]' (1023b28), and then distinguish elephants from other animals until we reach a more precise description applying only to elephants. But, although there is an analogy with other kinds of wholes, Aristotle is using it here to draw a conclusion specifically about *universal* wholes.

[37] Alternatively, if we take the 'confused things' at 184a21–2 more broadly and not as already implying wholes, Aristotle is taking up the old premise that we must begin from the things that are better known to us by sensation, and adding two new premises, that wholes are better known to sensation and that universals are wholes. The conclusion is the same.

Alexander endorses or at least entertains Reading B, saying that 'it is possible that by universals here he also means the axioms' such as 'that the ἀρχή must be either one or more [than one]', itself derived from the principle of excluded middle, and then 'that the ἀρχαί are several, and neither one nor infinitely many, then that there must be both a contrariety in them and something that underlies the contrariety', all of which Aristotle deduces before he says what the ἀρχαί themselves are (all quotes from Simplicius *In Physica* 19,21–33). Thomas also endorses Reading B when he says that Aristotle in this chapter shows the order in which we must proceed in physics, and proves two things, namely 'that we must begin from a study of the principles' and 'that among the principles we should begin from more universal principles' (or, presumably, from principles under more universal descriptions). By contrast, Simplicius and Averroes endorse Reading A, and Simplicius raises a difficulty against Reading B: Simplicius asks, if Alexander were right, 'how will this procedure send us up from composites to the ἀρχαί', and complains that the universals Alexander is talking about are 'observations *about* the ἀρχαί' rather than '*composed of* the ἀρχαί of natural things' (Simplicius 19,28–9 and 19,33–20,2).[38] One reply, on behalf of Reading B, might be that these universals are composed of the ἀρχαί in the broad sense that they are universally predicated of the ἀρχαί and, as Aristotle says here, the universal is ὅλον τι. Another reply might be that, when we know composites confusedly, we also know their ἀρχαί confusedly, and that we must proceed both from composites to ἀρχαί and from ἀρχαί known confusedly to ἀρχαί known distinctly. But, admittedly, Aristotle does not always make clear when he is talking about universals and when about other kinds of wholes, or when he is talking about knowing composites (or other derivative things) confusedly and when he is talking about knowing the ἀρχαί themselves confusedly.[39]

A further difficulty for Reading A would be in connecting what Aristotle says here about method with what he actually goes on to do in *Physics* I 2–9. From Reading A, we would expect that Aristotle would be interested in resolving

[38] Averroes, likewise, speaks repeatedly of going from composites to simples, and uses 'confused' and 'composite' as interchangeable.

[39] Perhaps to put it another way: when he speaks of things that are συγκεχυμένα, are these the mixtures (which are mixed and thus difficult to know clearly) or their ingredients (which are mixed with each other and thus difficult to know clearly)? Ross says that 'what Aristotle says of the actual method of physics is that it is one of analysis of the confused data of experience into their elements (a21–3), and at the same time one of coming to recognize the specific nature of that whose generic nature alone is at first recognized (a23–6); these two features not being clearly distinguished from each other' (Ross 1936, 458). I would prefer to say that the former is proceeding to the ἀρχαί (from 'confused' or compounded objects—Ross's 'confused data of experience' may be misleading, Aristotle isn't analysing *experience* as Kant or James Mill will do), while the latter is going from a vague to a more precise conception of the ἀρχαί.

wholes given in sense-experience into their constituents, and it is true in a broad sense that *Physics* I 2–9 does this. But we would not expect that Aristotle would start, immediately after our passage, with the *a priori* universal 'the ἀρχή must be either one or more than one' (I 2, 184b15), would spend several chapters discussing and criticizing the views of earlier philosophers on the ἀρχαί of all things, and would then discuss the contrariety and the number of these ἀρχαί, all in a generally 'top-down' rather than 'bottom-up' way, before giving his own analysis of coming-to-be in I 7. And we might have expected that, when he does start resolving wholes given in sense-experience into their constituents, he would begin from some richer empirical description, beyond the bare fact that natural things come-to-be (and that, say, humans come-to-be musical). As far as I can see, on Reading A, *Physics* I 1 will be a general description of the procedure of research in natural science, or of that phase of research in which we proceed from posterior things to their ἀρχαί, but it will not describe what Aristotle actually does in investigating the ἀρχαί in I 2–9. I 1 would have the look of a free-floating προοίμιον ready to be tacked on to any scientific work. (Perhaps a defender of Reading A should reply that *Physics* I 1 is not supposed to describe *Physics* I 2–9: rather, both I 1 and I 2–9 are supposed to describe the method of natural science, I 1 in a shorter and more general way, I 2–9 focusing on the explanation of coming-to-be, and conceptually clarifying the kinds of ἀρχαί that any explanation of coming-to-be will use. But is I 2–9 really about method?)

For Reading B, by contrast, it is entirely natural that Aristotle should begin with some extremely general descriptions of the ἀρχαί—they are either one or many; they are not one and unchanging; they are not one at all; but they are also not very many; they are or include contraries—and then proceed to give a more precise listing of how many and what the ἀρχαί are, and to distinguish matter from privation against those (especially Plato) who have confused them. The point is not that Aristotle, abandoning the method of beginning empirically, with what is most knowable to sensation, is instead seeking the ἀρχαί through dialectic, beginning with universal truths or endoxic plausible-truths, such as might be reflected in earlier philosophers' opinions about the ἀρχαί as well as in ordinary idioms for talking about coming-to-be. On the contrary, Aristotle is adhering to the method of beginning with the things that are most knowable to sensation, or rather most knowable to us while sense-perception is our standard of clear knowledge. But these will be unarticulated general observations, such as that things come-to-be out of their contraries (and therefore that the ἀρχαί are contraries), which are in need of dialectical scrutiny and articulation before they can pass as scientific first principles.

In favour of Reading B, and of an interpretation of 'universal' as meaning 'universal' and not just 'whole' or 'sensible composite', Aristotle later explicitly picks up on the idea that we should proceed from universals to particulars, and that this is a process that is happening within the argument of the *Physics*. *Physics* I 7 starts by saying that we should first speak 'about all coming-to-be: for it is according to nature after first speaking of what is common (τὰ κοινά) to then consider what is proper to each thing (τὰ περὶ ἕκαστον ἴδια θεωρεῖν)' (189b30–2). *Physics* III 1 says that, having undertaken to speak about nature (which is an ἀρχή of motion), we should now investigate what motion is, and also continuity, infinity, place, void, and time, because these are (or have a *prima facie* claim to be) necessary conditions of motion; and Aristotle adds that there is also another reason for investigating these things first, namely 'because they are common to all things and universal... for the consideration of particulars (ἡ περὶ τῶν ἰδίων θεωρία) is posterior to the [consideration] of common things' (200b21–5). And this seems to fit at least roughly the way that Aristotle actually proceeds in the physical treatises, with the *Physics* discussing things common to all natural beings, the *De Caelo* specializing to the simple bodies and the local motions natural to each of them, the *On Generation and Corruption* specializing to general and corruptible (thus sublunar) bodies and their generation (especially the inexhaustibility of their generation), and the *De Anima* and biological works specializing to living things and their souls and their characteristic actions and passions. (We might compare the argument in *Parts of Animals* I 1 that we should first explain attributes common to an entire genus before descending to explain attributes proper to each of its species, so that we will not have to repeat the same explanation many times over: e.g. we will explain breathing in all blooded land animals, rather than explaining it once for dogs, once for horses, and once for elephants. Thomas in the proemium to his *Physics* commentary apparently links the sequence of the physical treatises, the thesis of *Physics* I 1 that we should proceed from universals to particulars, and the argument from *Parts of Animals* I 1.) And this passage from more universal to more particular subject-genera (e.g. from all natural beings to corruptible ones) or from attributes that apply more universally to attributes that apply only to particular genera, also in at least some cases involves a passage from more universal ἀρχαί to more particular ἀρχαί, or from ἀρχαί under more universal descriptions to ἀρχαί under more particular descriptions. (As we will see, the formulation requires some care.) Thus in *Physics* I we talk about matter in general, but in *On Generation and Corruption*, in investigating the material ἀρχή of generation and especially of the inexhaustibility of generation, we reach the matter which is common to earth, water, air, and fire, and thus ultimately to all generable and corruptible things, but not common to them and the heavenly bodies. Likewise, in *Physics* I we talk about form in general, but in the *De Anima* we talk specifically about the soul, which is the formal (and efficient and final) ἀρχή of living things. Also, as we have seen, when *Physics* III–IV investigate motion, infinity, place, void, and time, the thought is not simply that these are *attributes* common to all natural things, but that they are, or have some claim to be, ἀρχαί in common of all natural things, existing prior to each particular natural thing.

So, at a first formulation, we can say that *Physics* I–IV, or I–VIII, discuss attributes common to all natural things, and the common ἀρχαί of all natural things as causes of those universal attributes, while subsequent physical treatises discuss attributes proper to some subclass of natural things, and more particular ἀρχαί as causes of those attributes. This formulation will need revising on several grounds. First, even within *Physics* I–IV there is a movement from more universal to more particular attributes and ἀρχαί. It is a striking oddity of *Physics* I that while its first sentence presupposes that we are pursuing a science of nature, and says that we should look for ἀρχαί of natural things, the concept of nature virtually drops out of the argument for the rest of the book. Aristotle does use the *word* φύσις, in marking out the domain of study and past thinkers who have studied it, in speaking of things that are 'more knowable by nature', and in redundant uses ('the nature of X' periphrastic for 'X', 'some nature' for 'something'), but the only *premise* he uses about nature is that 'some or all natural things are in motion' (I 2, 185a13), and thus that they come-to-be, or at least come-to-be *something*. Aristotle argues to the ἀρχαί from the fact of change, so they will be ἀρχαί of changeable things, but there is nothing here to distinguish natural from artificial things; it is only once he defines natural things in II 1 that he can recognize his ἀρχαί specifically as ἀρχαί of natural things, e.g. as the nature-as-form of natural things. (The ἀρχαί of natural things will be ἀρχαί of all changeable things, because all non-natural changeable things are posterior to natural things, but knowing them as ἀρχαί of changeable things is not the same as knowing them as ἀρχαί of natural things, and forms of artefacts will be included in the broader dragnet of book I but excluded from the narrower investigation of the ἀρχαί of natural things in book II.) So there is narrowing already from book I to book II, and Aristotle has not really delivered on the promise of an investigation of the ἀρχαί of natural things until book II. *Physics* I is complete and self-contained in that it gives as much of an account of the στοιχεῖα or constituent ἀρχαί of natural or changeable things as can be given without an account of what distinguishes natural things from other changeable things—that is, without an account of the ἀρχὴ κινήσεως. But because it clearly announces an investigation of the *ἀρχαί of natural things*, which cannot be completed without this investigation of nature, book I could never have been intended to exist on its own without book II.[40]

[40] But, as I suggested in a note above, it is entirely possible that *Physics* II initially existed on its own, and that *Physics* I was written later as a proemium to *Physics* II, or to a larger whole *Physics* I–IV.

As we have seen, I 7, taking up I 1's talk of proceeding from universals to particulars, says that we should speak first about what is common to all coming-to-be. What are the advantages of proceeding in this way? I would suggest the following: Aristotle frames his general account of coming-to-be as a response to an aporia against the possibility of any coming-to-be. But the aporia is much more acute against *substantial* coming-to-be, and most of the philosophers who rejected substantial change thought that apparent substantial changes were real qualitative changes or real combination and segregation (thus a local motion of the parts). None of these philosophers would have felt surprised or enlightened to learn that a thing-which-is-F can come-to-be when a persisting thing (or several persisting things) which had previously been not-F becomes (or become) F—hot air comes-to-be when air becomes hot, or a word comes-to-be when its letters become arranged in a certain way—although they might not have thought of putting the point in such abstract and general terms. But they would say that this does not help on the possibility of *substantial* coming-to-be, of γένεσις ἁπλῶς, since when S comes-to-be F the underlying thing is already there, and does not come-to-be but merely acquires a new attribute: γένεσις ἁπλῶς, they think, would have to be γένεσις ἐκ μὴ ὄντος, and this, they think, is impossible. Aristotle thinks that, once we have formulated the explanation of the possibility of coming-to-be in general terms, we can then specialize to the case of substantial coming-to-be. In substantial coming-to-be, as in other coming-to-be, there is a persisting subject, but this subject is not τόδε τι, and we should not be overly hasty in infering from the familiar cases of accidental change that *this* thing, τόδε, is always what persists through the change and never the result of the change.[41] The analogy with the familiar cases is supposed to lead us to recognize that, as in accidental change, so too in substantial change, there is something that preexists and persists through the change. But in accidental change the persisting thing is the οὐσία which 'this' refers to when we point to the result of the change, while in substantial change 'this' refers to the new thing that has come-to-be: and so what persists is not a τόδε, not what the thing that has come-to-be *is*.[42]

[41] The immediate target of the criticism is presumably the *Timaeus* account of the Receptacle, where if things change into each other, like fire and air or like triangles and squares shaped out of gold, then what perishes and what comes-to-be can at best be called τοιοῦτον, while 'that in which they each appear when they come to be, and out of which again they perish [namely the receptacle]—only this [should be] referred to using the words τοῦτο and τόδε' (49e7–50a2): so that, when we point to sensible fire or to a golden triangle, the gold or the Receptacle is what it *is*, while fiery or triangular is only what it is *like*. But Aristotle would probably think that Plato is just being unusually explicit about what most people implicitly assume.

[42] See Aristotle's criticism of the *Timaeus*' gold analogy at GC II 1, 329a13–21; and see *Metaphysics* Θ 7, where Aristotle insists (implicitly against the *Timaeus*) that when something comes-to-be by a substantial change out of matter X, it should be called 'X-en' (saying what it is *like*) rather than 'X' (saying what it *is*).

As Aristotle puts it in I 7, 'The underlying nature (i.e. the underlying whatever-it-may-be) is knowable by analogy: for as bronze is to a statue or wood to a bed, or as the material (ὕλη)[43] and formless, before taking on the form, is to any of the other things that possess a form, so *it* (sc. what underlies and persists through substantial change) is to οὐσία and τόδε τι and being' (191a7–12). And that sounds like as definitive a statement as we could hope for.[44] But a few lines later, after summarizing the results of I 5–7, Aristotle adds, 'but whether the form or the ὑποκείμενον is οὐσία is not yet clear, but it is clear that the ἀρχαί are three and how they are three, and what is their manner (i.e. how they are causes of the things that come-to-be)' (191a19–21). So Aristotle is deferring the issue, and with it the possibility of substantial coming-to-be. The 'not yet' clearly refers ahead to *Physics* II 1, where Aristotle argues that 'the φύσις and οὐσία of the things that are by nature' (193a9–10) is not (or not primarily) their matter but their form, which determines what a thing's motion is naturally directed towards: so that if (say) water is heated until it becomes air and thus acquires a new natural disposition towards upward motion, it has undergone a genuine substantial change. And the reason that in I 7 Aristotle defers the question whether the form or the ὑποκείμενον is οὐσία is that the answer is different in the case of artefacts (Antiphon is perfectly correct that the artisan has not added any new οὐσία to the bed) and in the case of natural things. And so Aristotle cannot answer the question until II 1, when he defines natural things and distinguishes them from artefacts, and so can make an argument about what their φύσις and οὐσία is.

So: the *Physics* I account of the ἀρχαί is 'universal' or 'general' in that it describes the ἀρχαί of coming-to-be in general, before *Physics* II narrows to ἀρχαί of natural things, and thus of substantial coming-to-be. We might also say with Alexander that *Physics* I itself proceeds from the more universal to the more particular, since Aristotle 'will show first that the ἀρχαί are several, and neither one nor infinitely many, then that there must be both a contrariety in them and something that underlies the contrariety. Then, making a transition from these [assertions], which are common, he will show what they (the ἀρχαί) are: for the person who knows these [more universal assertions] does not yet know

[43] I see no good reason for Ross's deletion (following Diels) of the words ἡ ὕλη καί, with Simplicius' quasi-quotation at *In Physica* 226,6–9. Ὕλη here means the raw material of craft-production or its analogue. This is only the fifth occurrence of ὕλη in the *Physics*; Aristotle does not start routinely using it as a name for one kind of ἀρχή until I 9. Ross is wrong to expect that ὕλη will be reserved for that use here.

[44] So too in the next sentence, 191a12–13, where 'this [sc. what plays the role of matter] is one ἀρχή, without being one, or being, in the way that τόδε τι is'. The syntax is compressed and the sense may be controversial, but I take the thought to be that this 'one' ἀρχή we've discovered ('one' in the sense that it and the form and the privation are three), is not like (say) the Receptacle of the *Timaeus*, a single τόδε underlying all natural things, nor indeed is it a τόδε in any case, whether a single τόδε for all or different τάδε for different natural things.

what the ἀρχαί are' (Alexander quoted by Simplicius 19,21–5). In narrowing down to what the ἀρχαί are, Aristotle starts with the most general question from the diairetic doxographical tradition, how many are the ἀρχαί (whether one or many, if many whether finitely or infinitely many, if finitely many how many exactly).[45] Indeed, he starts from the even more general question *whether* there are ἀρχαί, since he argues that if there were a single unchanging being there would be not one but zero ἀρχαί, since there would be nothing for it to be an ἀρχή *of* (185a3–5). In asking whether there are ἀρχαί and how many and what they are, Aristotle is broadly following the method of *Posterior Analytics* II, asking first εἰ ἔστι and then τί ἐστι. But we cannot proceed with the ἀρχαί as with an eclipse, first asking whether it is and then discovering what it is by investigating the causes of its existence, since the ἀρχαί have no further causes. Instead, by asking whether *coming-to-be* is and then what are the causes of *its* existence, we bring to light not only what coming-to-be is but also what its ἀρχαί are.[46] We can pursue these questions by investigating the aporiai against coming-to-be and seeing what kinds of ἀρχαί are involved in solving them, starting with the solutions given explicitly or implicitly in earlier philosophers' accounts of coming-to-be. This investigation will lead to the results, as Alexander says, 'that there must be both a contrariety in them and something that underlies the contrariety'. This is already more particular than saying that the ἀρχαί are more than one (or that they are just a few). And when Aristotle then distinguishes the material ἀρχή from the privation, and says that the matter as such is not contrary to form, he is further 'particularizing' as *Physics* I 1 recommends,

[45] The pre-Aristotelian doxography seems to have formulated the question as how many *beings* there are for each philosopher, but Aristotle argues that the question how many ἀρχαί there are amounts to the same (*Physics* I 2, 184b22–5). On the tradition of doxography classifying the philosophers by how many beings they posited, see Mansfeld 1990.

[46] Aristotle in the *Physics* is very much aware of the *Posterior Analytics* II methodology, proceeding from εἰ ἔστι to τί ἐστι, even where he deviates from it. (Thus at II 1, 193a3–4 he excuses himself from trying to show *that* nature exists, as the methodology seems to dictate, since this is too obvious to need showing (contrast chance, II 4, 195b35–6). At III 4, 202b35–6 he says that we must investigate about the infinite 'whether it is or not, and, if it is, what it is'; the same formulation about place (with an added 'and *how* it is') at the beginning of IV 1, about void at the beginning of IV 6, and a verbal variant about time at the beginning of IV 10. Even *Physics* VIII can be described this way: at *De Motu Animalium* 1, 698a10–11 he refers back to an investigation 'about eternal motion, whether it is or is not, and if it is, what it is'.) *Physics* II 1 gives a close parallel with the investigation of the ἀρχαί: rather than trying to give a definition that would specify a cause *of* nature, we investigate *natural things*, grasp what they have in common that differentiates them from non-natural things such as artefacts, and so bring to light what nature is by grasping it as a certain kind of cause (an internal principle of motion) present in natural things. The investigation of chance and spontaneity in II 4–6, and perhaps some of the investigations in III–IV, are similar; and grasping what eternal motion is, in VIII, also brings its cause to light, in the final chapter VIII 10.

distinguishing two things that were previously confused. ('Previously confused' not just in the sense that we had confused them until we draw the relevant distinctions in the course of the argument, but also in the sense that *Plato* had confused them, when he separates the material ἀρχή from all participation in the form and assumes that the resulting privative attributes are essential to the material ἀρχή.)[47]

But there is an important sense in which *Physics* I never says what the ἀρχαί are. To say that an ἀρχή is matter or form is not to say what thing this ἀρχή is, or even what *kind* of thing it is, but only to say what kind of *cause* it is, thus what *relation* it has to its effect. This is what Aristotle means in *Metaphysics* Λ 4–5 by saying that the matter and form of things in different genera are not numerically the same thing, or even the same in kind, but only the same *by analogy* (Λ 5, 1071a24–34, Λ 4, 1070a31–3). (Most obviously, things in different categories have no shared matter or form; but even within the category of substance, there is no common matter of sublunar and celestial things, Λ 2, 1069b24–6. And even within a single species the matters, forms, and movers of different individuals are numerically different, 'for the individual is the ἀρχή of the individuals: for human is (the ἀρχή) of human universally, but there is no [universal human who is the ἀρχή of humans universally], but rather Peleus of Achilles and your father of you', Λ 5, 1071a20–2.)[48] So it would be more accurate to say, not that *Physics* I discusses universal ἀρχαί, matter and form and privation, but that it discusses ἀρχαί under universal descriptions.

This is an important difference between Aristotle and his Platonist rivals. In *Physics* I 5 Aristotle says that his predecessors, constrained by the truth itself, have all posited pairs of contrary στοιχεῖα:

> But they differ from each other in that some of them take (i.e. assume or posit as ἀρχαί) prior things and others posterior things, and some of them [take] things that are more knowable by λόγος and others things that are more knowable by sensation (for some posit hot and cold, others moist and dry, others odd and even or strife and love as causes of coming-to-be, and these differ from each other in the aforesaid manner). So in one way they are saying the same things, in another way they are saying different

[47] This is not only what Aristotle says Plato does in *Physics* I 9 but also what Plato actually does, notably in the third Hypothesis of the *Parmenides*, where we discover the original nature of the others by stripping away all attributes which they acquire by participation in the one (including unity and finite multiplicity), concluding that 'there comes to be, in the things other than the one, from the combination of the one with themselves, something else that gives them a limit (πέρας) in relation to each other, whereas their own nature gives them unlimitedness (ἀπειρία) in themselves (καθ' ἑαυτά)' (158d3–6). With the Receptacle too, Plato separates it from everything it receives from the forms or from the demiurge, and assumes that what it would be when so separated is what it is by its essence.

[48] On the interpretation of this passage and its context see my discussion in Menn forthcoming (section IIIβ1 in the draft available on the internet) and in Menn 2011, 173 n. 13 and the text above it.

> things—different just as most people think, but the same by analogy... So in this way they speak both alike and differently, and better and worse, some [positing] things that are more knowable by λόγος, as we have said before, and others things that are more knowable by sensation (for the universal is knowable by λόγος, the particular (τὸ καθ' ἕκαστον) by sensation, for the λόγος is of the universal and sensation of the individual (τοῦ κατὰ μέρος)), e.g. the great and the small [are more knowable] by λόγος, the rare and the dense by sensation. (188b30–189a9)

This is in an obvious sense praising Plato (and some more or less kindred philosophers) over most Presocratic physicists: to identify all formal ἀρχαί with heat would be a childish mistake like calling all men 'papa'.[49] Aristotle's language here is reminiscent of the passage cited above from *Posterior Analytics* I 2, where 'the things that are closer to sensation', namely 'the particulars', are 'prior and more knowable to us', while 'the things that are more remote [from sensation]', namely 'the things that are most universal', are 'prior and more knowable without qualification' (72a1–5). But here in *Physics* I 5, without any explicit criticisms of the Platonists, Aristotle is implicitly distancing himself from them too, and reducing them to the same level as their opponents; while he says that the Platonists' more universal ἀρχαί are 'more knowable by λόγος' (and less knowable by sensation), he strikingly fails to say that they are 'more knowable by nature' or 'more knowable without qualification'. A more universal description of the ἀρχή is better than a too-restrictive particular description like 'the hot', but we should not think of it as a single universal ἀρχή. (And if, in trying to reduce all ἀρχαί filling the same analogical role to a single universal ἀρχή, we label all privative ἀρχαί as 'the great and the small' or 'the unequal', that may not be much better than labelling them all 'the cold'.) There may indeed be a single universal ἀρχή, i.e. an ἀρχή that is a cause of some attribute universally predicated of a very broad domain, while more particular ἀρχαί are causes of the distinctive attributes of its subdomains; but, if so, the ἀρχή-universally-described is *not* that single universal ἀρχή. We reach the first material ἀρχή of all sublunar bodies (but not of the heavens) by reasoning back from the manifest natural things first to their anhomoemerous parts, then to their homoeomerous parts, then to earth, water, air, and fire, then to what underlies the intertransformations of these four simple bodies—not by a single leap to 'matter' as what is common to all cases of material causality. Likewise, we reach the first *efficient* ἀρχή, not by grasping the universal concept of efficient cause, but by reasoning back step by step to the first efficient cause of all coming-to-be (the sun, or

[49] 'We too, as long as we discern things by following what is coarse and confused in sensory cognitions, are in no way unlike young children who call all the men they meet fathers' (Simplicius *In Physica* 17,18–21).

the mover of its yearly motion), or to the first efficient cause of all motion (the mover of the daily motion of the whole heaven).

Physics I 1 tells us that the ἀρχή-universally-described is the *starting point* of the path, more knowable to us, and not, as Plato thinks and as *Posterior Analytics* I 2 might suggest, the ending point, most knowable in itself.[50] The contrast between Aristotle's and his Platonist rivals' modes of argument is nicely illustrated by the text on the ἀρχαί which Sextus Empiricus (*AM* X 248–83, esp. 262–76) attributes to 'the Pythagoreans', but which he is clearly taking from an Old Academic or neo-Pythagorean (Old Academic-influenced) source. These 'Pythagoreans' start by arguing that everything falls into one of three classes, being said either *per se*, by contrariety, or correlatively. Then they argue that anything in each of these classes, if it is not the *summum genus* in that class, must be posterior to the *summum genus* in that class, because the species is destroyed when the genus is destroyed, but not vice versa. But these *summa genera*, which are prior to all other things, are either themselves the one (the *summum genus* of things said *per se*) or the dyad (excess and deficiency, the *summa genera* of correlatives), or else they are posterior to the one and the dyad (like the equal and the unequal, which are the *summa genera* of contraries). And so, the 'Pythagoreans' conclude, everything else will be posterior to the one and the dyad, and so the one and the dyad are the ἀρχαί of all things. These 'Pythagoreans', then, think that by taking the ἀρχαί under the most universal descriptions, as *summa genera*, they will reach the ἀρχαί that are the first of all things universally. And they think that they can formulate the οὐσία of these things, not merely their causal role in relation to other things but what they are in themselves, by calling them 'the one' and 'the dyad'. This is what Aristotle means by calling all men 'papa' and all women 'mama', and he is trying to educate us away from it in the *Physics*, and in all the physical treatises. Already in *Physics* I he has distinguished matter from privation, and distinguished the ἀρχαί of substantial change (and thus of those

50 Compare *De Anima* II, where we must *start* by 'trying to determine what soul is and what is its most universal λόγος' (II 1, 412a4–6); '*if we must* say something universal to every soul, it would be the first actuality of an instrumental natural body' (II 1, 412b4–6, cf. b10–11). But this is only a start: 'let it thus have been determined in outline and sketched about soul; but since from what is unclear (ἀσαφές) but more obvious (φανερώτερον) there arises what is clearer and more knowable according to λόγος' (II 1–2, 413a9–12), we need to go back again to say what soul is. The first definition is unclear partly because it does not state the cause (II 2, 413a13–20), but partly because it is so universal, like a universal definition applying to all polygonal figures (II 3, 414b20–5). To prove a geometrical proposition about all polygons (e.g. Euclid II 14, to square a given rectilinear figure) we need to prove it about triangles first and then about each successive kind of polygon, and to understand souls, or the living things of which they are ἀρχαί, we need to understand the different kinds of souls belonging to different kinds of living things. There is no ἀρχή 'soul in general': soul is like animal, where 'the universal animal is either nothing or posterior' (I 1, 402b5–9). ('Universal' is κοινόν in most of these passages, but καθόλου at I 1, 402b7 and I 2, 412b10, with no apparent difference in meaning.)

substances which come-to-be) from the ἀρχαί of non-substantial change, even if he is not yet in a position to give a criterion for when a change is substantial. And subsequent books, in the *Physics* and beyond, will have much more to say in making the ἀρχαί more precise, drawing further distinctions among them (for instance, distinguishing matter from place in *Physics* IV), critically examining inadequate accounts of the ἀρχαί or of things that claim to be ἀρχαί, and finally bringing us to the first causes of motion and coming-to-be.

2

Looking for a Starting Point—The Eleatic Paradox Put to Good Use

Physics I 2

MICHEL CRUBELLIER

After the general epistemological reflections in chapter 1, the opening lines of chapter 2 seem to mark the beginning of the inquiry of book I. The rest of the book shows an indisputable continuity; indeed, it consists mainly in one long argument in which an accurate and consistent model of natural change is built up step by step.

I have divided the chapter into five sections, which correspond to the paragraphs in Ross's edition except that in the last paragraph I take 185b5–25 and 185b25–186a3 to be two distinct units (that is, I do not see a particularly close relation between the mention of 'the more recent of the Ancients' and the part of the discussion preceding it immediately, about the many senses of 'one'). Actually, chapter 2, as it stands (i.e. as it has been articulated by Renaissance editors), is not a well-individuated unity. Its first section is linked to the subsequent chapters at least until I 5; the second is an introduction to the discussion of the

I thank the participants for all their objections and suggestions, especially Tim Clarke, who also read several versions of this chapter and greatly helped me to improve it; Marwan Rashed, with whom I could pursue the discussion about the doublet lines (see Appendix II), and Gabriela Rossi, who sent me written comments and called my attention to the important issues that emerge from the parallel with chapter 11 of the *SE*. Many thanks also to Henry Mendell, who sent me useful comments and information; to Zoé McConaughey, who revised and corrected the grammar and style, and to Claire Louguet who read the penultimate version.

immobilist monistic ('Eleatic') position, i.e. to chapters 2 and 3 taken together. Only the last three sections form a distinct unity (which nevertheless is closely related to chapter 3).

I will spend much more time on the first two sections than on the rest, on account of their strategical importance.

1. 184b15–25

1.1. *Where to start and where to head in order to get to the starting point(s)*

Chapter 2 begins by listing, in the form of a tree, all the possible options regarding the number of the principles, combined with some indications (the aim of which is not explicitly declared) about their properties (changing/unchanging) or the way they could differ from one another ('different in shape or species', 'contraries'):

> [184b15] Necessarily, the principle will be one or many, and if it is one, it will be either unchanging, as Parmenides and Melissus say, or changing, as the natural philosophers hold, some of them saying that air is the first principle, and others saying that it is water. And if there are many principles, they will be either limited or unlimited [in number]; if they are limited but more than one they will be two or three or four or some other number; if they are unlimited, either they will be of one and the same kind, as Democritus says, but different in shape or species, or they might also be contraries.[1]

[1] Some commentators felt the need to introduce a correction into the sentence about Democritus, at 184b20–2. All the witnesses agree (with two minor variations) on the following text:

> ἢ οὕτως ὥσπερ Δημόκριτος τὸ γένος ἕν, σχήματι δὲ ἢ εἴδει διαφερούσας ἢ καὶ ἐναντίας (δὲ is missing in mss. F and J and in Philoponus, and καὶ is missing in ms. I).

Ross prints it with an addition suggested by Torstrik:

> ἢ οὕτως ὥσπερ Δημόκριτος τὸ γένος ἕν, σχήματι δὲ [διαφερούσας], ἢ εἴδει διαφερούσας, ἢ καὶ ἐναντίας.

The effect of this addition (emphasized by Ross's punctuation) is to cut off ἢ εἴδει διαφερούσας ἢ καὶ ἐναντίας from the comment about Democritus. Actually, there is no particular difficulty in taking σχήματι ἢ εἴδει διαφερούσας (with or without the δέ) as one phrase, which could be attributed to Democritus (for instance), since the differences of shape between the atoms could be interpreted as differences in species. The notion that some types of atoms could be contrary to some other seems more difficult to fit in with what we know about Democritus; but actually (a) we will see Aristotle trying such a move at the beginning of chapter 5: 'All of them make their principles contraries: 'And Democritus [says that they are] the full and the empty, one of which he says is being and the other not-being; *he also [says that atoms differ] by their position, shape and order, and these are generic terms for contraries, [namely]* positions: up and down, before and behind, shape: with or without angles, straight and curved' (188a19–26). If one nevertheless (b) thinks that this is definitely too harsh or prefers to read ἢ καὶ ἐναντίας as a hint at Anaxagoras, that does not require changing the first part of the sentence; the ἢ καὶ (which should be conserved) is probably enough to indicate that the mention of contrariety refers to another philosophical position. I have tried to translate in a way that could be compatible with both readings (a) and (b).

Since this tree looks quite simple, and since it is well known that the final outcome of book I is that three principles are necessary (and to some extent sufficient) to account for physical change, it is all too natural to think that these lines indicate the question that will be addressed in book I and the steps through which the inquiry will proceed.

If this is so, it is far from being explicit. In fact, it is unlikely that the question of the nature of the principles and of the specific system that they form could be extracted from these lines by an innocent reader. Actually, the readers of book I will have to wait until the end of chapter 7, 191a20–2, to find a precise statement of the question the whole book deals with: 'how many principles there are, and what kind of principles they are'[2]—chapter 7, that is, the point at which an answer has first been reached. It is not that Aristotle is being careless; rather it is a consequence of the fact that in book I we are looking for principles. That has to be done according to some definite procedure, without which one could not claim to have carried out the task in a satisfactory, let alone demonstrative way; but no definite rules can be stipulated in advance, for that would amount to stating from the outset what the sought-after principles are. Thus it is plainly true, strange as it may seem, that one cannot get a clear understanding of the question itself before having reached and understood a sufficiently precise and complete answer.[3]

So we must start from a natural or informal notion of what a principle looks like or has to be, and try to refine it into a well-defined concept. This is the kind of situation that has been just described in chapter 1: we have to proceed from the awareness of something well known, but complex and unclear (*συγκεχυμένον*, cf. 184a22), to the precise knowledge of what makes it what it is. It is one thing to perceive and recognize a shape as 'round', it is another to be able to say what property makes it 'round', i.e. the equal distance its points have to a centre. For the sake of completeness, it should be added that since the aim of the book is to determine the first principles of natural philosophy, the search will have to lean on assumptions as to what 'nature' and 'natural processes' are. There is nothing on that topic in chapter 1, but such premises will be incorporated piecemeal in the course of the argument, as we shall see soon. The main task of a reader,

[2] Aristotle only said in chapter 1 (184a15–16) that we must 'try to settle first the matter of principles of the science of nature' (*τῆς περὶ φύσεως ἐπιστήμης πειρατέον διορίσασθαι πρῶτον τὰ περὶ τὰς ἀρχάς*).

[3] 'A question that you cannot properly understand until you have found the answer' is the definition of an enigma. Thus the question 'how to find the first principles' is inescapably enigmatic, that is to say, it cannot be solved by means of a straightforward given method. This specific difficulty of finding a convenient starting point for searching out the principles occurs again at two other places in chapter 2: at 184b25–185a20, where Aristotle ponders the advisability of discussing the basic thesis of immobilist monism even though it amounts to rejecting the notion of nature, and at 185a20–21, where he mentions the 'most suitable of all [possible] starting points' for the critical examination of this thesis (I will come back to these two points later on).

all along book I, is to check which properties of the principles are introduced at each step and to integrate them into a definite concept.

In chapter 1 Aristotle has given a few indications as to what a principle should look like and what purpose it has to fulfil. He described an ἀρχή (184a10–14) as what gives us real scientific knowledge of something, that is, in the terms of the *Posterior Analytics*, the principles let one know *why* things are like that, and that they *could not be otherwise*. Along the same lines, ἀρχή was associated to the notions of 'first cause' and 'element', a collection of three terms that recurs in many places of the *Metaphysics*, especially in passages that are considered parallel to *Physics* I, namely book A and the fourth chapter of book Λ. But it may be worth noticing that in *Physics* I Aristotle does not consider the different kinds of causes, as he does in *Metaphysics* A—maybe just because this distinction has to be set out and explained in book II—which gives to *Physics* I a quite different character, although it seems to deal with the same sort of topics as *Metaphysics* A and uses much the same doxographical material. In fact, it seems to me that the notion of a principle prevailing in chapter 1 is an implicit or, so to say, a grammatical one, which is expressed by the phrase ἐκ + genitive, i.e. the general notion of a starting point. To this we may add that since we are looking for first principles, there must be nothing more primary on which they could be said to depend, and they have to lead to an improved knowledge and understanding of that of which they are principles. By what means, then, and along what lines, does Aristotle intend to elaborate this general notion of a principle?

Let us take from chapter 1 the example of the circle (184b11–12): children very soon get spontaneously acquainted with the specific shape of 'round' objects (and from then on they are able to recognize and name this shape), but it will take more time before they are able to define it, that is, to distinguish what makes it round, i.e. this particular point, the centre, and its equal distance to any point of the circumference. Neither the centre nor the radius is given in the original experience of a 'round' shape: the principles are at a level of their own, distinct from that of everyday experience. Yet this does not mean that a divinatory insight is always needed to grasp the law according to which the circular form is generated. One may have observed a wheel, a sling, or a lathe at work and discover that a circular form is generated by the rotation of a line around one of its extremities. It seems to me that the doxographical material contained in book I is brought in for a similar purpose, i.e. in order to adduce some factual data that might help us form an adequate concept of a principle.

If the tree at lines 184b15–22 was meant to set the scene for the subsequent analysis, Aristotle would examine one after another, separately, the different positions he has listed there: he would first expound and discuss the monistic theories, then those who posit two, then three, four, etc. or infinitely many principles (possibly in a different order). But what he really does in the end is noticeably different.

He does begin with the Eleatic thesis, on which, as he himself says, he spends much more time than would be expected since he denounces it from the very beginning as alien to natural philosophy (and to which he will return once again in chapter 8).

Then, in chapter 4, he moves on to 'mobilist monists'. But there he introduces the pluralist theories, especially those of Empedocles and Anaxagoras, as if they were possible developments of a mobilist monistic position (187a12–23): some natural philosophers give their unique principle the capacity to turn into different kinds of stuff with different properties, while others claim that these different properties are actually present from the beginning within a unique entity and then become manifest through a process of separation.

After that, the choice between finite and infinite pluralism is settled first by referring to a sort of principle of parsimony: it is better (*βέλτιον*, I 4, 188a17–18 and I 6, 189a14–16) to have a limited and even a small number of principles, if they can explain as much as you would have explained by means of an infinite amount (I 6, 189a16–17); in fact this explanation through an infinite number of principles would just be a delusion, since it would give no real scientific knowledge (I 4, 187b7–13; I 6, 189a13). Another argument for a finite number of principles seems to be extracted from the fact that the principles (or at least some of them) must be contraries, if this is the meaning of the sentence at 189a13–14: *μία ἐναντίωσις ἐν παντὶ γένει ἐνί, ἡ δ' οὐσία ἕν τι γένος* ('there is only one contrariety in each genus and substance is one genus'). This (rather obscure) sentence might point to the possibility of reducing all the physical pairs of contraries to one fundamental contrariety; or Aristotle may have thought that the possibility of matching one pair of contraries with each genus rules out the possibility of an infinite number of primary pairs of contraries. Be that as it may, it appears that in the last part of the book Aristotle elaborates his thesis that there are three principles precisely on the basis of the assumption that some of them must be contraries.

It is thus clear that Aristotle's intention is not primarily to discuss and assess the theories of his predecessors. Although he mentions some of them by name at lines 184b15–22 and in the following chapters, the list he makes here is not a list of topics to be studied; rather it must be taken as the list of all possible answers to his own question about the principles of natural philosophy. I say 'the list of all possible answers' because it is meant to be exhaustive (cf. *ἀνάγκη δ' ἤτοι* etc., 184a15); but it is not complete since, as they stand, these answers are themselves incomplete: just saying that there are two or three principles does not solve the whole question. This is why the different positions are not discussed separately, but are rather introduced as successive steps in the course of

Aristotle's own inquiry, alongside the progressive elaboration of his own concept of a physical principle. The reader of *Physics* I has to observe how these options work, how they could, so to say, 'react' to some objections or survive a confrontation with rival theories.

It seems to me that the way Aristotle deals with the Eleatic position is indicative of this aspect of his project. While this position has been effectively refuted in chapters 2 and 3, it comes about once again in chapter 8, in which it is used as a touchstone in order to give a final confirmation of Aristotle's own theory, showing he is able to explain the rationale of that indefensible position and to defuse the most threatening point for natural philosophy.

The first section ends with a remark about 'those who try to find out how many beings there are':

> [184b22] Those who try to find out how many beings there are (*οἱ τὰ ὄντα ζητοῦντες πόσα*) inquire in the same way: for [they consider] first[4] that from which beings come about and they ask whether this is one thing or many and, if many, whether these things are limited or unlimited in number: thus they ask whether the principle or the element is one or many.

If *τὰ ὄντα* means 'the things there are' without qualification, the question seems naïve, even absurd ('Come on, let's find out how many things there are'). It makes better sense (1) if we are supposed to distinguish between things that really are and others that are just properties or appearances, and thus to distinguish between a phenomenal and an ontological level of reality; and still better (2) if one is able to set forth some criteria for counting something as a real being in this sense. Thus understood, such a remark may come from the ontological section of Plato's *Sophist* (242b–250d),[5] in which the question is first set out in the naïve form, just as 'stories they would tell to children' (242c), and later on rephrased in terms of criteria of reality (this is the contest between the 'Giants' and the Friends of Forms, 245dff.).

Even so, this remark may appear unnecessary. Aristotle says that those who inquire into the number of beings have to determine the number of the realities 'out of which' things come about (*ἐξ ὧν*, once again the vaguest possible notion of a 'principle'); which seems to mean that the two questions of the number of beings and of what the principles of natural beings are have one and the same structure and that, in some sense, this way of inquiring into the number of beings amounts to inquiring into first principles. So, why does he mention

[4] At 184b23, there is no reason to alter the received text: *ζητοῦσι πρῶτον* (or *πρῶτον ζητοῦσι*).

[5] As Mansfeld 1986, 1–59 (reprinted in Mansfeld 1990, 22–83) has shown, physical doxographies existed before Plato. But the explicit notions of 'things that really *are*' and of the search for a criterium in order to determine *which kind of things* really are most probably come from the *Sophist* and not from earlier classifications.

specifically the question of the number of beings and these (unnamed) people? Is it only for the sake of completeness? I venture two possible reasons. First, this might be an anticipation of what he will say about Anaxagoras and Empedocles in chapter 4, in which the notion of 'that out of which' will be more emphatically interpreted as a kind of material cause. Second, and most important, there is one case in which the distinction between 'the number of the principles' and 'the number of beings' may be crucial: the case of monism. In itself, the thesis that there is only one principle does not preclude the existence of many beings and of natural processes, as long as this unique principle is liable to change and develop into something else, as is the case with the natural philosophers discussed in chapter 4; but if it is not, then there will exist 'only one being' in the strictest possible sense of that phrase. This is what is at stake in the discussion of Parmenides and Melissus.

2. 184b25–185a20

2.1. *A path to principles: a slow-motion picture*

Although the general meaning of this section is clear, the progress of the argument seems curiously blurred and redundant[6] (in the following translation, I have divided it into short units, in order to make its analysis easier; the reason for grouping these units into three main parts will appear in the commentary).

> [*a*] [184b25] In fact, asking whether being is one and unchanging is not an inquiry about nature.
>
> [*b*] [185a] For just as the geometer cannot assert anything to those who reject his principles (this would rather belong to another science, or to a discipline common to all sciences), such is the case with the person[7] who inquires about the principles,

[6] The difficulty to get a clear picture of this passage is aggravated by the presence of three lines (185a9–12) in section [*e*] that are literally duplicated at the beginning of chapter 3 (186a7–10). This duplication occurred at an early date, since it is present in all our mss., in the Arabic translation and in Philoponus' commentary. I will deal with this textual problem separately (in Appendix II) because it seems to me that it has no real bearing on the argument. In fact, both passages could be read with or without these lines, without significantly affecting the general meaning; besides, the grammar and style of the period would not be made much rougher or smoother either. This situation makes the decision of ascribing these lines to one of the contexts difficult, but on the other hand it might take the sting out of the issue.

[7] τῷ περὶ ἀρχῶν, without any further specifications. There is no indication that this person should be the first philosopher, a dialectician, or even a natural philosopher inquiring into the principles of his own science.

[*c*] [and[8]] if there were just one thing in that sense, it would not be a principle any more, because a principle [has to be] the principle of some other thing or things.

[*d*] Then, to ask whether [being] is one in that way is like discussing any other claim that someone holds just for the sake of argument, for instance Heraclitus' claim (or if somebody claimed that what there is is just one human being),[9] or trying to defuse an eristic argument (*λύειν λόγον ἐριστικόν*)

[*e*]—which is precisely the case of both arguments, i.e. Melissus' and Parmenides'. For they start from false assumptions and they are not conclusive. Or rather, Melissus' argument is crude and does not show any real philosophical problem. In fact, once you have granted a first absurd proposition, the rest follows, and there is nothing difficult in that.

[*f*] But *we* will start from the assumption that things that are by nature, all or some of them, are subject to change; indeed, this is evidenced by induction.

[*g*]—However, 'resolving' (*λύειν*) is not ordinarily available (*οὐ . . . προσήκει*) for each and every argument, but only for the false conclusions that someone may have drawn from the [relevant] principles. Whenever this is not the case, it is not [available]:

[*h*] for instance, it is the geometer's business to solve the squaring [of the circle] by means of segments, but he is not concerned with Antiphon's squaring.

[*i*]—True; but[10] since the puzzles they bring forward, unnatural as they are, happen to be about nature, it might be worthwhile to make a short dialectical investigation about them; in fact, this investigation is philosophically relevant.

[*j*] The most suitable of all [possible] starting points is to ask . . . in which sense those who say that 'all [things] are one' mean it.[11]

Regardless of the complexity of its details, Aristotle's point is clearly indicated by the first [*a*] and the last [*i*] sentences of this section: does it make any sense to discuss the Eleatic position in a treatise of natural philosophy, given that this position seems to be alien to natural philosophy, and even to ruin the very notion of nature?

[8] 'And', because this *γάρ*- sentence, like the former one [*b*], is part of the explanation and justification of [*a*].

[9] The association of *ἢ εἴ τις φαίη ἄνθρωπον ἕνα τὸ ὂν εἶναι* with the mention of Heraclitus' paradoxes is somewhat strange, since the latter is a well-known example of a thesis held 'for the sake of the argument' (and mentioned as such by Aristotle in a different context, *Met.* Γ 3, 1005b25–6), while the former seems fictitious and might well come directly from line 185a24, where it does not figure as an example of an eristic thesis, but as an (assumedly absurd) interpretation of the thesis that 'all that there is is one'. I strongly suspect that it came into our text from a gloss, but since that cannot be proved, I have not excised it. It does not affect the meaning of the argument, anyway.

[10] 'True' stands for *οὐ μὴν* (which indicates the endorsement of the negative proposition contained in [*g*–*h*]).

[11] In fact, this sentence belongs to the next section (in which Aristotle discusses the monistic theories), but it brings an essential indication to complete the solution of the difficulty discussed in this section.

This doubt rests upon the general rule that you cannot carry out a refutation unless you are able to find some common ground on which you and your interlocutor agree and which can be used as a premise (or a set of premises) for the subsequent discussion. In the context of a science (which is what Aristotle's natural philosophy is supposed to be or to become), this rule is strengthened by the thesis that scientific knowledge has to proceed *ex præcognitis et præconcessis*, so that a science is characterized by a definite set of principles and cannot do anything outside this framework. This aspect is emphasized by calling on an analogy with geometry ([*b*, *h*]), which has an exact parallel in an important passage of the *Sophistical Refutations* (chapter 11), to which I will return.

One must add that the subjects under discussion are the principles themselves, which makes the situation even more problematic: how can one discuss first principles with opponents who seem to reject them altogether? [*b*, *g*–*h*.]

Yet Aristotle declares at the outset that he will nonetheless do so in this particular circumstance, because it is possible in a sense [*d*–*f*] and it will bring out some philosophically significant results [*i*].

The questions are: first, why does he feel entitled to break a rule that seems to be so crucial in his own philosophy; second, why does he takes so many little steps in formulating this objection and answering it? Is this sheer redundancy or what else?

I will now turn to the argument itself.

[*a*] states the objection: the examination of the monistic thesis does not belong to natural philosophy. It is followed by two sentences introduced by γάρ that must be put on the same level in the argument; I mean that [*c*] is not an explanation of [*b*], but should rather be taken as a kind of minor premise: after the statement of the general rule [*b*], [*c*] explains why a strict ontological monism is straightaway excluded from a discussion of principles. It is interesting to notice that Aristotle's objection is not (as one would perhaps be inclined to infer from [*f*] and from the occurrence of καὶ ἀκίνητον in [*a*], 184b26) that the Eleatic thesis is incompatible with the existence of motion or change; his claim is rather the much stronger one that this thesis is incompatible with any talk of principles whatsoever.

These lines introduce some important assumptions about the sought-after principles: first, that a principle has to be distinct from that of which it is a principle ([*c*]), a claim which immediately implies the existence of more than one being. Second, that *qua* principles of natural philosophy they must be compatible with the existence of change and changing objects (this is alluded to in [*a*] and Aristotle will make this claim explicit in [*f*]).

As it is expressed in [*b*], the analogy with geometry raises an intriguing and possibly important epistemological point, since it suggests that the principles of geometry could perhaps be established or defended by another prior mathematical science. More precisely, Aristotle offers an alternative between (1) another, presumably mathematical science bearing a specific relation to geometry, and (2) a science described as πασῶν κοινή (185a2–3), a phrase in which it is not clear whether πασῶν is meant to refer to all the sciences or only to all the mathematical sciences.

The notion that one science may be prior to another, i.e. that it be able to give better demonstrations (i.e. more universal) of its theorems, is a standard Aristotelian thesis (cf. the examples in *APo* I 13) and probably comes from Plato's *Republic*. The notion that this prior science would also be able to account for the principles of the subordinate science is perhaps never stated as distinctly as it is here, but it is not inconsistent with Aristotle's general picture of the sciences[12] (and by the way, it is most likely that Plato would have agreed with it; in fact, Aristotle's reluctance to assume that thesis without qualification is one of his points of disagreement with Plato). What is more surprising is to find it applied (even hypothetically) to geometry, since Aristotle never seems to make geometry dependent on a prior mathematical science. But there are indeed a few passages showing that he was acquainted with the Platonic concept of a 'universal mathematics' that would be able to state and prove the more general theorems and basic principles of arithmetic, geometry, and so on. In one celebrated passage (*Met.* E 1, 1026a23–7), he sketches a similar picture of the relations between mathematical sciences in order to shed light on the relation between first philosophy and a 'second philosophy' which happens to be physics:

> [1026a23] One could ask whether first philosophy is universal or specific to some genus or, so to say, to one definite nature. For there are different ways [of being 'prior'] in mathematics as well: while geometry and astronomy are concerned with a particular nature, universal mathematics is common to all (πασῶν κοινή).[13]

Now, if we take the analogy in its fullest and strictest sense, the mention of ἤτοι ἑτέρας ἐπιστήμης ἢ πασῶν κοινῆς hints at two distinct hypotheses that one may put forward as to the discipline that would be entitled to discuss immobilist monism and defend the (as yet not otherwise known) principles of natural philosophy: it must be either the science of some nature on which all natural beings depend or a science embracing every other science. These two descriptions

[12] Compare *Posterior Analytics* I 11, 77a26–35; I 13, 78b32–79a13.

[13] My interpretation of this passage is that geometry and astronomy are not given as two more or less equivalent examples of special sciences (both being opposed to the καθόλου one), but that the pair they form is an example of one science (namely, geometry) being prior to another (astronomy) while dealing with some particular nature, as opposed to the pair formed by any particular mathematical science and universal mathematics. I admit that this interpretation is heterodox. This is not a convenient place to argue for it, but I am inclined to take our passage of the *Physics* as a confirmation.

correspond to the main options about first philosophy;[14] the second one could fit dialectic as well.[15] The use of ἐπιστήμη at 185a2 and the mention of 'first philosophy' at the end of book I (192a34–b1) might be taken as an argument for first philosophy; on the other hand, the phrase 'to make a short dialectical investigation about them' (ἐπὶ μικρὸν διαλεχθῆναι περὶ αὐτῶν 185a19–20), which immediately introduces the discussion of the Eleatics, obviously refers to dialectic. It seems to me (as I will try to show) that this is the case, with the consequence that in the end, this distinction between a 'prior' and a 'universal' science will prove to have little relevance. In any case, I think it would be better to avoid loading section [*a–c*] with metaphysical implications. Aristotle's point is negative, he is just saying that the discussion that follows does not concern physics in the strict sense, but will come under another science, without further specification. Even 'science' should be taken in a loose sense here.

The most important point with regard to this first step of the argument is that it does not altogether rule out the possibility of discussing the Eleatic thesis, nor does it say that now is not the appropriate time to undertake such a discussion. It simply suggests that it would be an incursion into the field of another πραγματεία, whichever this may be.

The next move [*d–f*] is introduced by δή, which indicates precisely a step forward ('then...', or 'now...'), i.e. a new way of looking at the problem, which integrates what has been said hitherto while bringing in something new. This new element is that if we cannot manage to build a formal refutation (an ἔλεγχος) of the monist thesis, we may still recourse to another strategy, namely λύσις or 'resolution'. This type of dialectical operation is described and studied in the *SE*, in which it is clearly distinguished from refutation. While ἔλεγχος is directed against a thesis (and aims at demonstrating the contradictory of this thesis),[16] λύσις is directed against an argument. A resolution, Aristotle says, 'is the disclosure of a false deduction [by showing] what makes it false'.[17] It is thus suitable for dealing with the kind of theses that people hold 'just for the sake of argument' or with eristic claims.[18]

[14] Alexander (quoted by Simplicius, 19.20–3, 49.34–50.2) opts for first philosophy; Simplicius himself (17.4–50.2) and Philoponus (27.11) leave both alternatives open. Simplicius quotes (48.6–26) the first lines of Eudemus' *Physics*, which, he says, aptly addresses the issue, but postpones the solution to 'ulterior inquiries'.

[15] Cf. APo I 11, 77a26–35 and SE 11.

[16] It is defined as 'the deduction of the contradictory' of a given thesis (*SE* 1, 165a2; 9, 170b1–2).

[17] *SE* 24, 179b23–4; the same definition is given in Chap. 18, 176b29–30. On λύσις and its rules and uses, see Rossi 2006, 76–109, especially 91–7.

[18] It is not fully clear whether πρὸς ἄλλην θέσιν ὁποιανοῦν διαλέγεσθαι τῶν λόγου ἕνεκα λεγομένων (a5–6) and λύειν λόγον ἐριστικόν (a8) are meant to indicate two distinct cases, or are just two different descriptions of one and the same case. As Gabriela Rossi pointed out to me, holding a thesis just for the sake of the argument may occur in regular dialectical practice (cf. *Top.* I 11, 104b19–22 and VIII 5, 159b31, precisely with the examples of Heraclitus' and Melissus' theses). Nevertheless, the phrase πρὸς ἄλλην

This move implies that such honourable philosophers as Melissus and Parmenides himself be ranked among the eristics, although they certainly did not deserve the shameful epithets that Aristotle ordinarily associates with the name ἐριστικός.[19] Undoubtedly, Melissus and Parmenides aimed at finding and expressing the truth about what there is[20] and, in that sense, they should not be mixed up with typical ἐριστικοί. That is why Aristotle appears to be careful not to *call* the Eleatics themselves ἐριστικοί, but to say that *their arguments have* that eristic character (see ὅπερ... ἔχουσιν οἱ λόγοι 185a8 and ἐριστικῶς 186a6). He justifies this critical assessment by saying that 'they rest on false premises and are not conclusive' ([*e*]): these features are part of the standard dialectical definition of an eristic argument (see *Top.* I 1, 100b23–5, VIII 11, 162b3–15 and *SE* 2, 165b7–8).

In other words, his main motive in expressing these judgements is not to discredit his adversaries—if the outcome was just that these theses are not even worth being discussed, that would place us in a dead end—but rather to set up the frame of the immediately subsequent discussion.

A λύσις implies a weaker claim than an ἔλεγχος, since its success does not require completing a formal deduction, but only showing some important flaw in somebody else's argument. But Aristotle's aim is to prove that the thesis of the immobilist monists is false. Now, just showing that their premises are false and/or that their alleged deduction is not valid would not bring us any further forward. For as it is well known, a true conclusion can be derived from false premises (cf. *APr* II, 2–4). This is *a fortiori* the case if the inference is not drawn correctly, since the outcome of a piece of faulty reasoning is unpredictable.

This is why the crucial move in this section is [*f*]: ἡμῖν δ' ὑποκείσθω τὰ φύσει ἢ πάντα ἢ ἔνια κινούμενα εἶναι, by which Aristotle claims that the existence of change in nature must have the status of *prima facie* evidence, and thus lays the burden of the proof on the opposite party. From that moment on, the Eleatic arguments appear as a challenge to the claim he has just made; the aim of the game will be to show that these arguments are faulty. This will amount to establishing the claim he makes in [*f*]. In order to prove that their arguments are faulty, any one of the criticisms he has raised at [*e*] can work, i.e. it will suffice

θέσιν ὁποιανοῦν... τῶν λόγου ἕνεκα λεγομένων appears to imply that the Eleatic arguments are among the theses 'held just for the sake of the argument', and Aristotle describes them, immediately after, as having an eristical character. Thus it seems to me that in our text these are variants of what is basically the same situation.

[19] In *SE* 11, 171b24–6, the ἐριστικός is described as someone who wants nothing but to win at any costs in the current contest. Unlike the sophists, he does not even have the desire to gain a lasting reputation for wisdom and knowledge with his discourses.

[20] Although, as Aristotle does not fail to stress on some occasions, Parmenides appears to have had a problematic relation to truth: Aristotle blames the Eleatics for having preferred to follow their λόγοι rather than the facts (πράγματα, *Gen. et Corr.* I 8, 325a16ff.) or notes that Parmenides himself was eventually 'forced to follow the phenomena' (*Met.* A 5, 986b31ff.)

to show that they rest on false assumptions or that they are not logically valid. This is what he will set out to do in the rest of chapter 2 and in chapter 3.

At this point, one might raise two questions.

(1) Why is Aristotle entitled (or why does he claim to be entitled) to assume that at least some of the things that exist by nature are subject to change? His argument for this is laconic: 'this is evidenced by induction'. In other places, in fact, he is even more abrupt, just calling upon the consensus of all sensible human beings, for instance *Physics* II 1, 193a3–4: 'it would be ridiculous to set about proving that there is a nature, since it is manifest that there are many such beings', although, in the immediately following lines, he admits—and this is relevant for the general tone of the present discussion—that 'it may happen' (ἐνδέχεται) that one would reason about colours with a person born blind' (193a6–8). It has been doubted that the ἐπαγωγή mentioned here could be the explicit inferential procedure described and discussed in the *APr* (II 23) and alluded to in several passages of the Organon, so that it would more likely be the kind of spontaneous and silent process explained in the last chapter of the *APo*. This is plausible in view of the fact that the evidence for the assumption of the existence of natural change is omnipresent (and in view of the above-mentioned dictum of *Physics* II 1). But the outcome of the process of *APo* II 19 is a general notion, while what is needed here, according to the rules of dialectic, is a proposition. Still, the border between recognizing a notion and assuming a proposition is not necessarily clear-cut, especially when this proposition happens to be extremely general, as is the case here.

(2) Another question would be: if Aristotle feels free to assume this thesis, and with a very high degree of certainty, why does he need the long, intricate, and difficult discussion with the Eleatics that follows? Apart from the fact that it is always better to have the additional security that even the most clever opponents to one's thesis did not find a decisive argument against it, we have to remember that in *Physics* I Aristotle is not primarily concerned with the *existence* of change and multiplicity, but with the *principles* of nature and natural processes. Finding a convenient defence against the Eleatic paradox will help him determine what these principles have to be and, in the end, it will lead to the discovery of what they are. That is the reason why, in chapter 8, he insists that the general model of change he is advocating solves in a satisfactory way 'the puzzles of the Ancients'.[21]

[21] Of course, this research strategy is possible and fruitful, first, because Parmenides himself saw his own arguments as a matter of principles. He appears to have distinguished the level of 'true reality' (ἀλήθεια) from that of the phenomena, or at least Aristotle thinks he did (cf. *Met.* A 5). Second, the strategy is possible only if Aristotle admits that at least some of Parmenides' premises are true and

At this point, i.e. once the burden of the proof has been reversed, it seems that the way has been cleared and that we can start discussing the Eleatics' refutation. But here again comes something like an objection: ἅμα δ' οὐδὲ λύειν ἅπαντα προσήκει, etc. ([*g*]), which may seem to be more or less the repetition of the objection that was made in [*a*–*c*], especially since it is illustrated by a geometrical example which might be the kind of situation that was meant in section [*b*].

Some caution is needed here. First, there is some ambiguity in the phrase οὐ προσήκει: at first sight, it may mean either that one *cannot* put forth a λύσις unless the opponent's argument 'starts from the [relevant] principles' (ἐκ τῶν ἀρχῶν) and leads to a false conclusion, or simply that one *does not have to* look for a λύσις if this is not the case. In fact, we are not necessarily obliged to fully eliminate this ambiguity. To say that it is definitely impossible to resolve the Eleatic paradox would be too strong, since Aristotle does claim he can resolve it; but obviously he would not be satisfied with just moving on because he is not bound to find an answer. Rather, I am inclined to think that οὐ προσήκει etc. points to the fact that *under normal circumstances*, bringing forth a λύσις requires that the opponent's argument be inconclusive and that it 'start from the principles'.

'*The* principles' must mean some specification about the premises; the analogy with the different strategies for squaring the circle is clearly meant to illustrate that point. We learn, from the information given by ancient sources[22] on which it seems that we may safely rely and from chapter 11 of the *SE*, that Hippocrates' attempt at squaring the circle 'by means of segments' rested on a set of correctly demonstrated theorems (even though they were insufficient to support the claim that the circle could be squared). Antiphon's squaring, on the other hand, was rather a clever guess, which involved reasoning about the limit of an infinite sequence and thus did not allow a standard geometrical construction.

Since Aristotle says that geometry is concerned with the squaring by means of lunes, but not by Antiphon's conjecture, we may surmise that he meant that the premises had to be *relevant* or *appropriate* with regard to their subject matter. In *SE* 11, he says that a science rests on a set of true and acknowledged propositions specific to its subject matter, including principles or axioms, previously established theorems, and perhaps some rules of method. The possibility of working on the basis of such a set is characteristic of an established science; Aristotle refers to this condition with various phrases such as κατὰ τὸ πρᾶγμα, κατὰ τὴν τέχνην, κατὰ τὰ ὑπὸ τὴν τέχνην, or ἐκ τῶν (ἰδίων) ἀρχῶν, and so on.[23] He obviously

relevant for the inquiry into the principles: this is indeed the case with the principle of non-contradiction and with the claim that nothing can be generated out of nothing.

22 See Appendix I.

23 'Relevant' for the subject matter: κατὰ τὸ πρᾶγμα *SE* 11, 171b17–21; 'acknowledged' by the relevant interlocutors: πρὸς γεωμετρίαν (as opposed to πρὸς πολλούς), 172a5–6; κατὰ τὴν μέθοδον and κατὰ τὴν τέχνην (171b11–12) cover more or less both concepts. The long phrase ἐκ τῶν ἀρχῶν καὶ

thinks that none of these conditions is met by the premises on which the monistic theses rest, since he has made it clear from the start that they obliterate the very notion of a principle ([*c*]).

Thus, section [*g–i*] does not merely repeat [*a–c*]. The objection in [*a–c*] can be met by reversing the burden of the proof and looking for a simple λύσις instead of an ἔλεγχος, but [*g–i*] raises a new question, namely which premises—or kinds of premises—could be used as a starting point for the λύσις itself.

Readers may be disturbed by Aristotle's apparently casual attitude towards this difficulty, for he seems to declare quietly that, on view of the philosophical benefits he expects to draw from the discussion, he will not pay any heed to the objection. The only indication he gives for solving it is that the forthcoming discussion will be a form of 'dialectical' inquiry. But we can shed some more light on this point with the help of *SE* 11.

In this chapter, Aristotle intends to elucidate the difference that there is, in spite of some superficial similarities, between the dialectician and the eristic disputer. He considers a possible analogy with the proponents of incorrect geometrical solutions (ψευδογράφοι) in the realm of geometry:

> The eristic disputer is to the dialectician what the proponent of incorrect geometrical solutions is to the geometer: he builds his faulty arguments on the same basis as dialectic rests upon, just as the proponent of an incorrect solution [builds it on] geometry.[24]

As examples of 'incorrect geometrical solutions', he introduces the same attempts at squaring the circle by Hippocrates, Bryson, and Antiphon, and he immediately makes a distinction between two types of ψευδογραφήματα: Hippocrates' squaring, he says, is like an argument that rests on correct and appropriate premises but is not conclusive, while Antiphon's and Bryson's alleged solutions are faulty in that they rest on premises that are not appropriate, because they have a scope broader than geometry;[25] arguments of this second class are more genuinely eristic than those of the first type.[26]

συμπερασμάτων τῶν ὑπὸ τὴν τέχνην (171b38–172a1) might well be the paradigm, of which our ἐκ τῶν ἀρχῶν (185a15) would be a minimal counterpart. A similar phrase occurs in *APo* I 9, 76a13–15: φανερὸν ὅτι ἕκαστον ἀποδεῖξαι οὐκ ἔστιν ἀλλ' ἢ ἐκ τῶν ἑκάστου ἀρχῶν.

[24] *SE* 11, 171b34–7 (the text of the last sentence is difficult; I have paraphrased it according to the most obvious meaning).

[25] Nevertheless, this is a flaw in Aristotle's view, for a demonstration is supposed to show the cause of the fact that it demonstrates; and from premises which are too broad one cannot draw a complete account of the cause.

[26] *SE* 11 is somewhat uneven in that regard, since Hippocrates' squaring is labelled 'not eristic' (twice: 171b12–15 and 172a3–4), while it is clearly introduced as analogous to one of the two main types of eristic arguments. A possible solution could be that Hippocrates is absolved because he had a strong reputation as a serious scientist. Actually, *SE* 11 also defines eristic (in a short excursus at 171b32–4) by ethical criteria in addition to the logical and epistemological criteria that are its main topic. Alternatively,

Yet he strongly qualifies his own analogy a few lines later:

> If the eristic disputer was to the dialectician exactly (πάντῃ) what the proponent of incorrect solutions is to the geometer, there would be no eristic arguments on these topics [with which dialectic is concerned].[27]

The justification of the analogy is that the dialectician, just like the geometer and (we may assume) any other practitioner of a science, is described as 'considering the common facts in conformity to the subject matter' (*κατὰ τὸ πρᾶγμα θεωρῶν τὰ κοινά*, 171b6). But the reason for the difference lies in the fact that dialectic is not a science: it has no specific domain (*γένος*) of its own, with the consequence that it cannot build on a set of principles in the sense I have defined above (i.e. axioms, theorems, and methodological rules), for such a set must express the essence or nature of a given object or class of objects.

Still, the dialectician's practice must rest on something real, which differentiates him from the eristic disputer and enables him to carry out his specific function of testing other people's opinions and claims—that is, his 'peirastic' role, which requires some kind of access to truth, even without possessing specialized knowledge.

Since this 'something real' is not the nature or essence of an object or a class of objects, let us call it *facts*. These are the 'common facts' mentioned in the definition of the dialectician as *ὁ κατὰ τὸ πρᾶγμα θεωρῶν τὰ κοινά*. *SE* 11 gives the most detailed description of what forms the ultimate basis of dialectical practice:

> [172a21] Indeed, testing opinions (*ἡ πειραστική*) is not something like geometry, but rather the kind of knowledge that one has even when one does not know (*ἣν ἂν ἔχοι καὶ μὴ εἰδώς τις*). For even a person who is not an expert in the field can test another who does not know, if only he is granted (not on the basis of something he knows or of special properties of something he knows, but on the basis of what follows [from the proponent's claims]) whatever fact [meets the following conditions]: (1) one may know it without having an expert knowledge, while (2), not knowing it, one must remain someone who is ignorant in that matter.
>
> Such facts are the 'common facts', and laypersons know them no less [than experts], even when they seem to deviate too far from them.... And since there are many such facts about all sorts of objects—not the sort of facts that determine a nature and a domain of reality, but having rather the same status as negations have (*οὐ τοιαῦτα δ' ὥστε φύσιν τινὰ εἶναι καὶ γένος ἀλλ' οἷα αἱ ἀποφάσεις*), while the other sort of facts are rather specific [to their objects] –, it is possible to start from them in order to test

the fact that some *ψευδογραφήματα*, in the end, are *not* eristic, may be another reason why eristic arguments are not exactly like *ψευδογραφήματα* (but that may be seen as a mere tautology).

[27] *SE* 11, 172a9–11. The same qualification is recalled in the last lines of the chapter: 'That is why the eristic disputer is not absolutely (*πάντῃ*) in the same situation as the proponent of an incorrect geometrical solution' (172b2–3).

> opinions on every subject, and there can be [172b] some technical skill for that, but [it will] not be the same kind of discipline as those that demonstrate.[28]

These 'common facts' form the universal basis of dialectic.

Of course, in order to test an opinion, one must also have at the start some propositions specifically related to that opinion. This is what Aristotle does in the first lines of the next section, in which he says that 'the most suitable of all possible starting points' is to ask *what is meant* by the claim that 'all things are one'. Similarly, at lines 185b6–7, he begins the next part of his critique by inquiring into the possible meanings of 'one'. He does not explain why he thinks that the most suitable of all starting points is precisely this one, but it is just natural to think that when you have to start a discussion without knowing *a priori* if, and to what extent, you stand on a common ground with your interlocutor or opponent, the first thing to do is to try to agree with him on the exact meaning of his thesis. In fact, this is a common dialectical procedure, of which Plato's dialogues, and especially the so-called 'Socratic' dialogues, provide many examples. Thus, the difficulty raised in [*g*–*h*] gets an answer, or at least substantial elements of answer, in [*i*] and [*j*].

To finish on this section, I hope to have shown that, far from being just the garbled discussion of one single objection, the section as a whole gives a sharp slow-motion picture of the successive moves through which Aristotle reaches the possibility of inquiring into first principles. As we have seen, the whole process consists of three distinct steps, introduced and justified through the discussion of related ἀπορίαι:

- first [*d*–*f*], he rejects the burden of the proof and announces his own argument as a λύσις;
- second [*g*–*i*], he raises the question of the kind of principles that would be appropriate for such an argument;
- then [*j*], he asks what is meant by *being* and by *one*.

Seen in this way, our section provides a vivid and accurate illustration for a famous passage of *Topics* book I:

> [101a36] [The Topics] has yet another use, in relation to the most fundamental [notions] in each science. For it is impossible to say anything about them from the [principles] proper to the science in question, since the principles are prior to everything else, [101b] and one has to investigate them by means of the reputable opinions about

[28] *SE* 11, 172a21-27 and a32-b2. My translation is intentionally long and paraphrastic, so as to dispense with a special commentary on this dense and significant passage.

them. This task belongs properly, or most appropriately, to dialectic: for dialectic, being a process of criticism, *has a path to the principles of all inquiries* (πρὸς τὰς ἁπασῶν τῶν μεθόδων ἀρχὰς ὁδὸν ἔχει).[29]

3. 185a20–b5

3.1. The possible meanings of 'being'

Since διαλεχθῆναι περὶ αὐτῶν is most likely meant to cover the whole discussion, including chapter 3, the method and aim of the first part of this discussion are not explicitly announced by Aristotle. But at the beginning of chapter 3 (186a4–5), where he just calls it τὸν τρόπον τοῦτον, this tactics is opposed to ἐξ ὧν ἐπιδεικνύουσι [λύειν]. Thus the distinction is between a critical analysis aimed at showing the logical faults in an opponent's alleged refutation (chapter 3) and a preliminary discussion of the possible meaning of the opponent's main thesis, aimed at showing that one simply cannot make sense of it (ἀδύνατον φαίνεται τὰ ὄντα ἓν εἶναι, 186a4–5).[30]

> [185a20] The most convenient of all [possible] starting points is to ask, since 'being' is said in different ways, in which sense those who say that 'all [things] are one' mean it.

One might object that in doing so Aristotle is in fact introducing his own doctrine of the plurivocity of being, which straightforwardly contradicts the monist thesis, and that this is no better, or little better, than a *petitio principii*.

To this I would answer that the question, as we have seen, is primarily about 'being', not about being,[31] so that asking in which of its many possible senses the word τὸ ὄν has to be understood does not amount to straightforwardly contradicting the monist thesis. In some cases (the void, for instance), Aristotle does admit that there are uses of a word which have no adequate reference in the real world; so it is not *a priori* excluded that among the uses of the verb 'to be', there might only be one to which something corresponds at the level of 'true reality'. It is true that the case of τὸ ὄν and of the verb 'to be' is a peculiar case, since ὄν stands for the possibility of predicating something in general,

[29] *Top.* I 2, 101a36–b4.

[30] Simplicius and Philoponus think that the end of chapter 2 is a discussion of the monistic position 'in general' or 'for itself' (Simplicius 71.26, 102.20; Philoponus 71.23ff.) while chapter 3 intends to refute separately the theses of Melissus and Parmenides.

[31] What I am claiming here is that there are contexts (like this one) in which a list of categories does not need to be interpreted as a part of an ontological doctrine or even as implying some definite ontological claims.

which is a basic feature of human language (in this, Aristotle follows in Plato's footsteps, cf. *Sophist*, 251a–253a). Now, although one may feel immediately certain that the plurivocity of being will destroy the thesis of the unicity of being, this implication is not logically immediate. In fact, this is exactly what Aristotle has to spell out in lines 185a20–b5.

At 185a22–6 he asks two questions: first, to which type of predication ('category') is this unique being supposed to belong; second, could it be a single item in one of the categories?

> [185a22] Do they mean that all beings are 'substance', or of 'a certain quantity', 'of a certain quality'? Again, do they mean that it is just one substance, e.g. one human being or one horse or one soul; or that it is just one quality, [like] one white or [one] hot or some other item like these? All these options are quite different from one another, and none of them could be said.

I have translated literally ἀδύνατα λέγειν, which in other contexts could simply mean that these different variants of the thesis 'being is one' cannot be maintained, i.e. that they meet insuperable objections. But the discussion that follows shows that for Aristotle the root of their deficiency lies deeper. One could paraphrase: 'none of them can be articulated in a significant way'. This is shown by means of a complex dilemma:

> [185a27] For if they mean that there is[32] some substance and some quantity and some quality, whether apart from one another or not, there will be several beings; and if [they mean] that all there is a certain quality or a certain quantity, whether there will be some substance or not, this is strange, if I may call 'strange' what is impossible. For none of the other [types of being] is something separate beside substance: actually, all of them are said of a subject which is the substance. Now, Melissus says that being[33] is unlimited. Thus being has a certain quantity. For the unlimited belongs to quantity, while a substance, a certain quality or a property [185b] cannot be 'unlimited' except in virtue of a certain predication,[34] i.e. if at the same time they are 'of a certain quantity'; for the definition of unlimited refers to a quantity, but not to a substance or to a sort of thing. Then, if it is both a substance and a certain quantity, being will be two [distinct things] and not one; while if it is only a substance, it will not be unlimited, and it will have no size whatever (for [if it had,] there would be a certain quantity).

[32] Alternatively: 'if they mean that *being consists in* some substance and some quantity', etc.

[33] For the sake of homogeneity I had to translate τὸ ὄν by 'being' when it occurs in sentences that report Melissus' doctrines, although, as Tim Clarke pointed out to me, Melissus most probably meant by τὸ ὄν something much less abstract, which would be more aptly suggested by a phrase like 'what is' or 'what there is'.

[34] The standard translation would be 'accidentally' (κατὰ συμβεβηκός 185a1). This case of accidentality is parallel to that of 'the physician is building a house', in which the person who *is building a house* is 'a physician' only in virtue of a certain predication.

The structure of the dilemma can be reconstructed as follows: (*a*) if the proponent of the Eleatic thesis means that 'being' covers items belonging to different categories, then he will straightforwardly admit the plurality of being; (*b*) if he means that τὸ ὄν is a being belonging to one of the non-substance categories, then:

- either it will be 'of a certain sort' or 'of a certain magnitude' without being a quality or the magnitude of something, which is impossible (inconceivable);
- or it will be a property of something else, and then there will be several beings: that brings the opponent back to (*a*).

(*c*) If τὸ ὄν is some single substance, then:

- either it has a property—or properties—belonging to one of the other categories, and then again being will not be one;
- or it will be a substance without any property (*which would contradict Melissus' claim that being is infinite*).

The trouble with this dilemma is that Aristotle does not say why the last option must be rejected. He just says that it would contradict Melissus' claim. That would be unproblematic if his aim was just to show a contradiction between the opponent's thesis and another claim he has made, according to the standard rules of ἔλεγχος (one could even add that Parmenides' 'way of truth' is open to the same refutation, since it says that being is 'limited' and 'continuous'). Yet that would restrict the significance of the refutation.

According to the interpretation I am advocating, it is important that the outcome be that the thesis 'being is one', taken in itself, cannot be sustained because it is devoid of sense (cf. ἀδύνατα λέγειν 185a27 and ἀδύνατον 186a4). Admittedly, the fact that one may suspect that a more clever or more careful opponent could have escaped the refutation by avoiding saying that 'being' has any predicate whatsoever tends to weaken this interpretation (or, alternatively, in the way in which Aristotle implements his own argumentative tactics). A possible way out of this difficulty is to say that if nothing can be predicated of this unique 'being', then the claim itself cannot be uttered, so that, in a way, it simply does not exist (a literal interpretation of ἀδύνατα λέγειν). This is a stance that Aristotle might have taken (compare *Met.* Γ 4, 1006a12ff.) and Plato had already sketched a similar argument in the *Sophist* (244b–d)—a text which Aristotle had obviously in mind when he was writing these chapters of the *Physics*; but if this is what he intended, it is strange that the text contains no explicit indication to that effect.

4. 185b5–25

4.1. The possible meanings of 'one'

In these lines, Aristotle asks the same question about the meaning of the predicate 'one' in the Eleatic thesis. He considers three possible senses of the word, namely the continuous, the indivisible, and 'what falls under one and the same definition'. His argumentative tactics are not the same as in the previous section, and on the whole this part of the discussion seems to be less methodical and less fully worked out. His aim is simply to show that the thesis that 'being is one' cannot receive a meaning that would meet the requirements of the Eleatic thesis.

In fact, the lists of meanings of ἕν in the corpus are less constant and consistent than the lists of meanings of 'being'. This one, as compared to those in *Metaphysics* Δ 6 and I 1, is shorter. Among the senses that are missing here are, unsurprisingly, the different forms of accidental unity (cf. *Met.* Δ 6, 1015b16–36). We do not find either the different cases of identity (or partial identity) of form, i.e. things that are said to be 'one' by their species, their genus, or by an analogy—unless they should be included in the general notion of things that have the same *λόγος τοῦ τί ἦν εἶναι*. In any case, the list given here seems to be restricted to cases in which the object that is said to be 'one' is clearly one individual thing. Seen that way, the most interesting divergence between the list given in our chapter and that of *Met.* Δ 6 is that we do not find here the case of things that are said to be 'one' because they form 'a certain whole' (*ὅλον τι*), just as the different parts of a shoe are not 'one' in any disposition whatsoever, but only when they are put together according to the form 'shoe': for this notion of unity is a stronger variant of mere 'continuity' and in fact Aristotle's own solution of the puzzle about the parts and the whole mentioned at lines 185b11–16 rests precisely on this notion. But, as he himself says, the resolution of this puzzle is not relevant for his current purpose. Moreover, even in the case of an object endowed with a complete, fully developed form, the fact that it is continuous is enough to deny that it is 'one' in the strong sense required by the Eleatic thesis, since the definition of 'continuous' implies that it is made of parts.[35]

> [185b5] Moreover: since 'one' is said in different ways as well as 'being', we should ask in what sense they mean that the whole [of being] is one. Now, something is said to be 'one' when [*a*] it is continuous, or [*b*] indivisible, or [*c*] when the account of the essence is one and the same for several [terms], like 'cloak' and 'mantle'.

[35] 'I call something *continuous* when the limit at which two [elements] are in contact becomes one and the same [reality] and, as the name indicates, is held together (*καὶ ὥσπερ σημαίνει τοὔνομα συνέχηται*)' (*Physics* V 3, 227a11–12).

> [*a*] Then, if 'one' [means] [10] 'continuous', it will be many,[36] since what is continuous is indefinitely divisible.
>
> There is [also] a difficulty about the part and the whole, which is probably not relevant for our present argument, but rather in itself, that is: are the part and the whole one or several distinct objects, and in what sense are they 'one' or 'several'? and if they are several, in what way? ([The same questions may be asked] about non-continuous parts as well.) And if they are one with the whole in the sense of an indivisible unity, it follows that each one of them is one in that sense with the other part.
>
> [*b*] But if 'being' is one in the sense of an indivisible entity, nothing will have a certain quantity or be of a certain sort, and in fact 'being' will be neither unlimited, as Melissus says, nor limited, as Parmenides says (for the limit is indivisible, but not the limited).
>
> [*c*] Now if all the objects that 'are' are one in virtue of the definition [of their quiddity], like 'cloak' and 'mantle', then [the monists] will be committed to assume Heraclitus' paradoxes: the essence of 'good' and 'evil', of 'good' and 'not-good' will be the same, so that the same thing will be good and not-good, will be a human being and a horse, and the argument will no more say that all beings are one, but rather that they are nothing at all. The same holds for the essence of 'of such-and-such sort' or of 'of such-and-such quantity'.

It is interesting to notice that the first two arguments would not only work against the thesis that 'all that there is' is one and unchanging: they also entail that no being whatsoever, among those that are currently said to be 'one' in one of these senses, could be 'one' in that absolute sense. That means that in Aristotle's philosophy of nature at least, what is one can only be so against the background of some form or of some kind of multiplicity. The third argument, on the contrary, is especially directed at the thesis that 'all that there is is one'. But some form of multiplicity is presupposed in this case as well. In fact, the existence of several subjects, be they individuals or lesser extended classes, is analytically contained in the phrase 'the same account of the quiddity' (since 'A and B are one and the same' implies that there is an A and a B and that they can be distinguished from one another, at least by thought).

[*a*] The notion that what is continuous is multiple, since it is infinitely divisible, is assumed without further justification or elaboration; this is to be contrasted

[36] At 185b10, some mss. and Philoponus offer a variant reading: *πολλὰ τὸ ὄν vs. πολλὰ τὸ ἕν*. Since the aim of the passage is to raise objections against the thesis that 'being is one', one would expect *πολλὰ τὸ ὄν*, but it is safer to follow the majority of the witnesses. The variation does not really affect the meaning of the argument, for we might paraphrase this text, as Ross does: '*their* one', i.e. the unique reality which, according to them, 'being' is supposed to be.

with the extensive and accurate analyses of book VI, in connection with Zeno's paradoxes. Here, Aristotle mentions incidentally an *ἀπορία* about the parts and the whole, which is less sophisticated and which, as he himself points out, is not relevant for the present discussion and in fact not specific to continuous wholes.

[*b*] As to 'one' in the sense of 'indivisible', Aristotle's argument is that if being is one in that sense, it cannot have any quantity whatsoever. As such, the argument seems to be especially directed at Melissus and Parmenides since it points to an internal contradiction in their doctrines. Here again, we are confronted with the difficulty raised by the mention of Melissus' claim in the preceding section.

Nevertheless, there is something that might represent a move in the direction of a possible generalization, when Aristotle adds that if being is indivisible in that sense, 'it will not have a certain quantity *nor be of a certain sort*' (*οὐθὲν ἔσται ποσὸν οὐδὲ ποιόν*, 185b16–17). The assumption that something cannot have a qualitative property unless it has some kind of continuous extension is not otherwise explained. It might rest on the notion that a qualitative property is an 'affection' (*πάθος*), which should be conceived of as the result of a process. If this is what Aristotle had in mind, this part of his argument is circular, since that would presuppose that there are natural processes. Another way of putting the same explanation would be to refer to Aristotle's description of a qualitative property as a determined position in the range of a 'genus', i.e. somewhere between two contraries (for instance, a determined colour between white and black). But it would not be much better, since this description is the outcome of Aristotle's account of alteration, i.e. qualitative change.[37]

[*a* + *b*] Although Aristotle does not mention this possibility, these two arguments can easily be linked together into a dilemma, which applies to any given object: either it is continuous, and then it is divisible (and thus it is one and many at the same time); or it is indivisible (so that it will not be continuous) and therefore, according to [*b*], it cannot have any quantitative or qualitative property.

[*c*] As we have seen, the argument about 'one' in the sense of the identity of definition implicitly assumes the existence of a plurality of beings. This conception of the unity or unicity of 'being' appears to belong to a more sophisticated variant of immobilist monism, which would imply a distinction between the

[37] In the course of the discussion at Delphi, several participants expressed the opinion that such a subordination of physical qualities to magnitude or extension seems quite plausible in the context of Aristotle's natural philosophy. To this I do not object, but I do not see any text in the Corpus in which this doctrine is explicitly argued for, or even stated in a general form. Likewise, it does not seem to me that the standard accounts of the doctrine of categories explicitly indicate a priority of quantity over quality.

level of immediate experience and the level of true reality, not to say between the realms of the sensible and the intelligible. Parmenides may have held some version of this sophisticated monism; Aristotle at least credits him with such an attitude, and distinguishes him accordingly from Melissus' cruder version.[38] In itself, the idea that what has the same intelligible form is in some sense one and the same thing is quite defensible: it is the basis of Plato's doctrine of Forms and Aristotle maintained it to some extent; but, applied in a straightforward way to 'all that there is', it results in a nonsense 'close to madness', as he says in a parallel passage of *Generation and Corruption* (I 8):

> [325a17] On the side of argumentation, it seems that the consequences will be these; while with regard to facts, this opinion appears to be close to madness. Even among the insane, in fact, no one is so wild as to think that fire is one [and the same thing] with ice.

'The discourse of Heraclitus' (*τὸν Ἡρακλείτου λόγον*, 185b20–1) refers generically to the absurd consequences of the rejection of the principle of non-contradiction (PNC).[39] If Aristotle thought, as many modern commentators do, that Parmenides' monism rested on a strong realistic interpretation of the PNC, he might have felt a special feeling of satisfaction in showing that this position ends up in the same paradoxes as those who straightforwardly rejected the PNC.[40]

5. 185b25–186a2

5.1. *'The more recent among the Ancients'*

> [185b25] The more recent among the Ancients too were in trouble as to how they could avoid admitting that one and the same thing be one and many at the same time. Thus some of them eliminated 'is', as Lycophron did, and others changed the structure of the propositions: they would not say that a man 'is white' but that 'he has whitened', nor that 'he is walking', but that 'he walks', lest by introducing 'is' they would make the one many, as if 'one' and 'being' were always used to mean one and the same thing. Actually, the things that 'are' are multiple, either by their definition (for instance, the essence of 'white' and the essence of 'educated' are different, while the same subject is both; thus this 'one' [subject] is several things) or by division, like

[38] Cf. *Met.* A 5, 986b19–29: *Παρμενίδης μὲν γὰρ ἔοικε τοῦ κατὰ τὸν λόγον ἑνὸς ἅπτεσθαι*: 'Parmenides appears to have considered the one according to the notion', and a31–2: *τὸ ἓν μὲν κατὰ τὸν λόγον πλείω δὲ κατὰ τὴν αἴσθησιν ὑπολαμβάνων εἶναι*: 'he held that this one, [being one] according to its notion (*κατὰ τὸν λόγον*), was many according to sense-perception'.

[39] Compare *Met.* Γ 3, 1005b24–6; Γ 7, 1012a24–6, 1012a34–b2.

[40] More precisely, the positions of the Eleatics and Heraclitus are symmetrical: while Heraclitus seems to have derived his rejection of the PNC from the affirmation of universal change, the Eleatics denied change as being incompatible with the PNC.

> the whole and its parts. But at that point, [186a] they were perplexed and they admitted that the one was many, as if the same thing could not be one and many at the same time (but not the contradictory [one and many]: for 'one' can be in potentiality and in actuality as well).

The text does not tell us what Lycophron's preferred formulas were; the most plausible guess is that he recommended that we express predication through a nominal sentence (of the type ἄνθρωπος λευκός), a turn of phrase which is possible and relatively common in ancient Greek.

Given that at lines 185b26–7 the text says that the concern of 'the more recent among the Ancients' was that they might be forced to admit that one and the same thing would be 'one and many', one may be tempted to connect this remark exclusively or principally with the preceding section (I assume that this is why Ross divides the text as he does). But since the solutions that they adopted amounted to eliminating the verb 'to be' in its function as a copula from predicative sentences, it is clear that they thought that the problem lies in the existence of a special relationship between 'being' and unity conceived as identity. Now, what is the exact nature of this relationship? As I see it, they took 'is' to express strict identity, so that by saying 'S is P' one would say that one and the same thing is several things, that is, S itself, and P as well. This might be expressed either by ὡς μοναχῶς λεγομένου τοῦ ὄντος or by ὡς μοναχῶς λεγομένου τοῦ ἑνός (185b31–2): the first phrase would mean that for them 'is' had but one meaning (the strictest, i.e. identity), the second that they did not admit of weaker notions of unity that could fit in with the unity of the predicate and the subject. These puzzles are reminiscent of those of the νέοι and ὀψιμαθεῖς alluded to by the Eleatic Stranger at *Sophist*, 251a–c.

The dialectical structure of the last lines, from 185b32 (or 31) to 186a3, is not immediately apparent. As I see it, the proposition: 'the things that are multiple, either by their definition... or by division' expresses something that is a fact in Aristotle's opinion. To avoid a possible misunderstanding, it must be stressed that here, πολλά τα ὄντα does not mean only that there are many beings in the world, but that each of them must be considered as multiple in some regard, e.g. as having parts (or being part of a whole) or as being describable in different ways. These philosophers were not immobilist monists, although they may have been influenced by Eleatic arguments.

Aristotle says that the fact that beings are multiple was the starting point of their puzzlement (ἐνθαῦτα δὲ ἤδη ἠπόρουν). So they must have been aware of it, not necessarily (and most probably not) under the form of an explicit universal statement, but rather for particular things and on the occasion of particular experiences or particular assertions. This realization led them to state—this time explicitly and self-consciously—the proposition that 'X (whatever it may be) is one and many': ὡμολόγουν τὸ ἓν πολλὰ εἶναι, an acquiescence which made them feel embarrassed (ἠπόρουν). The reason for their perplexity is given at the end of the next sentence: 'as if the same thing could not be one and many at the same time, μὴ τἀντικείμενα δέ': here, as in the *Prior Analytics*, τὰ ἀντικείμενα must have the special meaning of 'contradictories'. If both 'one' and 'many' have just one meaning, these terms will necessarily be contradictories, and asserting that some object is one and many at the same time will be an intolerable violation of the PNC.

We are now in a position to reconstruct the progress of the last section of chapter 2. I am inclined to make it begin at line 31, even at the cost of admitting a loose syntax: that is, I suppose that after the general premise ὡς μοναχῶς λεγομένου τοῦ ἑνὸς ἢ τοῦ ὄντος, πολλὰ δὲ τὰ ὄντα etc. introduces a minor premise, the conclusion being that 'being' must be one and many at the same time. Be that as it may, the clause ὡς μοναχῶς λεγομένου etc. introduces us into the inner, and perhaps even the unconscious, reflections of these people:

- (*a*) they thought (b31–2) that 'one' (or 'being') is said in just one sense,
- (*b*) but in fact (b32–4) things are multiple, either by their definition or by division,
- (*c*) [They happened to realize this fact], so that they admitted that the same thing may be one and many; they felt this to be unacceptable [as a transgression of the PNC],
- while in fact (*d*) it is quite acceptable that one and the same thing be one and many,
- (*e*) provided the 'one' and the 'many' in question are not straightforwardly contradictory.

The very last sentence: 'for *one* can be in potentiality and in actuality as well' cannot be properly understood on the basis of what has been said since the beginning of book I. It is just a hint to the complete Aristotelian solution; in fact δυνάμει and ἐντελεχείᾳ do not appear elsewhere (as a pair of opposites)[41] in book I. The reader, if she or he is not already acquainted with Aristotle's doctrines, has to guess that what is 'in potentiality' may admit predications that would be plainly contradictory if they were thought of as fully achieved.

One difficulty that one may feel (among others) in reading these lines is that the same proposition—*a thing is one and many*—occurs thrice, each time with

[41] With one exception: at 192a1–2, Aristotle charges the Platonists with the (implicit) belief that what is numerically one is also 'one and not many'.

a different status: (*b*) as a plain fact, (*c*) as a statement made (reluctantly) by the 'more recent among the Ancients', and (*d–e*) as a sentence that has to be interpreted, and may be interpreted in two different ways.

The explanation of this is probably that in writing these lines Aristotle was not aiming at a formal refutation. Rather, he is trying, with a sort of equity, to explain the train of thoughts and experiences that has led to the theses he is discussing. This is not unusual on his part. See for instance *Met.* Γ 5–6: after having given strictly dialectical arguments to defend the PNC, he engages in a 'therapeutic' sequence, in which he turns to those who rejected the principle and tries to find out their possible motives. At the same time, such explanations will complete and corroborate the formal refutations. Here, dialectic grows into a sort of rhetoric.

6. Glancing Behind Instead of Concluding

It would not be easy, and it would make little sense, to write a conclusion on a portion of text that is made up of two different parts, each of which has to be considered as a preliminary step towards something else. The shorter first part (section 1, 184b15–25) begins the whole inquiry of book I, while the remaining sections (2 to 5, with the digression of section 6) are only the first part of the discussion of the Eleatic paradox, to be continued in chapter 3. In this respect, I can but express the hope that the landmarks I have located in this chapter will help readers find their way through the intricate discussions that follow.

There is, nevertheless, one important Aristotelian issue that shows up in both parts of chapter 2, on which each of them can throw some light in its own way, namely the well-known problem: how is it possible to argue about the principles, if there is nothing more primitive on which we could safely base an argument? The question is connected with the general picture of science that Aristotle has given in the *Posterior Analytics*; but here it receives a dramatic expression through the confrontation with the Eleatics, since they (especially Parmenides) appear to have deliberately rejected the most basic assumptions of natural philosophy. Aristotle answers that even if the possibility of another more fundamental principle is ruled out *a priori*, the natural philosopher cannot be sure that he has conceived his own principles with all the desirable precision and certainty, so that we still have something to learn by a further inner scrutiny: such is the instruction conveyed by chapter 1's comparisons with the definition of the circle *versus* the 'round' shape, or the fumblings of a child in the process of acquiring language. For this deeper scrutiny, we can gain significant—indeed essential—help from the critical discussion of the opinions of our forerunners.

Appendix I
On the Ancient Attempts at Squaring the Circle

There are two other places in the corpus[42] at which Aristotle mentions these attempts to square the circle. In both other passages he mentions the same names (Hippocrates of Chios and Antiphon, with the additional mention of Bryson, who seems to play the same role as Antiphon in the analogy) and illustrates points connected with the same issues that he raises in this section of *Physics* I 2. The most interesting is *Sophistical Refutations*, chapter 11, in which this issue is developed at length:

> [171b12] Incorrect geometrical solutions (*ψευδογραφήματα*) are not eristic, since the logical mistakes [that they contain] come within the subject matter studied by this science (*κατὰ γὰρ τὰ ὑπὸ τὴν τέχνην οἱ παραλογισμοί*), even when it is an incorrect proof about a true fact, *e.g.* Hippocrates' proof, or the squaring [of the circle] by means of lunes. While the way by which Bryson squared the circle, even if the circle is squared,[43] since it is not relevant for the subject matter, is therefore sophistical.[44]
>
> [172a2] For instance, the squaring by means of lunes is not eristic, but Bryson's squaring is eristic; and the one could not be transposed [to another subject matter] but holds only for geometry, because it rests on the specific principles of geometry; while the other is intended for the many who do not know what is possible and what is impossible in each [field], for it can adapt [to other, different matters].[45] Or [similarly] the way in which Antiphon squared; or if someone used Zeno's arguments in order to prove that it is not better to take a walk after dinner: that would not be a medical argument, for it would be [too] general.[46]

Posterior Analytics I 9 mentions the same topic only partly and shortly:

> [75b37] Since it is clear that it is not possible to demonstrate a specific fact except on the basis of its specific principles, if the demonstrated property belongs to it as such, scientific knowledge is not [just] this, i.e. proving something from premises that are true, indemonstrable and immediate. For it is possible to prove in that way, as Bryson

[42] Beside these passages, there is another mention of the squaring of the circle in the *Prior Analytics* (II 25, 69a29–34) without a name and in order to illustrate a different (though related) point, namely abduction: 'For instance, let D be *to square*, E *rectilinear figure* and F *circle*: if there is just one middle term in EF, namely *that a circle plus lunes is equal to a rectilinear figure*, then we will be near to knowing.'

[43] It should perhaps be pointed out that *εἰ καὶ τετραγωνίζεται ὁ κύκλος* does not mean *if the problem has been solved* (and still less *if Bryson did solve it*), but only *if there exists a polygon that has the same area as a given circle* ('if there exists' in the geometer's sense, that is *if this polygon can be regularly constructed*).

[44] *SE* 11, 171b12–18.

[45] Assuming that *ἁρμόσει γάρ* (a7) is parallel to *κοινὸς γάρ* (a9) and to the last line of the passage of the *APo* quoted below. Alternatively, it might mean that Bryson's alleged solution could 'fit in' *with the (limited) degree of geometrical instruction of the 'many'*.

[46] *SE* 11, 172a2–9.

> proved the squaring [of the circle]: arguments of that kind prove in virtue of a common feature which is the case also for something else. This is why such arguments [76a] [could] adapt to other objects of a different kind (*ἐπ' ἄλλων ἐφαρμόττουσι οἱ λόγοι οὐ συγγενῶν*).[47]

These mentions of different attempts at squaring the circle have not failed to arouse the interest of the historians of pre-Euclidean geometry. In fact, they had already interested the ancient Greek commentators of Aristotle, so that we have the good luck to have detailed and reliable information about the ancient attempts at squaring the circle in their commentaries to these passages.[48] In what follows, I will limit myself to the points that are relevant for my interpretation of *Physics* I 2, referring the reader to the specialized literature[49] for more detailed accounts from the point of view of the history of mathematics.

(1) Hippocrates squared three different lunes[50] using the standard geometrical methods of his time (commonly and aptly described as: *with the use of a ruler and compasses*). According to Alexander (quoted by Simplicius, 67.7ff.), he also proved that the total area of a circle plus a certain lune (other than the three he had squared) did equal the area of a rectilinear figure. All these demonstrations, as we can reconstruct them from ancient witnesses, resort to segments of circle (*τμήματα*) and to the proposition that *the areas of similar segments of circle are to*

[47] *APo* I 9, 75b37–76a4.

[48] The most important and valuable of these testimonies is Simplicius, *In Phys.* 54.12–69.34, which contains large excerpts from Eudemus' *History of Geometry* as well as quotations from Alexander of Aphrodisias. Evidence about Bryson's squaring (originating from Alexander and Proclus) is given by Philoponus, *In Anal. Post.* 111.3–114.17, cf. also Themistius, *In Anal Post. Paraphrasis,* 18.29–19.27. On our passage of the *Physics*, see also Philoponus *In Phys.* 31.1–32.3, and Themistius, *In Phys. Paraphrasis* 3.30–4.8. These texts can be read in English in the relevant volumes of the Duckworth 'Ancient Commentators on Aristotle' series, or in Thomas 1939, 234–53 (Hippocrates) and 310–17 (Antiphon and Bryson). Beyond this, we have little or nothing to learn from the Byzantine commentator Michael Ephesius (*In Soph. El.* 76.11–24, 90.1–25) who had no access to other sources than the texts that have come down to us.

Apart from some technicalities that are important for the history of early Greek geometry but not relevant for the current discussion, there is a general agreement among modern scholars (see next note) on the meaning of all these testimonies, which are consonant with one another and, taken together, give a fairly accurate picture of each one of the different attempts.

[49] The matter was set out at the turn of the nineteenth–twentieth centuries in several papers by Paul Tannery (Tannery 1876, 46–52; 1882, 168–77; 1883, 339–70; 1902, 119–30) and later on by F. Rudio (Rudio 1907). The best recent review and discussion of the topic that I know is given by M. Caveing (Caveing 1997, 88–131). There is another excellent presentation by Henry Mendell (no date), including translations of all relevant ancient Greek texts. See also the relevant chapters in Heath 1921, 182–200 (Hippocrates) and 220–5 (Antiphon and Bryson); and Netz 2004, 243–86 (with new and attractive insights into what could have been the literary and graphic form of Hippocrates' original text).

[50] A lune is a plane figure bounded by two circular arcs. There are lunes of different shapes, depending on the ratio between the diameters of the two circular arcs.

each other in the same ratio as the squares of their bases (the chords that subtend them). He is said to have engaged in the squaring of lunes 'because of its affinity (οἰκειότης) with the squaring of the circle' (Simplicius *In Physica* 61.2–3) and he may have interpreted his first achievements in this field as a sign that there existed a solution for the squaring of the circle.

(2a) In a quite different manner, the sophist Antiphon suggested that a circle could be assimilated to a regular polygon with a very large (possibly infinite) number of sides. Although we cannot, from the evidence available to us, figure out precisely how Antiphon would have implemented this suggestion, it is clearly on the way to equalling the area of a circle to that of a rectilinear figure.

(2b) The solution put forth by the third man, Bryson, may be seen as a more accurate variant of Antiphon's concept. According to Alexander, he said (a) that the area of the circle is greater than that of any inscribed polygon and lesser than that of any circumscribed polygon; if (b) there is a polygon which is smaller than all the circumscribed polygons and greater than all the inscribed polygons, then, since (c) magnitudes that are respectively greater and smaller than the same magnitudes are equal, then there is a polygon which has the same area as the circle.

(2c) One might ask why Aristotle mentions both Antiphon's and Bryson's solutions, if they are but two varieties of the same argument. Apart from the contingent fact that he and his audience knew of two well-known philosophers who made such claims, there might be a minor but significant difference between their arguments. Antiphon is said[51] to have rejected some basic principles of geometry, while Bryson, according to the ancient testimonies, did not: the assumption that the circle is larger than any one of the inscribed polygons and lesser than any one of the circumscribed polygons is perfectly compatible with standard Euclidean geometry.[52] Moreover, in *APo* I, 9 (quoted above), Aristotle says, apparently without reservation, that Bryson's premises were or, at least, might well be true and that it might be possible to 'prove' (δεῖξαι) something along these lines, although it would not be a real scientific proof (ἀπόδειξις). On the other hand, he does not say in so many words that Antiphon did reject geometrical theses, nor does he seem to make use of this point in the development of the analogy in *SE* 11. What he says, in all occur-

[51] Simplicius, *In Physica* 55.13–17, 55.22–3; Philoponus, *In Physica* 31.27–8; Themistius, *In Physica paraphrasis* 4.7–8. The assumptions which amounted to rejecting some fundamentals of geometry may have been (1) the claim that a section of a curve (an arc of a circle) could coincide with a straight line segment (Themistius 4.3–8; Simplicius 55.13–18; Philoponus 31.24; Alexander, quoted by Simplicius 55.13–15, suggests that the rejected proposition was the theorem that a straight line touches the circumference in one single point) and possibly also (2) the claim that one could achieve the division of a geometrical length into minimal parts.

[52] Admittedly, the story will get worse if Bryson intended to suppose a polygon that would be the largest of *all* inscribed polygons or the least of *all* the circumscribed polygons.

rences, about eristic or sophistical arguments is just that they rest on premises which are 'common', i.e. that they apply, or could apply, to things of a quite different kind from the specific object of the alleged proof. Nevertheless, he may have had this difference in mind when he mentioned two 'infinitist' squarings and not just one.[53]

A prejudicial question may be raised for the correct interpretation of these testimonies: namely, whether these authors really claimed to have solved the problem of squaring the circle and whether Aristotle could have believed they had made such a claim.

It seems to me that this does not result from the texts. To have actually solved the problem would mean to be able to exhibit some defined polygon, of which one could prove that its area is equal to that of a given circle. Now, it does not appear in the testimonies of the commentators that either Antiphon or Bryson made such a claim.[54]

This is even more clear in the case of Hippocrates. The results ascribed to him are regular constructions that could be made with ruler and compasses in perfectly determined cases. Eudemus tells us he conceived his work about the lunes as preliminary research that was likely to open the way to a solution for the squaring of the circle. The fact that he squared the sum of a circle plus a lune strongly suggests that he looked for a resolution in that direction:[55] if you can square a circle plus a certain lune, and if you are able to square that lune, then

[53] Some caution is needed here, because the notion that an 'eristic' squaring involves rejecting some basic geometrical theses may well have come from our chapter of the *Physics*, in which the topic of squaring the circle shows up in connection with the question: 'is it convenient to argue with people who straightforwardly deny the basic principle of natural philosophy?'. More generally, one must be aware that all the evidence we have about these examples was ultimately triggered by Aristotle's texts, and especially by our chapter. Of course, it would be interesting to see if something in this information about Hippocrates and Antiphon (or Bryson) could lead us to complete or to rectify something in the interpretation that is generally given of the first part of [*b*] above (185a1–3). But it appears that in their conjectures about the persons and the facts to which Aristotle alludes in [*h*], the ancient commentators were led by the assumption that [*h*] was an illustration of [*b*], so that we cannot hope to extract from them something new and useful for our purpose. Anyway, the fact that this line of interpretation of Antiphon's squaring starts with Eudemus (cf. Simplicius 55.23–4), who was a close associate of Aristotle, may suggest that it was in keeping with Aristotle's teachings.

[54] The strongest conclusion one can draw from their arguments is that *there is* such a polygon, but these arguments give no hint of how it could be constructed or how one could find out the construction. Apparently, they did not even take the step that Archimedes was to make a century after Bryson, i.e. show that the area of their hypothetical polygon (and thus that of the circle) was equal to that of a triangle having its basis equal to the circumference of the circle and its height equal to the radius (Archimedes, *The Measurement of the Circle*).

[55] Thus it is but natural to suppose that Hippocrates was meant by the example of *apagoge* given in *APr* I 25.

you will have solved the problem. But there is no reason to suppose that he was sure that the problem could be solved in this way.[56]

Thus, although Aristotle freely speaks of 'Antiphon's squaring', of 'the squaring by means of segments' or 'the way in which Bryson squared', it seems to me that we have to interpret such phrases in a loose or weak sense, as meaning either the claim that there is a solution (Antiphon–Bryson) or the claim to have carried out the first step of an abductive resolution (Hippocrates). It could reflect a state of the art in which the problem of squaring the circle was no more seen (naïvely, so to speak) as not yet solved, but was growingly suspected of being insolvable.[57] In this situation, arguments tending to establish positively that its resolution was possible could be seen as milestones in the debate. Such arguments could be grounded either on real geometrical findings (Hippocrates) or on more general principles, as is the case with Antiphon and Bryson.

Appendix II
The Doublet at 185a9–12 (= chapter 3, 186a7–10)

> For they [= Melissus' and Parmenides' arguments] start from false assumptions and are not conclusive. Or rather, Melissus' argument is crude and does not show any real philosophical problem. In fact, once you have granted a first absurd proposition, the rest follows, and there is nothing difficult in that.

These lines would fit into both contexts, because in each one they would be immediately preceded by a remark about the 'eristic' character of Melissus' and Parmenides' arguments and followed by a refutation in which Melissus is mentioned first.

Now, in chapter 2, as we will see, the refutation does not deal separately and successively with each philosopher: it considers first the possible meaning of

[56] Eudemus (Simplicius, *In Physica* 67.3–6) suggests a seeming argument to that effect, when he notes that Hippocrates had squared 'every [*kind of*] lune, since he squared the semi-circular lune, the one that has its external limit (περιφέρεια) larger than a semicircle and the one that has this limit shorter'. Of course, the sentence πάντα μηνίσκον ἐτετραγώνισεν cannot be taken literally. Even if we take it to mean 'every *kind of* lune', it gives at best an incentive to trying the abductive approach. If Hippocrates had given some credit to such an argument, he would not have escaped being blamed as an eristic. But, as Caveing 1997, 125–8 rightly reminds us, he was a skilled geometer. His work on the lunes appears so extensive and accurate that he had most certainly realized that only certain lunes could be squared. In the same direction, in a 1987 paper, Geoffrey Lloyd pointed out that, on close scrutiny, neither the Aristotelian texts nor the additional information given by the Greek commentators imply beyond any reasonable doubt that Aristotle meant Hippocrates himself when he mentioned a *pseudographema* or a fallacious claim to have solved the squaring of the circle (Lloyd 1987).

[57] That might explain the fact that Aristotle is careful to point out the indecisive status of the question: 'even if the circle is squared', *SE* 11, 171b16–17, and probably also: 'even when it is an incorrect proof about a true fact', 171b14. See also *Categories* 7, 7b31–3.

'being' and second the possible meaning of 'one' in the thesis ἓν τὸ ὄν. That would plead in favour of assigning these lines to chapter 3, in which we find first two separate refutations, first of Melissus, then of Parmenides. Moreover, these refutations develop the two charges mentioned in the doublet: (1) they assume false premises, and (2) their arguments are not valid deductions—while the discussion in chapter 2 would only bear on (1), and that only if one considers that giving a certain meaning to a word is equivalent to a propositional assumption. One may doubt that Aristotle would have made this move, since he writes in the *Posterior Analytics* that terms (ὅροι) are not ὑποθέσεις, because 'one has just to understand them and this is not making a supposition, unless someone would say that even hearing is an ὑπόθεσις' (εἰ μὴ καὶ τὸ ἀκούειν ὑπόθεσίν τις εἶναι φήσει, 76b37–8).

Another reason for excising these lines in chapter 2 would be that this might attenuate the impression of muddle and redundancy in lines 184b25–185a20.[58]

On the other hand, Ross has an impressive argument in favour of chapter 2. He remarks that the clause ἀσυλλόγιστοί εἰσιν has an obvious subject in this context (namely ἀμφότεροι οἱ λόγοι 185a8–9) while it has none in chapter 3, except in ms. E, which adds: καὶ ἀσυλλογιστοί εἰσιν [αὐτῶν οἱ λόγοι]. Since it seems implausible that just these words could have been lost before the lines were (say) copied in the margin of chapter 2, it is certainly safer to follow Ross and keep the passage in our chapter.

If this is the case, the passage was probably meant to introduce the whole discussion of the Eleatic position, including chapter 3. Since a λύσις must be directed at a θέσις, i.e. at a philosophical stance expressed in a proposition, it seems to me that the 'resolution' announced at lines 185a7–8 and a14–15 is not carried out before chapter 3, in which the monistic position is discussed in the form of a full-fledged thesis, while the discussions of chapter 2 about the possible or plausible meaning(s) of the terms 'being' and 'one' in this claim, should be seen just as a preparatory step.

Note

During the discussion at the Symposium, Marwan Rashed put forth an alternative option for solving that difficulty. It rests on the principle that since these lines fit well in both contexts, it may be pointless to attempt to determine from which one it should be excised. Besides, Rashed remarks that the two discussions of the Eleatic position, that of chapter 2 and that of chapter 3, not only have (1) the same 'introduction' (i.e. these three doublet lines with some additional sentences), but also follow along parallel lines, with (2) a short criticism of Melissus,

[58] See Section 2.

(3) a more accurate discussion of Parmenides, and (4) a remark about some thinkers who reacted to the Eleatic paradoxes and a criticism of the solutions they advanced in order to avoid them (although these philosophers are not the same in both passages). From this parallelism, Rashed suggests that 185a5–186a3 and 186a4–187a11 (that is, chapter 3 as a whole) might have been two successive versions of the same part of Aristotle's plan for *Physics* I.

This is a clever and attractive guess, which would make a significant contribution to our understanding of how the text of the *Physics* was established. Nevertheless, I must say I am not fully convinced, mainly because[59] it seems to me that there is nothing in chapter 2 that could match exactly the phrases ψευδῆ λαμβάνουσιν and ἀσυλλογιστοί εἶσιν. Furthermore, the discussion in chapter 2 does not consider separately each one of the immobilist monists, although it refers occasionally to some specific theses of Melissus or Parmenides—that is, I do not see Rashed's steps (2) and (3) in chapter 2.[60] There is, on the other hand, a striking parallelism between the two sections (4); but it has to be stressed that they introduce quite different people and different solutions: in chapter 2, these are philosophers who tried to avoid the paradoxes by means of some logical or linguistic *ad hoc* rules, while those in chapter 3 do assert some theses about what things are (the indivisible magnitudes or the existence of something else besides αὐτὸ τὸ ὄν). This difference corresponds to what is, in my view, the distinct character of each of these two discussions, the first being concerned with questions about *meaning* and the second with *reality* itself, be it from a physical or an ontological point of view.

Appendix III
'Unnatural Puzzles About Nature': On the Punctuation of Lines 185a18–19

I will add a few observations on a philological problem about the first sentence of [*i*]. I do not think that this is a crucial issue, although it has attracted much attention among commentators since antiquity.

[59] These are only 'negative' reasons, i.e. the reasons for my reservations about Rashed's hypothesis. But I also have positive reasons, namely my general interpretation of chapters 2 and 3 as one and the same refutation of Eleatic theses (on which see above). On the whole, a decision on that point must be based on an appreciation of the strengths and weaknesses of both interpretations. I have to remember that the main weakness of my own interpretation lies precisely in its inability to account for the existence of the doublet.

[60] See Sections 3 and 4. To this Rashed objected, very sensibly, that his hypothesis does not require strict parallelism; on the contrary, if we have here two ways of presenting the same point, one has to expect some (possibly high) degree of similarity, but not necessarily a full term-to-term mapping, otherwise it would not be two *different* ways.

We are told by Simplicius (pp. 70–1) that the interpretation of the first part of this sentence was disputed in antiquity because it can be punctuated either with a comma after *μέν* or (as printed by Ross) after *οὔ*:

- ἐπειδὴ περὶ φύσεως μέν, οὐ φυσικὰς δὲ ἀπορίας συμβαίνει λέγειν αὐτοῖς
- ἐπειδὴ περὶ φύσεως μὲν οὔ, φυσικὰς δὲ ἀπορίας συμβαίνει λέγειν αὐτοῖς

According to Simplicius, the first option was favoured by Porphyry, while Alexander himself preferred the second.[61] Many medieval authors followed Porphyry, but the modern editors follow Alexander. It appears that both punctuations seemed tolerable to the ancient commentators, and in both cases the sentence means that the Eleatic position lies at the margin of natural philosophy, because its proponents did not do exactly what they were supposed to do; this circumstance is indicated in the text by the phrase *συμβαίνει λέγειν αὐτοῖς*. The question is whether 'what they are supposed to have done' means 'what they intended to do' or 'what an intelligent critic might (perhaps hastily) have supposed them to have done'. I assume that this last meaning is what Ross's punctuation implies: after having written ([*a*]) that the Eleatics' main thesis was not about nature, Aristotle would concede at [*h*] that, after all, the difficulties they raised were in some sense about nature.

I am not sure that we should put much weight on the quasi-repetition (in one case) of the phrase *οὐ περὶ φύσεως* or (in the other case) on the apparent contradiction between *οὐ περὶ φύσεως* at line 184b26 and *περὶ φύσεως μέν* at line 185a18. In fact, *οὐ περὶ φύσεώς ἐστι σκοπεῖν* might as well be taken over by *οὐ φυσικὰς δὲ ἀπορίας* in the second passage (in other words, I do not see a significant difference between the phrase *περὶ* + genitive and the adjective in *-ικός*).

So it seems to me that the decision depends on two points:

- if Ross's punctuation is correct, what are the 'physical difficulties' raised by the Eleatics? Alexander, quoted by Simplicius (70.32–71.4) mentions motion, the infinite and the void.[62] But although Aristotle deals with these problems later in the *Physics*, the following pages (which must be the *ἐπὶ μικρόν* mentioned at line 19) mention only the infinite, in connexion with Melissus' theses; and

[61] Alexander puts forward three different arguments: (1) someone who speaks 'about *x*' must refer to what *x is*, which is not the case of the Eleatics; (2) that punctuation would be more in keeping with 184b26 ([*a*]); (3) he adduces parallel sentences from the beginning of *SE* 164b21–2 *τὰ δ' ἔστι μεν οὔ, φαίνεται δὲ κατὰ τὴν αἴσθησιν*, and 165a3–4 *οἱ δὲ τοῦτο ποιοῦσι μὲν οὔ, δοκοῦσι δὲ διὰ πολλάς αἰτίας*. Porphyry is not quoted so extensively: the one argument that Simplicius reports is that the title of Melissus' treatise was *On Nature*.

[62] See also Ross 1936, 467.

what Aristotle says about it does not look like the discussion of a 'physical ἀπορία'; rather he appears to stress that Melissus' concept of the infinite is inconsistent and confused;

- Porphyry's solution makes these lines one sentence, in which the μὲν and the δὲ function like adverbs or accents enhancing an opposition. Indeed, we could as well read the sentence without these particles and it would be lucid and smooth:

  περὶ φύσεως οὐ φυσικὰς ἀπορίας συμβαίνει λέγειν αὐτοῖς

  'they bring forward non-physical puzzles about nature'.

With Ross's punctuation, we would have to assume two distinct propositions:

(1) περὶ φύσεως μὲν οὐ λέγουσι (or συμβαίνει λέγειν αὐτοῖς),

(2) φυσικὰς δὲ ἀπορίας συμβαίνει λέγειν αὐτοῖς.

The first is a mere repetition of [*a*], while the second seems to come out of nowhere without being otherwise justified (although it is far from being self-evident).

That is why I opt for Porphyry. I would paraphrase the whole sentence as follows:

> They raise difficulties which impact on physics (περὶ φύσεως μέν) though they are alien to physics (οὐ φυσικὰς δὲ); that justifies straying for a while from the field of natural philosophy (and venturing out into the field of dialectic, διαλεχθῆναι) in order to remove these difficulties.

3

Towards the Principles—Resolving the Eleatics' Arguments for Absolute Monism

Physics I 3

DIANA QUARANTOTTO

1. The Role of Chapter 3 in *Physics* I: An Introduction

In the second and third chapters of *Physics* I Aristotle provides a criticism of Eleatic monism. Chapter 2 contains a refutation (ἔλεγχος) of the Eleatics' view that 'all things are one', whereas chapter 3 gives a resolution (λύσις)[1] of some of their arguments, most of which concern the unity of the things that are.[2]

In the first two sections of chapter 3 (186a4–22, 22–32) Aristotle resolves some arguments by Melissus and one argument by Parmenides, which he describes

I wish to thank all the participants at the conference in Delphi, in particular Francesco Ademollo, Andreas Anagnostopoulos, Sarah Broadie, John Cooper, Carlo Natali, and Gabriela Rossi for their generous and helpful comments on previous drafts. My gratitude goes also to the friends and colleagues of the 'Seminario itinerante di storia della filosofia antica' with whom I discussed *Physics* I in 2013, and to the participants at two conferences on *Physics* I, held in Rome in 2013 and 2015. My thanks also go to the anonymous readers. The translations of Aristotle's *Physics* I are taken from Quarantotto (2018), sometimes slightly modified. Emphasis is always mine.

[1] Cf. *Top.* VIII 10, 160b33–4, *SE* 16. On Aristotle's distinction between the refutation of a thesis (ἔλεγχος) and the resolution of an argument (λύσις), cf. Rossi 2014.

[2] The focus of chapter 3 is clearly on Eleatic monism. Eleatic immobilism is dealt with in chapter 8. However, in chapter 3 Aristotle also tackles some arguments by Melissus for the immobility of what is, arguments that he describes as lacking ἀπορία (185a10–12).

as eristic: they assume some falsity and their conclusions do not follow (185a8–10, 186a6–8). In the third section (186a32–b14) Aristotle puts forward a strengthened version of Parmenides' argument and criticizes it. Then (186b14–35), building on this criticism, he formulates an argument for the internal division of what is (conceived of as what is just being). Lastly, he tackles the view of some post-Eleatic philosophers who gave in to two Eleatic arguments. The whole criticism of Eleatic monism is concluded in chapter 3 with the following words: '[i]t is therefore clear that it is impossible that what is is one in this way' (187a10–11).

The role of this criticism within the inquiry into principles that Aristotle conducts in *Physics* I is not immediately clear. This is mainly due to the fact that this inquiry is carried out according to various schemes, some of which operate together in complete or partial overlap.[3]

At the beginning of chapter 2 (184b15–16), the Eleatic theory that what is is one and changeless is the first mentioned in the classification of views about the nature and number of the principle(s). This classification is used to provide a scheme for the subsequent inquiry into the principles, albeit with some modifications. One of these modifications is that Aristotle does not start his inquiry by tackling the whole Eleatic theory. Rather, his treatment of it is divided into two parts, which bracket most of the rest of the book: a criticism of Eleatic monism in chapters 2–3 and a criticism of Eleatic immobilism (i.e. a resolution of an Eleatic puzzle about coming to be) in chapter 8. The first criticism occurs before the formulation of Aristotle's theory of principles, whereas the second occurs after it and is accomplished by means of it. Why does Aristotle break down the treatment of the Eleatic theory in this way? In chapter 8 he claims that his resolution of the puzzle about coming to be is the only possible or effective one (191a23–4). This suggests that he *needs* to solve that puzzle by means of his theory of principles. But why does he criticize Eleatic monism before formulating his theory of principles? Is it only because he does *not* need his theory of principles to challenge Eleatic monism and to resolve the Eleatics' arguments for the unity of what is?[4] Or is it also because he has some reason for *needing* to do so in order to construct his theory of principles?

To answer these questions, we must consider several factors. First of all, we must consider a passage from chapter 2 where Aristotle introduces and sets out his criticism of the Eleatic theory (184b25–185a20).[5]

In this passage, Aristotle claims that the Eleatic theory that what is is one and changeless is not a theory of principles, and that it does not concern nature, despite the fact that he listed it in the opening classification. It is not a theory of

[3] In Quarantotto 2018, 9–28 I tackle the issue of the various schemes used in *Physics* I more extensively.

[4] This hypothesis is put forward by Clarke 2018, 62.

[5] On this passage see also Crubellier in this volume and Clarke 2018, 62–5.

principles because it, in fact, denies the existence of principles: the existence of principles (and even of one single principle) calls for a pluralistic ontology (i.e. a distinction between a principle and what it is a principle of). Moreover, the Eleatics' theory does not concern nature, since it denies the fundamental presuppositions of natural science: the existence of both plurality and change. Hence, an examination of it falls beyond the scope of natural science.

That said, from 184b20 Aristotle starts his criticism of the Eleatics' theory (by focusing first of all on Eleatic monism), on the grounds that the Eleatics raise 'physical difficulties'[6] and that the inquiry into their theory 'contains some philosophy'.[7] Arguably, the difficulties mentioned here are the puzzles about plurality and change that Aristotle tackles respectively in *Physics* I 2–3 and in I 8. These difficulties are physical in the sense that they are pertinent to and significant for natural science. They are so for at least two reasons. First, because they concern the basic presuppositions of natural science, and because the Eleatics' failure to solve these difficulties led them to dispute these presuppositions. Thereby, the examination of the Eleatic theory is presented as a pre-physical and philosophical (or philosophically relevant) inquiry that will be especially focused on the physical difficulties raised by them. Aristotle's aim is to address and to solve these difficulties, in order to answer the Eleatics' challenge to natural science. Second, accomplishing this task is also important because the Eleatics' arguments informed the post-Eleatic physical tradition, in particular the post-Eleatic theories of principles.[8] In Aristotle's view, none of his post-Eleatic predecessors had been able to solve properly the physical difficulties raised by the Eleatics. As a consequence, the post-Eleatic theories of principles are ill founded.

This brings us to consider another scheme of construction of *Physics* I, which is particularly important to clarify further the role of chapter 3 within the

[6] I follow Ross's punctuation: *περὶ φύσεως μὲν οὔ, φυσικὰς δὲ ἀπορίας*. For a detailed discussion of this line and a defence of an alternative punctuation (with the comma after *μὲν*), see Crubellier in this volume. For other authors who use this alternative punctuation, see also, e.g., Simpl. *In Physica* 70.5–17, Mansion 1913, 66 n. 48, Pellegrin 1994, 126–8.

[7] These grounds mitigate the harsh assessment of the Eleatics' theses and arguments provided at 185a5–9. There Aristotle claimed that examining the Eleatic theory 'is like arguing dialectically against any other thesis put forward for the sake of argument . . . or like solving an eristic argument'. Aristotle's mitigation of this previous assessment can be formulated as follows. The Eleatics' theses are paradoxical, but they are not put forward merely for argument's sake. The Eleatics' arguments are eristic insofar as they are incorrect, but not insofar as the intention that motivates them is concerned (*Met.* Γ 2, 1004b25). Rather, the Eleatics were among the first to inquire into truth and the nature of things (191a24–5). They arrived at false conclusions because of real difficulties that they met and that they were unable to solve properly due to their inexperience (191a26–9). Therefore, they deserve a different treatment to those who put forward theses merely for argument's sake (*Met.* Γ 5, 1009a16–22), although their treatment does not fall to the natural philosopher.

[8] Cf. *Physics* I 2, 185b25–186a3; I 3, 187a1–10; I 9, 191b35–192a1; *Met.* N 2, 1088b35ff., *Gen. et Corr.* I 8, 324b35–325b5.

inquiry into principles: the scheme of the path and the progressive solution of difficulties along this path.

The metaphor of the path is introduced at the beginning of the inquiry to describe the method for discovering the principles of natural things (184a16–26). Then it is used again several times in crucial passages of the book (184a16, 188b26–30, 191a24–7, 191b31–4). This metaphor is a traditional one, which Parmenides introduced into the philosophical domain and made famous, and which Plato developed further.[9] It is therefore plausible that Aristotle chose to use it in *Physics* I to place his inquiry and achievements in the context of Parmenides' investigation, especially given Parmenides' profound influence on the tradition of natural science.[10]

In *Physics* I, the metaphor of the path expresses the idea that the search for principles is like a natural development that is both cognitive, i.e. in accordance with our human nature, and objective, i.e. in accordance with how things are.[11] The path has a fixed course: it is characterized by a starting point (184a16–17), an end point (184a17–18), some intermediate steps (188b26–30), a distinctive procedure (184a21–6), and specific difficulties that must be overcome (184b3–5).[12] Given that the path is an objective feature of the natural world and of human nature, this scheme is used to articulate not only Aristotle's own search for principles, but also his reconstruction of the search conducted by the preceding Greek tradition of natural philosophy.[13]

The path goes from what is more knowable to us (i.e. natural changeable substances as they appear at the beginning of the inquiry) to what is more knowable by nature (i.e. the principles of natural substances), and does so by way of dividing a confounded whole (i.e. a natural changeable substance as it appears at the beginning of the inquiry) into its parts (i.e. the principles of natural substances), and by proceeding from the καθόλου to the καθ' ἕκαστα. The end point of the path—insofar as *Physics* I is concerned—corresponds to Aristotle's principles: matter and form. These are the final result of a correct application of the method

[9] *NE* I 4, 1095a30–b5. [10] I defend and develop this hypothesis in Quarantotto 2016.

[11] The path is a cognitive natural development, since it starts from how things appear to us (184a16), i.e. from how things are perceived by us (184a24–5), and since its course amounts to a transformation of our view on things: what was at first obscure to us becomes clear to us. The cognitive character of the path is also suggested by the fact that the path is compared to the ontogenetic development of linguistic abilities in human beings (184b3–5). Moreover, the path is objective, since it is sometimes engendered by truth (188b29–30) and the things themselves (191b31–4).

[12] I take it that the naturalness and necessity of the path attaches not only to its general direction from what is more knowable to us to what is more knowable by nature, but also to the specific route to be followed, i.e. the division of a confounded whole into its parts and the passage (which is said to be necessary) from the καθόλου to the καθ' ἕκαστα. This is suggested also by the presence of a deviation from the path (191a24–7, 191b31–4).

[13] However, the scheme of the path is used in *Physics* I without that the doxographical discussion is in the form of an historical narrative. Of course, this does not exclude the underground presence in *Physics* I of an historical narrative and the possibility of reconstructing one out of *Physics* I's indications, but eventually it complicates the understanding of its sense and relevance.

prescribed in chapter 1. Instead, the theories of principles of Aristotle's predecessors are placed at earlier points of the path. Moreover, among these theories, a special position is granted to Eleatic monism and immobilism. The Eleatic theory is described as a side track of the natural path (191a24–6, 191b10, 31–3). Arguably this description applies primarily and somewhat ironically to Parmenides' arguments for the unity and immobility of what is, i.e. to his Way of Truth. *Physics* I 8 provides some clues as to how the Eleatic deviation from the natural path is viewed by Aristotle. There, after including the Eleatics among the first to inquire 'philosophically into truth and the nature of things' (191a24–5), Aristotle tackles the Eleatic ἀπορία about coming to be and describes it as the initial part of an argument that led them to deny first coming to be and then plurality, and to claim that only what is itself exists (191a27–33). This view also fits with the account of the Eleatics provided in *Met.* A 3. At 984a27–b1 the Eleatics are described as natural scientists that, just like the Ionians, posited one principle but that, unlike the Ionians, recognized a puzzle about change.[14] Since they were unable to solve this puzzle, they concluded that this unitary entity is completely changeless (i.e. completely changeless and absolutely one), thereby crossing the boundaries of natural science and deviating from the natural path of research.

This argumentative order (from the ἀπορία about coming to be to the conclusion that what is is changeless and then to the further conclusion that what is is absolutely one), however, is not followed in *Physics* I. There—as already noted—Aristotle first resolves Melissus' and Parmenides' arguments for the unity of what is (chapter 3) and then the Eleatic argument for its immobility (chapter 8). Some clues to explain why, in *Physics* I, Aristotle changed the order in which he tackles the Eleatics' arguments are provided by his description of the natural path and the method for discovering the principles.

Following the path amounts to articulating a confounded whole into its parts. As Aristotle's result of the application of this method in *Physics* I (i.e. the distinction between matter and form) shows, the articulation at issue here is not (or is not primarily) a physical division but a conceptual and ontological division (i.e. a distinction in account and in being).[15] So, the existence of principles of natural things requires a pluralistic ontology. Put more carefully, the existence of principles requires a special kind of pluralistic ontology, i.e. one in which what is can be both one in number (the whole) and many in account and in being (the parts). Indeed, the principles of natural things—which Aristotle presents as the final result of the correct application of the method prescribed in chapter 1—are constituents (στοιχεῖα) of natural substances (189b16). Moreover, the internal division of natural

[14] I assume that the puzzle about change mentioned in *Met.* A 3 and the one tackled in *Physics* I 8 are, if not the same, at least linked to each other.

[15] Later I shall suggest that the distinction in account and in being is Aristotle's primary focus in *Physics* I 3, too.

beings into parts that are their principles is presented as the ontological structure whereby natural beings are subjected to change (190b11, 20–3).

This suggests that, from a heuristic point of view that takes into account the path of research followed by Aristotle's predecessors, it is necessary first to argue for a pluralistic ontology that makes it possible to articulate what is into parts, and then to tackle the problem of change and to solve the ἀπορία about change, by showing how it is possible for what is (conceived of as a compound entity) to undergo change. Hence, from this point of view, resolving the arguments for the absolute unity of what is has priority over resolving the argument for its immobility. Moreover, resolving the arguments for the absolute unity of what is is particularly appropriate if the result of the resolution is not only that the things that are are numerically many, but also that what is (i.e. a substance) is both one in number and many in account and in being, i.e. a whole articulated into parts. This is one of the hypotheses that I shall put forward in the following sections.

Broadly, my hypothesis goes as follows. In *Physics* I 3, Aristotle's focus is on Parmenides' argument for the unity of what is. This argument tackles the entity that, in chapter 1, Aristotle describes as the starting point of the natural path (what is more knowable to us) and as a confounded whole. Parmenides' argument deduces that this entity is in fact absolutely unitary: there is only one entity and this entity is only one, i.e. undifferentiated (in being and in account). Aristotle's resolution amounts to showing that this argument is not able to exclude that there are many (numerically distinct) entities, each of which is both one in number and many in being and in account, i.e. each of which is a unity articulated into parts. His resolution is aimed at blocking the Eleatic deviation from the natural path, by removing the fundamental mistake from which that deviation stems, and at establishing the basis for a successful search for the principles.

In sum, Aristotle's employment of the scheme of the path to argue for his theory of principles suggests that his aim in *Physics* I is not simply to propose a theory that replaces those of his predecessors. More precisely, his aim is to propose a theory that, unlike those of his predecessors, is *fully correct*. This is because, unlike those of his predecessors, this theory is the result of a correct application of the method prescribed in chapter 1 and is built by solving progressively the problems that are found along the path and that were left unresolved (or not satisfactorily resolved) by his predecessors.[16] In particular, from both a heuristic and a historical point of view, the resolution of Parmenides' argument for the absolute unity of what is corresponds to the initial (pre-physical) segment of the path that leads to the principles.[17]

[16] For a similar approach, cf. *Physics* IV 4, 211a7–11.

[17] Of course, this does not imply that the solution of the problems met by his predecessors and the resolution of their arguments is viewed by Aristotle as necessary without qualification to construct his theory of principles. Rather, it is necessary only given the special approach of the book, i.e. given that in *Physics* I the construction of Aristotle's theory is done by retracing the steps made by his predecessors along the path.

In the following sections, I shall provide an analysis of Aristotle's resolution of Parmenides' argument for the unity of what is and of the subsequent sections of the chapter. I shall skip the resolutions of Melissus' arguments for reasons of space and because Melissus' arguments are explicitly described as lacking ἀπορία (185a10–12): they simply derive conclusions from the assumption of an absurdity. So, the working hypothesis, which I shall argue for in the following sections, is that, in Aristotle's view, the physical puzzle about plurality that is especially relevant to natural science and the decisive mistake lie in Parmenides' argument for the (absolute) unity of what is.

2. Aristotle's Resolution of Parmenides' Argument for the Unity of What Is (186a22–32)

Having argued against the Eleatic view that 'all things are one' in chapter 2, and having resolved some arguments by Melissus in the first part of chapter 3 (186a10–22), Aristotle provides a resolution of Parmenides' argument for the unity of what is, at 186a22–32. He does not set out Parmenides' argument and, while resolving it, uses the term 'pale' in place of 'being'.[18] The resolution is introduced by the claim that Parmenides made a false assumption and that the conclusion does not follow (186a23–5). The false assumption is that ὂν λέγεται ἁπλῶς: this is false since ὂν λέγεται πολλαχῶς. Arguably, Aristotle means that Parmenides did not distinguish the different senses of 'being' and that he instead assumed that 'being' has only one sense.[19] The invalidity of the argument amounts to Parmenides' inferring that 'pale things are one' from 'only pale things are' and 'the pale σημαίνει ἕν'. Aristotle disputes the conclusion that 'pale things are one' in two ways: pale things are one neither by continuity nor in account and in being.[20]

[18] I translate ὄν sometimes with 'what is' and others with 'being', for stylistic reasons only.

[19] Insofar as *Physics* I 3 is concerned, I share the traditional interpretation that the expressions ὂν λέγεται ἁπλῶς and ὂν λέγεται πολλαχῶς are primarily about the different senses of 'being' and, derivatively, about the different kinds of being. I shall defend this hypothesis here and in the following sections. Further clues in favour of it can be found in *SE* 33, 182b22–7 and *Physics* I 2, 185b25–186a3. The first passage seems to refer to the kind of resolution of Parmenides' argument that Aristotle provides in *Physics* I 3. The significant fact is that this resolution is included among those that depend upon homonymy. In the second passage, Aristotle tackles the worries about unity and plurality, which some later philosophers shared with the Eleatics, and describes these worries as concerned with the senses of 'being' and 'one'.

[20] By tackling unity by continuity, Aristotle is arguably referring to Parmenides' claim that what is is ἕν, συνεχές (B8.6, 22–5). Instead, by tackling unity in account, Aristotle may be referring to Parmenides' claim that what is is μουνογενές (B8.4), a hypothesis that is suggested by Clarke 2018, 77.

Concerning Parmenides' argument, the text mentions five assertions:

(1) 'Being' is said without qualification (ὂν λέγεται ἁπλῶς)

(2) Only pale things are

(3) The pale signifies one thing (σημαίνει ἕν)

(4) Pale things are one

Assertion (2) is rephrased at 186a29–30 as follows: 'there is nothing separate from the pale'.[21]

The corresponding assertions for being would be:

(1^1) 'Being' is said without qualification

(2^1) Only beings are = there is nothing separate from being

(3^1) Being signifies one thing

(4^1) Beings are one

These lines pose several problems. Next, I shall tackle the following: what is the respective meaning of assertions (1)–(4)? Do (1) and (3) have the same meaning? Moreover: are (1)–(3) all premises of Parmenides' argument? Are (1)–(3) the only premises of the argument or are there further premises not mentioned by Aristotle at 186a22–32? And finally: why does Aristotle use 'pale' in place of 'being'?

Let us start by considering assertions (2) and (3), which are clearly premises of Parmenides' argument. I shall reconstruct their meaning by analysing Aristotle's resolution.

Premise (2) is 'only pale things are'. This premise means or implies that 'there is nothing separate from the pale' (186a29–30), i.e. it states that the relation between the things that are and the pale is of non-separation: what is not separate from the pale is, whereas what is separate from the pale is not. As we shall see shortly, the subsequent lines suggest that for something to be not-separate-from-the-pale means that it is a substrate that receives the attribute pale.[22]

Premise (3) claims that 'the pale σημαίνει ἕν'. By itself, it may be interpreted in several ways. Aristotle's resolution of the argument—especially the part

[21] 'There is nothing separate from the pale' can be conceived of as the meaning or an implication of 'only pale things are'.

[22] On this sense of χωριστόν, and cognate words see *Physics* I 2, 185a27–32 (ἀπολελυμένα); I 4, 188a6 (ἀχώριστα); *Met.* N 2, 1089b25–6 (χωριστά). Instead, later in chapter 3 (186b22, 28), Aristotle employs χωριζόμενον and χωριστόν to describe the *status* of attributes that are not included in the definition of a subject and of attributes that do not include the subject in their definition.

concerning the pale's lack of unity in account—however, helps us to understand how it should be intended.

At 186a28–39, Aristotle claims that the pale is not one in account, since the being of the pale and the being of what receives it are different. Arguably, the latter claim means that the essence of the colour pale and the essence of the subject of which 'pale' is predicated are different: the essence of the colour pale is, e.g., penetrating colour,[23] whereas the essence of its subject, e.g. the essence of the human being, is, e.g., animal biped; moreover, the essence of the colour pale is the essence of an accident, whereas the essence of its subject is the essence of something that receives it, i.e. of a substrate or substance.[24] If this is correct, then the claim that 'the pale is not one in account' means that the pale substance (e.g. a pale human being)—which is the result of the predication 'x is pale' and which is thereby signified by 'the pale'—is not one in account: it comprises two parts that are different in account.[25]

Then, at 186a29–31, Aristotle specifies that the distinction between the being of the pale and the being of its substrate does not imply that there is something separate from the pale (186a30–1), i.e. it does not contradict the claim that 'only pale things are'. Arguably, this means that the pale and what receives it, albeit different in account and in being, form a unity. This unity is a numerical and accidental unity: when the predicate 'pale' is said of a subject, the result of the predication—which, therefore, can be called 'pale' or 'the pale'—is something numerically and accidentally unitary (e.g. a pale human being).[26] This—I suggest—is what premise (3) means: '"the pale" σημαίνει ἕν' means that the predicate 'pale' (when it is said of a subject: 186a32–3) signifies a numerical and accidental unity.

If this hypothesis is correct, ὂν σημαίνει ἕν does not mean ὂν λέγεται ἁπλῶς, i.e. it does not mean that 'being' has only one sense. Rather, ὂν σημαίνει ἕν expresses what sense 'being' has when it is predicated as an accident of a subject: 'being' signifies a numerical and accidental unity. This is the sense of 'being' that

[23] *Met.* I 7, 1057b2–34.

[24] This, if correct, implies that the kind of predication that Aristotle attributes to Parmenides' argument is intercategorial accidental predication (ὂν κατὰ συμβεβηκός: *Met.* Δ 7, 1017a7–22). This hypothesis, as we shall see shortly, is also suggested by Aristotle's distinction between σημαίνειν ἕν and σημαίνειν ὅπερ ὂν and ὅπερ ἕν (186a23–34), and by ll. 186a34–b1.

[25] For a different interpretation, see Ross 1936, *ad loc.*, Bostock 2006, 108–9, Castelli 2018, 86–93. According to these commentators, the point at issue is not the internal composition of 'the pale', but the ambiguity of the term 'pale', i.e. the fact that it can mean either a substance or a quality. Of course, in my interpretation, the fact that 'pale' can mean either a substance or an attribute is the reason why 'the pale' (i.e. the result of the predication 'x is pale') has two parts that differ in account.

[26] In Aristotle's view, accidental being and accidental unity are signified by the same kind of sentences (*Met.* Δ 6, 1015b16–34; *Met.* Δ 7, 1017a7–22): the pale (i.e. the pale substance) is both an accidental being and an accidental unity.

(in Aristotle's view) Parmenides employed in his argument and that he wrongly considered as the only one.[27]

In sum, premises (2) and (3) of Parmenides' argument can be explained as follows:

(2) only pale things are = only things that receive the accidental predicate 'pale' are = nothing is separate from the predicate 'pale'

(3) the predicate 'pale' (when it is said of a subject: 'x is pale') signifies a numerical and accidental unity (i.e. a pale substance)

One part of Aristotle's resolution amounts to showing that Parmenides' premises do not imply that the pale (a pale substance) is absolutely one, namely undifferentiated in being and in account.[28] Even if only the things that receive the accidental predicate 'pale' are, and even if 'the pale' signifies a unity, this does not imply that this unity is undifferentiated, i.e. that the pale is *only* one. Rather, the pale is *both one* (numerically and accidentally) *and many* (in being and in account): it is a numerical and accidental unity, which comprises two parts that are different in being and in account.[29]

This interpretation of Aristotle's resolution can help us answer other questions among those previously raised: why does Aristotle use the term 'pale' in place of 'being'? Given that ὂν σημαίνει ἕν and ὂν λέγεται ἁπλῶς do not have the same meaning, is ὂν λέγεται ἁπλῶς conceived of by Aristotle as a premise of

[27] For other interpretations of 'the pale σημαίνει ἕν', see, e.g., Bostock 2006, 107–8 and Castelli 2018, 87–93. Bostock maintains that 'the pale σημαίνει ἕν' (which he takes as equivalent to ὂν λέγεται ἁπλῶς) means that the term 'pale' has one meaning, because 'pale', unlike 'is', falls within the categories. However, he also maintains that the sense of 'being' that Aristotle attributes to Parmenides is that in which 'being' applies to substances (109). Instead, Castelli argues that σημαίνειν is used here in a sense that is meant to mirror Parmenides' signs (σήματα) in B8: 'the pale σημαίνει ἕν' means that the sign 'pale' signifies something (i.e. a substance) that also happens to be one, i.e. to be signified by the sign 'one' (89).

[28] More radically—although this is not made explicit—this part of Aristotle's resolution amounts to showing that Parmenides' premises *require* the difference in being and in account between subject and predicate, i.e. the internal articulation of the pale. Otherwise, 'x is pale' would not mean 'x receives the colour pale', but 'x is the colour pale'.

[29] The distinction between unity in number and unity in being and in account has a central role in *Physics* I. It is first alluded to in chapter 2 within the criticism of Melissus' claim that what is is unlimited (185a32–5). Then it is mentioned, again in chapter 2, within the criticism of the 'more recent of the earliest thinkers' (who are described as people that, like the Eleatics, were 'troubled lest the same thing should turn out for them to be at the same time both one and many' (185b25–7)). One of the examples provided there is very close to the one employed in chapter 3. At 185b31–4 Aristotle claims that the pale and the educated are both one and many (πολλὰ . . . τὸ ἕν), since they are the same (insofar as they are accidents of the same thing and hence comprise a numerical unity) but differ in account. Finally, the distinction between unity in number and unity in being or account is employed in chapter 7 when Aristotle puts forward his theory of principles: at 190a14–17, he claims that what comes to be (the uneducated) is one in number but many in account.

Parmenides' argument? Moreover, this interpretation can clarify the relation between 'only pale things are' and '"the pale" σημαίνει ἕν'.

If the claim '"the pale" σημαίνει ἕν' refers to the numerical and accidental unity of subject and predicate signified by 'the pale' (when 'pale' is said of a subject), and if this unity comprises parts that are different in being and in account, then the kind of predication that Aristotle attributes to Parmenides is intercategorial accidental predication: a predication in which the subject is a substance and the predicate is an accident.[30] This is the kind of predication that corresponds to the sense of 'being' that, in *Met.* Δ 7 (1017a7–8), Aristotle labels as ὂν κατὰ συμβεβηκός and distinguishes from ὂν καθ' αὑτό (intracategorial essential predication). Only intercategorial accidental predication is such that the subject and the predicate are both numerically and accidentally one and different in account. Intracategorial essential predication would not work. 'Socrates is a human being' or 'this colour is pale' are such that the subject and the predicate do not differ in account and in being: 'human being' expresses the being of Socrates; 'pale' expresses the being of this colour. Similarly, 'the human being is an animal' is such that 'animal' expresses part of the being of 'human being'. Therefore, it would be inaccurate to say that 'human being' and 'animal' differ in account and in being. Moreover, an assertion like 'the human being is an animal' exemplifies a kind of predication (i.e. σημαίνειν ὅπερ *x*) that Aristotle distinguishes from σημαίνειν ἕν (186a32–3) and that—as we shall see in section 3—corresponds to intracategorial essential predication (i.e. ὂν καθ' αὑτό).[31]

These considerations both explain why Aristotle uses 'pale' in place of 'being' and clarify the relation between 'only pale things are' and '"the pale" σημαίνει ἕν'. The premise 'only pale things are' is formulated in terms of pale things, since 'pale'—insofar as it is an accident that is predicated of substances—is suited to express the sense of 'being' that Aristotle attributes to Parmenides. In other words, the premise 'only pale things are' expresses not only the idea that there is nothing that is separate from the pale but also the idea that 'being'—insofar as it is equivalent to 'pale'—is the accidental predicate of the things that are. If this is correct, '"the pale" σημαίνει ἕν' is a specification, making explicit what is implied

[30] Arguably, the predicate 'pale' is here conceived of as a *per se* accident (*Met.* Δ 30, 1025a30–4), since it is always predicated of the subject.

[31] Generally speaking it is no surprise that Aristotle employs his own concepts and distinctions to report and deal with his predecessors' theories and arguments. In *Physics* I this is also a consequence of the general approach and aim of the book. As suggested in section 1, the search for principles conducted in *Physics* I proceeds along a path that Aristotle considers natural and necessary. Moreover, he clearly believes that his own concepts and distinctions are fully objective and that they correspond to how things really are (cf., for instance, *Physics* I 5, 188b29–30; I 8, 191b33–6).

by 'only pale things are'. This is arguably the reason why '"the pale" σημαίνει ἕν' appears in a different grammatical construction (a genitive absolute: 186a26–7).[32]

Let us consider the argumentative role of the claim that ὂν λέγεται ἁπλῶς: given that ὂν σημαίνει ἕν and ὂν λέγεται ἁπλῶς do not have the same meaning, is ὂν λέγεται ἁπλῶς conceived of by Aristotle as a premise of Parmenides' argument? Aristotle's resolution is based on the distinction between the sense that 'pale' has when it is used as an accidental predicate of a subject and the sense that 'pale' has when it is considered from the point of view of its essence (and the sense that the subject, which receives the predicate 'pale', has when it is considered from the point of view of its essence). In the first case, 'pale' means something's being accidentally pale, i.e. that a substrate receives the colour pale. In the second case, 'pale' means its own being or essence, which is expressed by the account of 'pale': pale is a certain kind of colour, e.g. a penetrating colour. These are clearly two different senses of 'pale' (and of 'being'): the first is accidental being (ὂν κατὰ συμβεβηκός), whereas the second is essential being (ὂν καθ' αὑτό).[33] From this—and from the hypothesis that ὂν λέγεται ἁπλῶς means that '"being" has only one sense'—it follows that the claim that ὂν λέγεται ἁπλῶς is not a premise of Parmenides' argument. If it were a premise of Parmenides' argument, Aristotle's resolution would be a *petitio principii*.[34] Charity urges us to suppose that the claim that ὂν λέγεται ἁπλῶς is rather a tacit assumption of Parmenides' argument.[35] It is the kind of assumption that in *SE* 8 Aristotle calls a 'false validating premise': if it were used in the argument as an explicit premise, it would make the argument valid, albeit incorrect.[36]

In sum, in his argument Parmenides used a unique sense of 'being' (ὂν σημαίνει ἕν, which is ὂν κατὰ συμβεβηκός) and assumed tacitly that this sense is the only one. Thus, the argument that Aristotle resolves is:

(1^2) Only beings are = there is nothing separate from being

(2^2) 'Being' signifies one thing (ὂν σημαίνει ἕν)

(3^2) Beings are one

[32] Later I shall suggest that this is also the reason why this premise does not occur in the version of Parmenides' argument referred to in *Met.* A 5, 986b28–30.

[33] Cf. *Met.* Z 6, 1031b23.

[34] Other reasons in favour of this hypothesis are the following: (1) Aristotle believes that he is the first to raise the issue of the different senses of 'being' (*Met.* N 2, 1089a35ff.; *Met.* A 9, 992b22–6; *SE* 33); (2) unlike σημαίνειν ἕν, which occurs inside the resolution of the argument, which is mentioned again at the end of it (186a32–3), and which is arguably used in one of its premises, the expression λέγεται ἁπλῶς occurs only in the lines that introduce the resolution; (3) in the last section of the chapter (187a1–11), while referring again to Parmenides' argument, Aristotle mentions the claim that ὂν σημαίνει ἕν but not the claim that ὂν λέγεται ἁπλῶς.

[35] For a defence of the hypothesis that ὂν λέγεται ἁπλῶς is a premise of Parmenides' argument, see Ross 1936, *ad loc.*, Bostock 2006, 107–8, Castelli 2018, 86–93.

[36] On the notion of 'false validating premise', see Fait 2013.

Since Aristotle does not set out Parmenides' argument, one may wonder whether (1^2) and (2^2) are its only premises or whether there are additional premises. This question arises also because later in the chapter (at 187a1–11), while referring again to an argument by Parmenides for the unity of what is, Aristotle mentions other premises. The argument referred to at 187a1–10 can be reconstructed as follows.

(1^3) There is nothing besides (i.e. separate from) being (187a6–7)[37]

(2^3) 'Being' signifies one thing (ὂν σημαίνει ἕν) (187a2, 4)

(3^3) Contradiction is impossible (187a4–5)

(4^3) There is no not-being (187a5)

(5^3) All things are one (187a1–2, 7–8)

To answer the question whether (1^2) and (2^2) are the only premises of Parmenides' argument, useful information is provided at 186a30–2 and 186a34–b1. At 186a30–2, Aristotle underlies what Parmenides overlooked: he overlooked that to claim that the being of the colour pale is *different* (ἕτερον) from the being of what receives it does not amount to claiming that they are *separate* (χωριστόν). In other words, he overlooked that *what is not pale* is not only what is separate from the attribute pale (e.g. a substance whose colour is not pale), but also what is not separate from the attribute pale (i.e. a substance that receives the colour pale). This—if applied to being—means that Parmenides overlooked that *what is not* is not only what is not absolutely (i.e. what is separate from being), but also what is not something, while being something else (i.e. what receives the attribute being, while having an essence that is different from the essence of the attribute being): if 'pale/being' is predicated of a subject 'x' and if 'x' is thereby not separate from 'pale/being' (i.e. 'pale/being' σημαίνει ἕν), it is true to say both that 'x is pale/being' (for 'x' receives 'pale/being' as a predicate) and that 'x is not pale/being' (for the essence of 'x' is different from the essence of 'pale/being'). These two claims—unlike 'absolute not-being is'—are not contradictory, since 'pale/being' is used in two different senses: x is pale = x receives the colour pale; x is not pale = x is not the colour pale; x is being = x receives the attribute 'being'; x is not being = x is not what the attribute 'being' is (i.e. the essence of x is not the essence of 'being').[38]

[37] I take it that the παρά at 187a7 has the same sense as the παρά the at 186a30.

[38] This interpretation is also suggested by 186a34–b1. These lines—introduced by γάρ—serve to explain why the premise that ὂν σημαίνει ἕν is not sufficient to conclude that 'the things that are are (absolutely) one'. The explanation is the following: an accident is said of a substrate, so that the substrate of which 'being' is an accident, since it is different from 'being', *is something* that is not (τι... οὐκ ὄν: 186b1). In this passage Aristotle contrasts this sense of 'being' (σημαίνειν ἕν) with the sense in which ὂν

These considerations suggest that Aristotle's resolution of Parmenides' argument amounts to showing that from Parmenides' premises it does not follow not only that being is absolutely one but also that there is no not-being. At 186a25–32, the claim that there is no not-being is an unmentioned intermediate conclusion from (1^2) and (2^2), whose derivation from (1^2) and (2^2) is challenged by Aristotle. By challenging this intermediate conclusion, he also challenges the final one, i.e. that what is is (absolutely) one (3^2).

Thus, the argument that Aristotle resolves is:

(1^4) Only beings are = there is nothing separate from being

(2^4) 'Being' signifies one thing (*ὂν σημαίνει ἕν*)

(3^4) Contradiction is impossible

(4^4) There is no not-being

(5^4) Beings are one

This is the same argument that Aristotle tackles again at 187a1–10. There, Aristotle's main aim is to criticize the way in which other philosophers tried to avoid Parmenides' monistic conclusion. Since these philosophers were unable to see that Parmenides' argument is a *non sequitur*, they thought that plurality and change could be accounted for by supposing that there is absolute not-being (187a2).[39] Thereby, they violated the principle of non-contradiction. Aristotle shows that it is unnecessary to violate the principle of non-contradiction in order to avoid the Eleatic conclusion. At 186b3–6, he stresses that from Parmenides' premises it does not follow that there is no not-being, for nothing excludes that there is a kind of not-being, which is something that is not something else.

Having reconstructed the argument that Aristotle resolves at 186a25–32, and before analysing the other part of his resolution, let us consider the historical *status* of this argument. First of all, it is worth tackling the relation between this argument and the one Aristotle attributes to Parmenides in *Met.* A 5 (986b28–30): *παρὰ γὰρ τὸ ὂν τὸ μὴ ὂν οὐθὲν ἀξιῶν εἶναι, ἐξ ἀνάγκης ἓν οἴεται εἶναι, τὸ ὄν, καὶ ἄλλο οὐθέν.* This argument can be reconstructed as follows:

(1^5) What is besides being is not (= only beings are = there is nothing separate from being)

σημαίνει ὅπερ ὂν καὶ ὅπερ ἕν. As we shall see in section 3, in this latter case, if 'being' were predicated of a subject or if it were the subject of a predicate, what is different from being would be absolute not-being.

[39] The philosophers Aristotle mentions here also gave in an Eleatic argument from dichotomy, thereby concluding that there are indivisible magnitudes. I shall come back to this passage in section 5.

(2^5) There is no not-being

(3^5) Being is one[40]

Supposing that the premise about contradiction (3^4) is implicit in this version of Parmenides' argument, the main difference between it and the one resolved in *Physics* I 3 is the claim that ὂν σημαίνει ἕν: this claim is a premise of the argument resolved in *Physics* I 3 but is absent from the argument referred to in *Met.* A 5. Moreover, whereas the other premises find some correspondence in Parmenides' poem, this does not hold for the claim that ὂν σημαίνει ἕν.[41]

Parmenides did not explicitly tackle the issue of the senses of 'being'. So, on what grounds does Aristotle attribute to him the claim that ὂν σημαίνει ἕν? A possible answer is the following. What Aristotle sees as Parmenides' main premise concerning 'being' is the claim that 'there is nothing besides being'. He interprets this claim in the sense that 'there is nothing separate from being' (i.e. nothing that does not receive the predicate 'being'). Thus intended, the claim states that the relation between the things that are and being is of non-separation: what is not separate from being is, whereas what is separate from being is not (i.e. absolute not-being). In Aristotle's view, this kind of relation is the one that holds between an accidental predicate, e.g. 'pale', and the subject of which it is predicated, e.g. a 'human being' conceived of as a pale thing. On these grounds, he takes it that Parmenides' claim that 'there is nothing besides being' implies that ὂν σημαίνει ἕν, and that this is its only sense. Finally, the reason why Aristotle interprets Parmenides' claim that 'there is nothing besides being' in terms of non-separation and attributes to Parmenides the implicit assumption that ὂν λέγεται ἁπλῶς is that this interpretation—together with the recognition of what Parmenides overlooked—makes sense of his conclusion that 'the things that are are one'.[42]

40 Formulated thus, this argument is identical to Theophrastus's version (according to Simplicius (*In Physica* 115, 11–14)): τὸ παρὰ τὸ ὂν οὐκ ὄν· τὸ οὐκ ὂν οὐδέν· ἓν ἄρα τὸ ὄν.

41 B 8.36–7: οὐδὲν γὰρ [ἢ] ἔστιν ἢ ἔσται ἄλλο πάρεξ τοῦ ἐόντος (= 1^4, 1^5); B 6.2: μηδὲν δ' οὐκ ἔστιν; B 7.1: οὐ γὰρ μήποτε τοῦτο δαμῆι εἶναι μὴ ἐόντα (= 4^4, 2^5); B 6.6–9: οἱ δὲ φοροῦνται . . . οἷς τὸ πέλειν τε καὶ οὐκ εἶναι ταὐτὸν νενόμισται κοὐ ταὐτόν, πάντων δὲ παλίντροπός ἐστι κέλευθος (= 3^4).

42 A further clue in this direction can be found in *Met.* B 4 (1001a4–b25). There, Aristotle claims that his predecessors conceived of 'being' either as a predicate that is said of something different or as a term that expresses its own essence. Since the second sense of 'being' is attributed only to Plato and the Pythagoreans, we can infer that all Aristotle's other predecessors (that he labels as 'physicists'), the Eleatics included, used 'being' in the first sense. This hypothesis fits well with the fact that, in *Physics* I 3 (186a21–b12), Aristotle uses the second sense of 'being' to formulate a stronger version of Parmenides' argument (cf. section 3). The inclusion of the Eleatics among the physicists is consistent with Aristotle's view in *Met.* A 3, 984a16ff., where the Eleatics' inquiry is described as an investigation that started from physical issues and from a Ionian-style account of principles but ended up with the claim that what is is one and changeless (cf. also *Physics* I 8, 191a24–7, and section 1 above).

Let us now move to the other part of Aristotle's resolution. At 186a28, Aristotle contends that from Parmenides' premises it does not follow that the pale is one by continuity, i.e. that pale things comprise a continuous whole, which is the only thing that is. In the light of the considerations made so far, this part of the resolution can be explained as follows: even if all things receive the accidental attribute pale/being, and even if there is nothing separate from pale/being, this does not imply that all things are a single continuous thing. Many things can have the same accidental attribute without them being a single continuous thing. They would be a single continuous thing, if numerical plurality were impossible because of the non-existence of any kind of not-being. But Parmenides' argument does not imply that there is no not-being. A subject that receives the predicate 'being' is such that it is true to say both that it is (since it receives the predicate 'being') and that it is not (since its essence is different from the essence of 'being'). Likewise, x and y—two numerically distinct substances that receive the predicate 'being'—are such that it is true to say both that they are (since they receive the predicate 'being') and that they are not, and this (that they are not) both in the sense that they are not 'being' (since their essence is different from the essence of 'being') and in the sense that x is not y (since their respective essences are different).[43]

Thereby, Aristotle has challenged two senses in which the unity of Parmenides' being can be intended: unity in account and continuity.[44] From the point of view of the 'first' part of the resolution, the meaning of Parmenides' conclusion is that 'being is only one, i.e. undifferentiated in being and in account'.[45] From the point of view of the 'second' part of the resolution, the meaning of Parmenides' conclusion is that 'there is only one (token) being'.[46] Aristotle's resolution amounts to showing that Parmenides' argument is not able to exclude that there are many (numerically distinct) beings, each of which is both one (numerically and accidentally) and many (in account and in being).

2.1. *Parmenides' argument and the puzzle about plurality*

If the physical puzzles mentioned in *Physics* I 2 (185a18) include the puzzle about plurality tackled in *Physics* I 3, Aristotle's resolution of Parmenides'

[43] *Met.* N 2, 1089a16–19.

[44] As already noted (cf. n. 20), by tackling unity by continuity and unity in account, Aristotle may be referring to Parmenides' claims that what is is ἕν, συνεχές (B8.6, 22–5) and μουνογενές (B8.4).

[45] The view that what is is one in account (i.e. that all things have the same account) is also tackled by Aristotle in chapter 2, at 185b19–25.

[46] The view that what is is one by continuity is also tackled by Aristotle in chapter 2, at 185b9–11. There, however, he does not dispute the claim that what is is continuous, but shows that, if it is continuous, it is (also) many, since it is divisible to infinity.

argument for the unity of what is shows that the puzzle about plurality has two aspects, which can be expressed as follows: how can there be many numerically distinct things that are? How can something that is be at the same time both one and many?[47]

Both kinds of plurality are impossible, if there is no not-being. Moreover, there cannot be the kind of not-being needed for the second kind of plurality, if 'being' has only one sense, for the existence of not-being would imply contradiction.

By assuming that 'being' has only one sense, Parmenides concluded that there is no not-being and that being is absolutely one (a single, i.e. token, entity that is undifferentiated in being and in account). The post-Eleatic thinkers, who before Aristotle tried to account for plurality and change, had not been able to resolve Parmenides' argument and to give a proper solution of the puzzle(s) about plurality implied by it.[48] Rather, they accepted Parmenides' premises and assumptions and they thought that a principle distinct from being (i.e. not-being) is necessary to account for plurality and change,[49] even if this implies contradiction.[50]

Aristotle's resolution consists in denying Parmenides' implicit assumption that 'being' has only one sense and, thereby, in showing that his argument is a *non sequitur*.[51] The thesis that 'being' has many senses—in particular the distinction between ὂν κατὰ συμβεβηκός and ὂν καθ' αὑτό, and within ὂν καθ' αὑτό between substance and accidents[52]—safely makes room (i.e. without any violation of the principle of non-contradiction) for several kinds of not-being[53] and hence for several kinds of plurality.

[47] The second aspect of the puzzle has already been mentioned in chapter 2, when Aristotle mentions some later philosophers who, like the Eleatics, were 'troubled lest the same thing should turn out for them to be at the same time both one and many' (185b25–7).

[48] In *Met.* N 2, Aristotle complains that his predecessors focused only on the numerical plurality of substances and neglected the issue of the many categories of being (1089a31–5, b20–4). Moreover, he stresses that the latter issue has priority over the former: it is by tackling the latter that also the former can be resolved (1089b2–4, 24–7). This order of inquiry seems to be followed in *Physics* I–II: *Physics* I tackles the internal composition of substances (which is built on the distinction between substance and accident), whereas *Physics* II, besides inquiring further into the constituents of natural substances (matter and form), tackles the other two causes (mover and goal) and, thereby, the numerical plurality of substances.

[49] *Met.* N 2, 1088b35–1089a6; *Physics* I 2, 185b34–186a3; I 3, 187a1–2; I 9, 191b36–192a1.

[50] *Physics* I 2, 185b34–186a3; I 3, 187a1–2; *Met.* Γ 5, 1009a22–36. It is unclear, however, how Aristotle judges Plato's view on the *status* of not-being (conceived of as 'different') and of the dyad, insofar as contradiction is concerned.

[51] Moreover, Aristotle's resolution is based on the assumption that the essence of things is expressed by an account (*logos*). This assumption is tackled, later in chapter 3, at 186b14–187a1 (cf. section 4).

[52] According to *Met.* Δ 7 ὂν καθ' αὑτό comprises the different categorial senses of 'being'. Each ὂν καθ' αὑτό is either a substance or an accident and, if an accident, it is either a quality or a quantity etc.

[53] *Met.* N 2, 1089a16–19.

2.2. *The resolution of Parmenides' argument: its role in Physics I*

By resolving Parmenides' argument, Aristotle solves a puzzle that lies at the heart of the post-Eleatics' inquiry into nature, for their inability to solve it properly undermined their theories of principles. Moreover, by resolving Parmenides' argument, Aristotle sets the basis for developing a correct theory of principles. The aspect of Aristotle's resolution that is most relevant to the inquiry into principles conducted in *Physics* I is that concerning the articulation of being (conceived of as a unity) into parts, which are different in account and in being both from the whole and from each other. This is suggested, first and foremost, by the way in which Aristotle's inquiry proceeds from chapter 4 onwards.

In the passage from chapter 3 to chapter 4 Aristotle introduces a new division of the theories of principles, which is different from that presented at the beginning of chapter 2.[54] He distinguishes between the Eleatic being and the physicists' being: the first is an absolute (i.e. undifferentiated in being and in account) unity, whereas the second is a unity comprising parts; moreover, the first—insofar as it is undifferentiated—is unchangeable, i.e. a unitary being from which nothing comes to be, whereas the second—insofar as it is a unity comprising parts—is conceived of as changeable and as a principle from which other things come to be.

The internal articulation of what is, the different ways in which this articulation (as well as the generation of other things from what is/what is one) was intended by Aristotle's predecessors, and the way in which, in Aristotle's view, it should be intended is the focus of the subsequent inquiry. By approaching the inquiry in this way, Aristotle employs the method prescribed in *Physics* I 1 (the division of a confounded whole into parts) and shows where and why his predecessors failed to apply it properly. In particular, in chapter 4 Aristotle criticizes Anaxagoras' theory that what is/what is one is a mixture of everything. Then, in chapter 5 he starts a positive inquiry that will lead, in chapter 6, to a result that is an intermediate version of his own theory and that is similar, in some relevant senses, both to the Ionian and the Platonic theory of principles (189a34–b16).[55]

[54] In Quarantotto 2018, 9–28 I tackle the relation between these two divisions more extensively. On this point see also Cerami 2018, 107–13 and Betegh in this volume.

[55] The Ionian and the Platonic theories of principles are presented at the beginning of chapter 4 as the first physical way of understanding what is/what is one (187a12–20). Both in chapter 4 (187a12–13, 19) and in chapter 6 (189b5–7) the principles of the first way of understanding are described as ὑποκείμενον (in the sense of 'what underlies'), on one hand, and as forms and differences, on the other. Moreover, in chapter 6 (189b1–16) Aristotle mentions a general version of the first physical way of understanding (the one, excess and defect: 187a16–20, 189b11–13) to corroborate the preliminary results of the positive inquiry conducted from chapter 5 onwards.

The argumentative path followed by Aristotle in this positive inquiry shows the role that his resolution of Parmenides' argument for the unity of what is has within the inquiry into principles: as we are now going to see, the path followed from chapter 5 to the end of chapter 6 corresponds to the path of Aristotle's resolution, i.e. it moves from ὂν κατὰ συμβεβηκός to ὂν καθ' αὑτό (by articulating ὂν κατὰ συμβεβηκός into two ὄντα καθ' αὑτά) and to different ὄντα καθ' αὑτά (by distinguishing between substance or substrate and accident).

In chapter 5 Aristotle begins the positive part of his inquiry from an ἔνδοξον shared by all his predecessors: opposites are principles (188a19–27). By itself, this general ἔνδοξον only serves to suggest that the opposites are principles, but it is not sufficient to determine the specific role played by the opposites, how many opposites are principles (which opposites are principles, if not all of them), and if besides the opposites there are other kinds of principles. The subsequent argumentation in chapters 5–6 is aimed, among other things, at sketching the role played by the opposites (i.e. as constituents and as termini of change: 188a27–30, 188a30–188b26) and at introducing a third principle that plays a different role from the opposites (189a21–189b16). The introduction of this third principle amounts to the distinction between the being of the opposites, conceived of as affections (πάθη), and the being of the substrate (189a20–b16, esp. a32–4). In this connection, Aristotle emphasizes that his predecessors—in particular the Ionian philosophers—did not make this distinction and instead conceived of the opposites and the substrate as intertwined (189b4–7), i.e. as comprising a confounded whole (184a21–2): just like Parmenides, they did not distinguish, from the point of view of their respective being, between an accidental property (e.g. hot) and the bearer of the property (e.g. fire: a/the hot thing).[56] So, this shows that, in this respect, the starting points of both the resolution provided in chapter 3 and of the argumentative path followed in chapters 5–6 are the same, as are their respective end points:[57] from ὂν κατὰ συμβεβηκός to ὂν καθ' αὑτό, which in turn is divided into substrate and accident.[58] The result achieved in chapters 5–6 is then employed

[56] 188a20–2.

[57] The main difference consists in the fact that from chapter 4 onwards the inquiry becomes a physical one, i.e. an inquiry into the principles of the things that change and come to be.

[58] This correspondence suggests that the senses of 'being' are presented as they are in *Met.* Δ 7 because Aristotle has followed a heuristic path and not simply conceived of them as a list. The inquiry into the senses of 'being' starts from ὂν κατὰ συμβεβηκός, since this is what is more knowable to us. It is the sense of 'being' that is expressed by most propositions of ordinary language: e.g. 'Socrates is pale', 'Plato is tall', and the like. This kind of predication expresses what accidents something has and not what something is essentially. Starting from this kind of being, the next step is to find out its 'principles', i.e. the ὄντα καθ' αὑτά on which ὂν κατὰ συμβεβηκός is based, and to articulate these ὄντα καθ' αὑτά into the different categories. Indeed, it is by asking what the subject is and what the predicate is of propositions such as 'Socrates is pale' that one arrives at ὂν καθ' αὑτό: e.g. what is Socrates? Socrates is a human being. What is a human being? A human being is an animal so and so, e.g. animal biped; What is pale? Pale is a colour so and so, i.e. a penetrating colour. Moreover, by pushing the question 'what is it?' in the direction of further

in chapter 7 to argue that what comes to be (e.g. the uneducated) is one in number but many in account (190a13–17).

This suggests that Aristotle's resolution of Parmenides' argument for the unity of what is is a pre-physical segment of the path towards the principles: it is the first phase of the division (at the level of being and account) of what is (conceived of as a unity or as a confounded whole) into parts.[59] In *Physics* I this resolution serves to block the Eleatic deviation from the natural path, to rectify the path followed by the physicists and, thereby, to move towards a correct theory of principles.

In Aristotle's view, the only effective way to proceed along the natural path towards the principles, i.e. to articulate the confounded whole into its parts, is to start with a distinction that none of his predecessors drew (or drew and applied correctly)[60], i.e. the distinction between the different senses of 'being' (in particular between substance and accidents): 'if we search for the elements (στοιχεῖα) of the things that are without distinguishing the many senses in which things are said to be, we cannot succeed' (*Met.* A 9, 992b18–20).

Of course, the distinction between substance and accidents is only the first step towards Aristotle's principles (matter and form). It is both the starting point for and the model on which Aristotle elaborates the distinction between matter and form.[61] Aristotle could have argued for his theory of principles by relying immediately on the distinction between substrate (or substance) and accidents, which is tackled also elsewhere in the corpus, especially in the *Categories* and in the *Topics*. In other words, he could have started his inquiry from chapter 7. However, the aim of the book is not simply to argue for his theory of principles. More specifically, it is to arrive at his theory of principles by following the natural path and by retracing and rectifying the steps already made by his predecessors along this path. This is why the resolution of Parmenides' argument for the unity of what is—i.e. the elimination of the Eleatic deviation—has an important role here.

generality one arrives at the distinction between the categories, or better, one arrives at one or another of the categories: e.g. What is Socrates? Socrates is a human being. What is a human being? A human being is an animal. What is an animal? An animal is a living thing. What is a living thing? A living thing is a *substance*; What is pale? Pale is a colour. What is colour? Colour is a *quality*. In this sense, as Aristotle stresses in *Met.* Δ 7, ὂν καθ' αὑτό comprises the different categories. Hence, from this heuristic point of view, ὂν κατὰ συμβεβηκός has priority over ὂν καθ' αὑτό, and in turn ὂν καθ' αὑτό has priority over the distinction between the categories.

[59] On the importance that a division done at the level of being and account has for Aristotle's theory of principles, see Quarantotto 2018.

[60] Insofar as his theory of principles is concerned, Plato approached the notion of substrate and the distinction between substrate and form, but wrongly identified the substrate with what is not, without grasping that the substrate is a composite entity comprising a matter-component and a privation component (192a1–12).

[61] This is shown by the fact that the inquiry of chapter 7 starts from an 'uneducated human being', i.e. from ὂν κατὰ συμβεβηκός.

3. A Strengthening of Parmenides' Argument and Aristotle's Criticism of It: ὂν σημαίνει ὅπερ ὂν καὶ ὅπερ ἕν (186a32–b12)

After concluding his resolution of Parmenides' argument and after stressing what Parmenides overlooked, Aristotle claims that: 'it is necessary ... to assume not only that "being" signifies one thing (*σημαίνει ἕν*), of whatever it is predicated, but also that [it signifies] what is just being (*ὅπερ ὄν*) and what is just one (*ὅπερ ἕν*)' (186a32–4).

Thereby—I take it—Aristotle means that, to avoid the objections made at 186a25–32 and to try to arrive at Parmenides' monistic conclusion, a stronger premise is needed: 'being' should not be conceived of as signifying simply one thing but what is just being and what is just one. If the interpretation advanced so far is on the right track, the distinction between *σημαίνειν ἕν* and *σημαίνειν ὅπερ ὂν καὶ ὅπερ ἕν* is a distinction between signifying something that is and that is one accidentally and signifying something that is and that is one essentially (i.e. a distinction between *ὂν κατὰ συμβεβηκός* and *ἓν κατὰ συμβεβηκός*, on one hand, and *ὂν καθ' αὑτό* and *ἓν καθ' αὑτό*, on the other).[62] This stronger premise is such that it rules out the existence of what is not something else: what is different from being is only absolute not-being. Hence, this stronger premise leaves no room for the difference in account and in being between subject and predicate, nor does it leave room for the numerical difference between distinct beings or for the physical division of what is into parts.

[62] In *Physics* I Aristotle does not explain the meaning of the expression *σημαίνειν ὅπερ ὂν καὶ ὅπερ ἕν*. However, useful indications on the distinction between *σημαίνειν* x and *σημαίνειν ὅπερ* x are provided by other passages of the corpus. Particularly helpful are *APo* I 22 and *Met.* Γ 4, 1007a31–3. These passages suggest that the expression *ὅπερ* x concerns the essence and the essential predicates of x (cf. also *Met.* Γ 2, 1003b32–3 and *Physics* I 3, 186b14–16), and that the contrast between *σημαίνειν ὅπερ* x and *σημαίνειν* x is the contrast between the use of a term to signify its essence or to predicate its essence of a subject (intracategorial essential predication) and the use of a term as an accident of a subject, i.e. as a predicate that belongs to a non-substantial category and that is said of a subject that instead belongs to the category of substance (intercategorial accidental predication). In sum, *ὅπερ x* is the content of a reply to the question 'what is x?', and it is distinguished from the content of a reply to the question 'what accident does x have?'. A translation of *ὅπερ x* that is consistent with this interpretation would be 'what is just x' with the meaning 'what x is essentially' (*Met.* Z 4, 1030a3: *ὅπερ γάρ τί ἐστι τὸ τί ἦν εἶναι*). The distinction between signifying the essence (intracategorial essential predication) and signifying the accident of a substance (intercategorial accidental predication) is expressed in the corpus with variable terminology: as a distinction between *σημαίνειν ὅπερ x* and *σημαίνειν* x; as a distinction between *σημαίνειν οὐσίαν* or *τὰ ἐν τῇ οὐσίᾳ* and *σημαίνειν ποιόν τι ἢ ποσόν τι ἤ τι τῶν τοιούτων* (*APo* I 22, 83a21–32, 83b13–24); as a distinction between 'being predicated of another as a subject' and 'being in a subject': subject (*Cat.* 2a19–34); as a distinction between 'being said *καθ' αὑτό*' and 'being said *κατὰ συμβεβηκός*' (*Met.* Z 6; *Met.* Δ 7, 1017a7–8).

In what follows, I shall argue that at 186a34–b12 Aristotle shows and criticizes the consequences that follow from the hypotheses that 'being' signifies ὅπερ ὄν and ὅπερ ἕν and that it is either a predicate or a subject. The argument is a *reductio ad absurdum*. To avoid the absurdity it is necessary to assume that 'being' means several things (186b2–3, 11–12).[63]

Now, it is not immediately clear what kind of predication is at issue here. If 'being' signifies only the essence of being and the essence of one, any kind of predication seems excluded, at least according to Aristotle's conception of predication. However, the terminology employed to describe the relations between 'being' and its subject, on the one hand, and 'being' and its predicate, on the other, seems to concur with the hypothesis that these relations are accidental ones.[64] I shall suggest that Aristotle tackles ὅπερ ὄν from the point of view of accidental predication at 186a34–186b14, and then, at 186b14–35, from the point of view of essential predication.[65]

From 186a34 to 186b4 Aristotle tackles the hypothesis that ὅπερ ὄν/ἕν belongs to something else. The main purpose of the passage is to rule out this possibility: if ὅπερ ὄν/ἕν belonged to something else, and if ὅπερ ὄν/ἕν were the only sense of 'being', the subject of ὅπερ ὄν/ἕν would not be a being (ὄν τι: 186b2), but it would be not-being (i.e. absolute not-being). This result is achieved after a preliminary remark, introduced by γάρ (186a34). At 186a34–b1 Aristotle claims that an accident is said of a substrate, so that the subject of which 'being' is an accident, since it is different from 'being', is something that is not (τι ἄρα οὐκ ὄν: 186b1). This remark has a double function. On the one hand, it serves to explain why it is necessary to assume the premise that 'being' signifies ὅπερ ὄν and ὅπερ ἕν rather than ἕν. The explanation is that the hypothesis that

[63] Another argument that is based on the premise that being and one are conceived of as being itself and one itself (i.e. as an entity whose essence is to be and to be one) can be found in *Met.* B 4, 1001a26-b1. There, Aristotle claims that if being and one are conceived of (as Plato and the Pythagoreans did) as being itself and one itself, it is very difficult (πολλὴ ἀπορία) to explain how there will be anything else besides being itself and one itself (πῶς ἔσται τι παρὰ ταῦτα ἕτερον), i.e. how beings will be more than one (πῶς ἔσται πλείω ἑνὸς τὰ ὄντα), for what is different from being is not (τὸ γὰρ ἕτερον τοῦ ὄντος οὐκ ἔστιν). He concludes by pointing out that from this hypothesis it follows necessarily Parmenides' conclusion that all beings are one and that this is being (*ὥστε κατὰ τὸν Παρμενίδου συμβαίνειν ἀνάγκη λόγον ἓν ἅπαντα εἶναι τὰ ὄντα καὶ τοῦτο εἶναι τὸ ὄν*). I take it that the main difference between this argument and the one formulated in *Physics* I 3 (186a32–b12) is that in the latter Aristotle assumes that 'being' is either a predicate or a subject, thereby providing a *reductio ad absurdum* of the hypothesis that being signifies ὅπερ ὂν καὶ ὅπερ ἕν. This difference is due to the fact that the argument in *Met.* B 4 is part of an ἀπορία that Aristotle leaves unresolved, whereas the argument in *Physics* I 3 is part of a criticism of a monistic position. I shall come back to this point shortly.

[64] To describe these relations Aristotle employs the expressions ἄλλῳ ὑπάρχον (186b1–2) and συμβέβηκεν (186b4–5), both of which are used to mean ὂν κατὰ συμβεβηκός, respectively at 186a31 and at 186a35. Moreover, in the rest of the chapter (especially from 186b14 to 186b35) συμβεβηκός and συμβέβηκεν are used in the sense of accidental predicate and accidental predication.

[65] On ll. 186b14–35 see section 4.

'being' signifies ἕν does not imply that being is absolutely one (in being and in account): it does not exclude that the subject of which 'being' is predicated *is something* that is not (ἔσται τι ἄρα οὐκ ὄν: 186b1). On the other hand, the remark serves to stress the issue of the relation between 'being' and what is different from 'being', an issue that Aristotle will show to be fatal for the hypothesis that 'being', conceived of as signifying ὅπερ ὄν/ἕν, is either a predicate or a subject. Indeed, while in the case of 'being' conceived of as signifying an accident of something else (ὂν κατὰ συμβεβηκός) the subject *is something* that is not (τι . . . οὐκ ὄν: 186b1), in the case of 'being' conceived of as signifying ὅπερ ὄν/ἕν it is impossible for the subject to be *something that is* (ὄν τι: 186b2).[66] In order for a subject of (accidental) predication to be something that is, it is necessary that 'being' signifies many things so that each is something (εἶναί τι ἕκαστον: 186b3).

After ruling out the hypothesis that ὅπερ ὄν/ἕν belongs to something else, Aristotle considers the hypothesis that ὅπερ ὄν/ἕν is the subject of something else, for instance that ὅπερ ὄν/ἕν is pale. The argument against this hypothesis goes as follows. First, Aristotle claims that the being of 'pale' is not ὅπερ ὄν (186b7). This is arguably because pale is different from ὅπερ ὄν, so that its essence is not the essence of being. Moreover, since pale is a non-substantial item, the fact that its essence is not the essence of being is true *a fortiori*, if ὅπερ ὄν is instead a substance. In addition, Aristotle stresses that being does not belong to pale accidentally either, since what is not ὅπερ ὄν is not (186b7–8). Hence pale is not-being, not simply in the sense that it is something that is not something else (τι μὴ ὄν: 189b9), but in the sense that it is absolute not-being (ὅλως μὴ ὄν: 186b9–10).[67] From this and from the hypothesis that ὅπερ ὄν is pale it follows that ὅπερ ὄν is absolute not-being.

Aristotle does not explicitly label this conclusion as impossible or absurd. However, it seems clear that neither he nor the Eleatics would have found it acceptable. If this is true and if it is also true that 'being' must be either a predicate or a subject, the argument is a *reductio ad absurdum*.[68] Its conclusion is that, to avoid the absurdity that being is (absolute) not-being, it is necessary to acknowledge that also 'pale' signifies ὅπερ ὄν (186b11–12), i.e. to deny that 'pale' signifies ὅλως

[66] Cf. also the contrast between τι μὴ ὄν and ὅλως μὴ ὄν at 186a9–10.

[67] For a similar use of ὅλως, cf. for instance *Met.* Z 3, 1029b5–6.

[68] This is also suggested by 186b4–6, where Aristotle asks '[i]f, then, what is just being does not belong as an accident to anything, but other things belong to it, why should it be the case that what is just being signifies what is rather than what is not?' I take it that this sentence sums up the part of the argument formulated at 186b4–12, and that the question 'why should it be the case that what is just being signifies what is rather than what is not?' expresses the conclusion of that part of the argument (186b10), i.e. if what is just being is pale, what is just being is (absolute) not-being. Since any property of what is just being would be (absolute) not-being, it follows that 'what is just being' signifies not-being rather than being (or that it is unclear whether 'what is just being' signifies being rather than not-being).

μὴ ὄν or οὐκ ὄν (186b9–10, 11). Arguably this means that also pale has an essence, and hence that 'being' signifies many different things or ὅπερ ὄντα.[69]

What is the meaning of the argument? The assumption that 'being' must be either a subject or a predicate is problematic, since it is not clear whether true monists would accept it.[70] Arguably, this assumption is shared by the Eleatics, since they attribute various properties to being. However, the position tackled at 186a32–b14 is a strengthened version of Eleatic monism. A defender of this stronger version could support her monism by claiming that being is a self-subsistent entity (ὅπερ ὄν/ἕν), which has no property whatever but its own absolutely unitary essence, which has no relations with other things (like the οὐσία μόνον mentioned in chapter 2, 185b4), and which is such that its existence implies the annihilation of everything else. In this view, the argument can be read as a *reductio ad absurdum* (which a monist would endorse) of the hypothesis that ὅπερ ὄν/ἕν is either the predicate or the subject of something else,[71] rather than as a *reductio ad absurdum* of a monistic position.

However, some clues—both inside and outside *Physics* I 3—suggest that this is not what Aristotle intends. First, nothing in the argument suggests that the assumption that ὅπερ ὄν/ἕν is a subject of predication is denied and hence that the conclusion of the argument is that ὅπερ ὄν/ἕν is not a subject of predication.

Second, after concluding the argument put forward at 186a32–b14, Aristotle starts a new argument, which is aimed at showing that ὅπερ ὄν breaks down into parts that are themselves ὅπερ ὄν τι. In introducing this argument, Aristotle claims that the division of ὅπερ ὄν into ὅπερ ὄν τι ἄλλο is also (καί: 186b15) evident τῷ λόγῳ. The καί suggests that, like this new argument, the preceding one is aimed at showing that being cannot be an undifferentiated unity.[72]

Third, at 187a6–9, Aristotle goes back to the view that 'being' means ὅπερ ὄν (187a8–9). He first labels as absurd the idea that 'if there is nothing besides being

[69] If the only meaning of 'being' is ὅπερ ὄν, this implies that the claim 'ὅπερ ὄν is pale' means 'ὅπερ ὄν *is* ὅπερ λευκόν (i.e. the colour pale)'. Alternatively, the former claim means 'ὅπερ ὄν *is* accidentally pale', which requires that 'being' means both ὂν καθ' αὑτό and ὂν κατὰ συμβεβηκός. This second reading is needed, if ὅπερ ὄν is conceived of as a subject of accidental predication. Moreover, this second reading would be correct even if the claim that 'also the pale signifies ὅπερ ὄν' means that the phrase 'the pale' signifies the accidental unity comprised by the subject ὅπερ ὄν and the predicate 'pale'. Lastly, there is a further possible reading of the conclusion of this argument, i.e. that 'pale' is another name for ὅπερ ὄν (just like 'jail' and 'prison'), so that ὅπερ ὄν would have two different names: 'being' and 'pale'. I find this reading less likely than the others, since the point at issue in this passage seems to be the difference in being between 'pale' and ὅπερ ὄν (186b7). Moreover, the issue of the two names could also have been raised about 'being' and 'one' (i.e. the fact that 'being' means both ὅπερ ὄν and ὅπερ ἕν). However, it is not raised.

[70] On this problem, see Bostock 2006, 111.

[71] For a defence of this interpretation, see Castelli 2018, 93–7.

[72] The argument provided at 186a32–b12 plus b12–14, as I suggest above, shows that what is cannot be an undifferentiated unity from the point of view of accidental being, whereas the argument provided at 186b14–35, as I will suggest in section 4, shows that what is cannot be an undifferentiated unity from the point of view of essential being.

itself (αὐτὸ τὸ ὄν), all things will be one'. Thereby he probably means that it is absurd to believe that the only way to avoid Eleatic monism is to claim that there is absolute not-being. But why is it absurd? Clearly it is absurd since it implies contradiction and since contradiction should and can be avoided. However, the next sentence seems aimed at stressing a further point: 'For who understands (μανθάνει) being itself (αὐτὸ τὸ ὄν), if what is just being is not a certain thing (εἰ μὴ τὸ ὅπερ ὄν τι εἶναι)?' (187a8–9). This probably means that, even if we grant that there is absolute not-being (and, hence, even if we grant that there are two distinct principles: being and not-being), it will be impossible to understand being. For in order to understand being, it is necessary for being (ὅπερ ὄν) to be something (τι), i.e. to have properties and, primarily, to have an essence. If this is so (i.e. if *ὅπερ ὄν is ὄν τι*)—concludes Aristotle—'then nothing prevents things that are from being many, as it has been said'.[73]

This passage suggests that, in Aristotle's view, the hypothesis that being is a self-subsistent entity, which has no property whatever but its own absolutely unitary essence and which has no relations with other things, is untenable. This conception of being would allow only the claim that 'being is being'. In this connection it is worth recalling that, in *Met.* Z 17, while dealing with οὐσία conceived of as principle and cause, Aristotle maintains that inquiring into 'why a thing is itself' (αὐτό ἐστιν αὐτό) amounts to inquiring into nothing (1041a14–15). Similarly, in *Met.* Γ 4, 1006a21–32, he emphasizes that for a term to signify at all is to signify something determined (τι . . . ὡρισμένον: 1006a24–5): a name (e.g. 'human being') signifies 'being or not-being *this*' (e.g. animal biped) (1006a29–32). Moreover, the passage where Aristotle claims that, if ὅπερ ὄν is ὄν τι, 'then nothing prevents things that are from being many, *as it has been said*' (187a9–10), suggests that Aristotle has already argued for this point. The phrase 'as it has been said' (187a10) must refer to one or both of the two arguments about ὅπερ ὂν provided in *Physics* I 3 (186a32–b14, 186b14–35). If the reference includes the argument put forward at 186a32–b14, then this argument is a *reductio ad absurdum* of a monistic position.

It may be objected that these considerations, as well as Aristotle's claim at 187a8–9, concern essential predication, whereas the argument put forward at 186a32–b12 concerns accidental predication. Whether or not this objection is correct, however, it can be argued that the hypothesis that all that is is a self-subsistent entity is also at odds with Aristotle's view on Eleatic monism from a second point of view, which takes accidental predication into account, too.

Eleatic monism is presented by Aristotle as a theory about the being of things, namely about being conceived of as the principle of all the things that

[73] οὐδὲν ὅμως κωλύει πολλὰ εἶναι τὰ ὄντα (187a9–10) can be read as follows: there is nothing else that prevents that there are many beings.

are, and not as the postulation of a self-subsistent entity that does not have any relation with the things of our ordinary experience or whose existence even implies the annihilation of everything else.[74] Moreover, the Parmenidean argument that being is absolutely one is presented as an argument that starts from what is more knowable to us, i.e from ὂν κατὰ συμβεβηκός, and deduces the fundamental and true unity of what is (apparently) many. So, if the Eleatic being (or the stronger version of it) is intended to be the principle of all things or the true being of all things, the hypothesis that this being is a self-subsistent entity is untenable not only for Aristotle but also for the Eleatics themselves (and indeed for an Eleatic-style monist) at least in Aristotle's view.[75]

Therefore, this line of thought suggests that the meaning and the final conclusion of the argument put forward at 186a32–b12 is that it is necessary to acknowledge that 'being' signifies many things. Aristotle's argument can be seen as aimed at showing that Parmenides' Way of Truth, if travelled consistently and from a more suitable starting point than Parmenides chose, does not lead to an absolutely unitary being but to contradiction (being is absolute not-being). His Way of Truth is both wrong and a side track of the natural path and leads exactly where he did not want to go.[76]

3.1. A further consequence of the hypothesis that 'being' signifies ὅπερ ὄν: what is just being lacks magnitude (186b12–15)

After concluding, at 186b12, that 'being' signifies many things, at 186b12–14 Aristotle continues to examine the consequences of the hypothesis that 'being' signifies only ὅπερ ὄν.[77] He stresses that this hypothesis implies that being lacks magnitude. This is because, if being had magnitude, the being of each of the

[74] A reference to this approach can also be found at 186b35, where Aristotle mentions the hypothesis that the universe (τὸ πᾶν) is composed of indivisibles, for here being seems to be conceived of as *all* that is. Cf. also *Physics* I 2, 184b15–16, 184b25–185a5, 185a22; *Met.* A 5, 986b10ff.; *Met.* B 4, 1001a5–6. This is also Plato's view on Eleatic monism (cf. *Soph.* 251d5–9, e8–d2).

[75] Moreover, note that, contrary to the implications of the hypothesis that being is a self-subsistent entity, in Parmenides' view the plurality of beings belongs to the Way of opinion and not to the impassable Way of absolute not-being (B6).

[76] Tellingly, this conclusion is similar to the one that Aristotle draws in chapter 2 (185b19–25) while tackling the hypothesis that what is is one in account. There, he deduces that being is not-being (i.e. all things are nothing) from the hypothesis that all the things that are have the same essence and from the implicit assumption that these things are. In chapter 3 (186a32–b12), instead, he deduces that being is not-being from the assumptions that what is different from being is absolute not-being and that 'being' must be either a subject or a predicate.

[77] In his formulation of this assumption Aristotle speaks only of ὅπερ ὄν (186b13), but he arguably means both ὅπερ ὄν and ὅπερ ἕν.

parts would be different, thereby contradicting the assumption that 'being' signifies only *ὅπερ ὄν*.

Aristotle has in mind the division of something into two parts that have a different essence from each other. It is not immediately clear what kind of division is at issue. There are at least two possibilities:

(a) A dichotomic division of being into quantitative parts that have an essence and whose respective essences are different from each other. As Ross emphasizes, if this is the case, Aristotle would probably have in mind Zeno's proof that, if being is extended, it is subject to *διχοτομία ad infinitum*: each of the two parts resulting from each successive dichotomy will necessarily have a different being from the other part.

(b) A logical distinction between being, conceived of as a substrate, and magnitude, conceived of as an attribute of a substrate.

These two possibilities need not be incompatible, insofar as the general sense of the passage is concerned.[78] Rather, in the preceding section Aristotle has shown that, if 'being' signifies only *ὅπερ ὄν/ἕν*, it can be neither an attribute (which belongs to a substance) nor a substrate (which receives attributes). From this it follows that being cannot have magnitude (i.e. any quantitative property would amount to absolute not-being), since magnitude falls into the category of quantity, which is different from the category of substance.[79] Further, if being lacks magnitude, it cannot be divided into quantitative parts, each endowed with a different being from each other (and, arguably, from the whole). Thus, being is something indivisible.[80]

Interpreting the passage in this way means viewing it as a compressed and elliptic argument. Interpreted thus, however, the passage serves nicely as a link between the preceding section (186a32–b12) and the following one (186b14–35):[81] the preceding section deals with the distinction between substrate and accidental attributes, whereas the following one deals with the internal division of *ὅπερ ὄν*,

[78] Cf. Castelli 2018, 96–7.

[79] Cf. *Physics* I 2, 185 a32–b5.

[80] Cf. *Physics* I 2, 185b16–19.

[81] This is also suggested both by the way in which the passage is introduced and connected to the previous part and the way in which the passage is linked to the subsequent part. The passage is introduced by *οὐ τοίνυν οὐδὲ*. *Τοίνυν* may have either a weak logical or a transitional sense, but the distinction between these senses is not very sharp (Denniston 1954: 568–80). The subsequent part of the chapter is linked to this passage by *καί*. At 186b14–15, Aristotle makes the following claim: that *ὅπερ ὄν* breaks down into a different *ὅπερ ὄν τι* is also (*καί*) clear by definition. Then he tackles the composition of the definition of what is just a certain being (*ὅπερ ὄν τι*). The *καί* may (but need not) be read as implying that also the previous passage (186b12–14) concerns the division of *ὅπερ ὄν* considered by itself and hence that the parts mentioned at 186b14 are the parts that result from a dichotomic division of *ὅπερ ὄν*.

considered by itself from the point of view of definition (i.e. not as a substrate of accidental attributes), into parts that are themselves ὅπερ ὄν τι; moreover, the following section contrasts the idea that ὅπερ ὄν breaks down into such parts with the hypothesis that the universe is composed of indivisible entities (186b35).

As we have seen, Aristotle did not label the conclusion of the preceding argument (186b10–11) as absurd, impossible, or unacceptable, nor does he do so here. However, the conclusion drawn at 186b12–14 is in conflict with Parmenides' and Melissus' claims that being is (respectively) limited and unlimited, for both these claims imply that being has magnitude.[82] Further, this conclusion is not simply at odds with the plain facts of experience,[83] but more radically it implies the annihilation of everything else. Similarly to the preceding argument (186a32–b12), it implies that the things that are are nothing.[84]

4. The Argument from Definition (186b14–35): What Is Just a Certain Being (ὅπερ ὄν τι) Breaks Down into Parts

In this section Aristotle argues that ὅπερ ὄν is articulated into parts that are themselves ὅπερ ὄν τι. The ὅπερ ὄν tackled here is not the ὅπερ ὄν and ὅπερ ἕν dealt with in the preceding two sections.[85] Rather, it is a determined substance, which is described as 'what is just a certain being' (ὅπερ ὄν τι: 186b15–16), since substances are the only beings that are what they are without being something else (*APo* I 4, 73b3–8).

To argue that substances are articulated into parts Aristotle uses the example of the human being and considers the human being from the point of view of its essence and as the subject of definition: 'the human being is an animal biped'. The purpose is to show—from the point of view of the structure of definitions (τῷ λόγῳ: 186b15)[86]—that what is just a certain being, e.g. the human being, breaks down into something else that is what is just a certain being, e.g. animal biped (186b14–15). This does not simply mean that animal and biped comprise

[82] Cf. *Physics* I 2, 185b16–19, where Aristotle disputes the hypothesis that what is is indivisible, by showing that it conflicts with Parmenides' and Melissus' claims that what is is, respectively, limited and unlimited.

[83] Cf. Ross 1936, *ad loc.*

[84] According to an alternative interpretation (Castelli 2018, 96–7), the conclusion of this argument describes the kind of being that genuine monists want or should want: an entity that lacks magnitude.

[85] Arguably, this is because the hypothesis that 'being' means only ὅπερ ὄν and ὅπερ ἕν has been refuted in the preceding two sections (186a34–b14).

[86] For other interpretations of τῷ λόγῳ, cf. Philop. *In Physica* 75, 20–5, Bostock 2006, 113.

the essence of the human being, but that they comprise the essence of the human being and thereby have the same ontological status as the human being. So, the purpose of this section is to argue that a substance is composed of various substances: the essential properties of substances are themselves substances.[87] Moreover, claiming this amounts to claiming that, not only at the level of ὂν κατὰ συμβεβηκός but also at the level of ὂν καθ' αὑτό, being is both one and many.[88]

Arguably, here ὅπερ ἕν means 'substance' in the sense of the *Categories* and not in the sense used, for instance, in *Met.* Z. If ὅπερ ὄν means 'substance' in the sense of the *Categories*, then both the genus 'animal' and the difference 'biped' can be labelled as ὅπερ ὄν τι.[89] 'Animal' is clearly a substance and 'biped' is a substance as well, at least if it is considered together with animal, i.e. as meaning 'biped animal' rather than the property of having two feet.[90] Instead, if ὅπερ ὄν means 'substance' in the sense of *Met.* Z, then neither the genus 'animal' nor the difference 'biped' (at least if considered by itself, i.e. as meaning the property of having two feet) is a substance (*Met.* Z 13, 1038b15–16, 23–5). Besides, in *Met.* Z 13, Aristotle clearly maintains that it is impossible that a substance is composed of substances that are present in it in actuality, since two things that are in actuality cannot comprise a unitary thing (1039a3–8). So, according to *Met.* Z 13, animal and biped cannot be conceived of as two substances.

The issue tackled at 186b14–35 is connected with important points of Aristotle's inquiry in chapters 6–9. In chapter 6 (189a29–34), Aristotle claims that a substance cannot derive from, or be composed of, things that are not substances, i.e. its essence cannot be identical to or made of the essence of the opposites, since opposites are affections of the substance and are therefore posterior to it.[91] Moreover, one of the main achievements of chapter 7—which

[87] Cf. *APo* I 22, 83a38–b1: ἢ γάρ τοι ὡς οὐσία κατηγορηθήσεται, οἷον ἢ γένος ὂν ἢ διαφορὰ τοῦ κατηγορουμένου.

[88] The issue of the articulated unity of the *definiens* is dealt with by Aristotle several times in the *Metaphysics*, particularly in *Met.* Z 12–13 (cf. also *Met.* H 6, *De Int.* 5, 17a13–15, *APo* II 6, 92a29–33). In the light of this, the argument from definition (186b14–35) may be connected with the example of the 'circle' that Aristotle uses in chapter 1 to illustrate the search for principles as the articulation of a whole into parts. One possible reading of this example is the following: the term 'circle' expresses its meaning in a unitary (ὅλον: 184b2), confounded and unarticulated way (ἀδιορίστως: 184b2), whereas its *definiens* articulates this meaning into its parts (διαιρεῖ: 184b3).

[89] Actually, at 186b14–15 Aristotle does not say that ὅπερ ὄν breaks down into many ἅπερ ὄντα (plural), but that it breaks down into another ὅπερ ὄν τι (singular). Moreover, in the whole section animal and biped are treated mostly as a unity, i.e. as a unitary but complex structure, and not as two distinct elements. However, that in *Physics* I 3 each of the elements comprising the definition of the 'human being' is conceived of as ὅπερ ὄν τι is suggested by 186b31–3. This is arguably because 'biped' is conceived of as 'biped animal' and not as the property of having two feet.

[90] Cf. *Top.* I 4, 101b18–19; *Met.* Z 12, 1037b29–33, 1038a5–9, 19–20, 25–6.

[91] The construction ἐκ plus genitive is used both in this section of chapter 3 (186b35) and in chapter 6 (189a33) to mean what a substance is essentially composed of.

is then employed in chapters 8 and 9—is that privation is not the essence of matter: matter 'is near substantial being and is substantial being in a way, while the other [sc. privation] is in no way at all substantial being'; hence, matter 'is what is not incidentally, while privation is in itself what is not' (*Physics* I 9, 192a3–6). In the light of these connections, the argument from definition can also be read as preparing the ground for the subsequent claims about the difference in being and account between substance and opposite, and between matter and privation.

Let us now consider the train of thought that leads to the argument from definition. In dealing with the structure of definitions Aristotle is continuing the investigatory path travelled so far in chapter 3: after focusing, at 186a25–32, on the difference in account and being between subject and predicate, Aristotle now considers the internal structure of the account that expresses the being or essence of things. Here, his assumption is that the essence of things (in particular, the essence of substances) is expressed by a definition, which is a unitary but composite linguistic structure comprising a genus and a difference.

Aristotle's argument is set out as a dilemma. It assumes that 'animal' and 'biped' are either what is just a certain being or accidents. Thereby, it contends that, if they are not accidents, they must be what is just a certain being. It shows, by means of a *reductio ad absurdum*, that they cannot be accidents and hence concludes that they are what is just a certain being.

The *reductio ad absurdum* goes as follows:

(1) If animal and biped were accidents, they would be either (a) accidents of the human being or (b) of something else (186b17–18).

(2) If animal and biped were accidents of the human being, i.e. (a), they would be either (a^1) attributes that can be separated from (i.e. denied of) the human being or (a^2) attributes that include the human being in their own definition (186b18–23).

(3) It is impossible that animal and biped are accidents of the human being in sense (a^1), since it is impossible for the human being not to be animal and biped (186b26–30).

(4) It is impossible that animal and biped are accidents of the human being in sense (a^2), since they are part of the definition of the human being and what is part of the definition of something does not include the *definiendum* in its own definition (186b23–6, 29–31).[92]

[92] Ross suggests that this latter hypothesis is ruled out since otherwise definitions would be circular. This is arguably a point at issue. However, it may not be the only one. If a term A, which is present in the definition of another term B, includes the term B in its own definition, this implies that the term A

(5) It is impossible that animal and biped are accidents of a subject different from the human being, i.e. (b), because also the human being would be an accident of this subject (186b31–2). This is impossible since the human being, insofar as it is ὅπερ ὄν τι (i.e. a substance), is not the accident of anything (186b33–4).

The *reductio ad absurdum* of the hypothesis that animal and biped are accidents implies therefore that a substance (ὅπερ ὄν τι) like the human being is articulated into parts that are themselves substances (ὅπερ ὄν τι). This means—as already noted—that not only ὂν κατὰ συμβεβηκός but also ὂν καθ' αὑτό and, in particular, substantial being is both one and many: it is a unity articulated into parts.

In the concluding line of this section Aristotle says: ἐξ ἀδιαιρέτων ἄρα τὸ πᾶν (186b35). The meaning and argumentative function of this sentence is controversial. It seems unlikely that ἐξ ἀδιαιρέτων ἄρα τὸ πᾶν is the conclusion of the whole argument: from the fact that a ὅπερ ὄν τι breaks down into parts that are themselves ὅπερ ὄν τι it does not follow that the universe is composed of indivisible entities.[93] Nothing excludes the possibility that the parts of an ὅπερ ὄν τι, can themselves divide into further ὅπερ ὄν τι. For instance, the argument from definition can arguably be applied to animal.

Another possibility is to consider ἐξ ἀδιαιρέτων ἄρα τὸ πᾶν as the conclusion of the lines that immediately precede it (186b34–5): καὶ καθ' οὗ ἄμφω ... λεγέσθω. The meaning would be that, if animal and biped were accidents of a subject different from the human being and if, therefore, the human being were an accident of this subject, the universe would be composed of indivisible entities. Aristotle might be referring to a conception according to which the fundamental beings are indivisible in being and account, and all the other beings are accidents of them. This would be a kind of monism that kept absolute unity (in being and account) and the status of ὅπερ ὄν only for the ultimate constituents of reality. A possible objection to this interpretation is that the argument is a *non sequitur*, since from its premises it is also possible to conclude that everything is accidental. Moreover, καὶ καθ' οὗ ἄμφω ... λεγέσθω is preceded by, and

has two distinct and incompatible relations to term B: insofar as it is present in the definition of term B, it is part of the expression of the essence of term B; insofar as it includes term B in its own definition, it is a necessary accident of term B. In the first case, the relation is such that every B is A (e.g. every human being is an animal), but not the reverse, whereas in the second case the relation is such that every A is B (e.g. every male is an animal), but not the reverse. Moreover, if animal and biped are treated as distinct accidental properties rather than as a unity, the hypothesis implies that every animal is a human being and every biped is a human being, which is patently not the case.

[93] Also, if τὸ πᾶν is taken to mean the whole (e.g. animal biped), rather than the universe, the conclusion still does not follow.

connected to, ἀλλὰ . . . συμβεβηκός, i.e. by an adversative claim that states Aristotle's assumption that the human being is a ὅπερ ὄν τι, rather than a συμβεβηκός. In the light of this, it seems absurd that the statement of Aristotle's assumption (ἀλλὰ . . . λεγέσθω) should comprise an argument for its own negation. Rather, καὶ καθ' οὗ ἄμφω . . . λεγέσθω can be interpreted as stating what 'animal biped' and 'human being', insofar as they are ὅπερ ὄν τι, are correctly said of, i.e. a particular human being. Of course, one may reply that the argument is formulated in a compressed and elliptic way, so that these objections are not fatal.[94]

Ross advances a further interpretation, which fits better with the syntactical construction of the preceding lines. He takes it that the main argument ends at b33 with ἑτέρῳ, and that ἀλλὰ . . . τὸ πᾶν adds a reflection of Aristotle's own. Moreover, following Pacius, he suggests that ἐξ ἀδιαιρέτων ἄρα τὸ πᾶν is a question rather than a statement, and that we should read ἆρα instead of ἄρα: 'is the all composed of indivisible entities, *then*?' The question would be a rhetorical one that is introduced here to present an alternative view to Aristotle's in a rather dismissive way: 'what would an opponent say to avoid the conclusion of this argument? Would she deny its most basic assumption, i.e. that the essence of things is expressed by a λόγος (i.e. a composite phrase)? Would she claim that the universe is composed of indivisible entities?' The hypothesis that the universe is composed of indivisible entities is alternative to Aristotle's view, since the latter implies that substances (as well as accidents and the essential predicates of substances and accidents) can be analysed (i.e. defined).[95] The assumption that the essence of things is expressed by a λόγος is clearly in place since the beginning of Aristotle's resolution of Parmenides' argument, for the resolution is based, among other things, on the idea that the difference in being is a difference in account (186a28–31). Moreover, the introduction of a hypothesis on the structure of reality that challenges the most basic assumption of Aristotle's argument (i.e. that the essence of things is expressed by a λόγος) fits well with the preceding lines (186b33–5), where Aristotle presents two other assumptions of his argument.[96]

[94] In reply to the first objection, one may say that the inference from the premises to the conclusion is based on an implicit argument for the existence of indivisible entities (the argument from dichotomy, mentioned at 185a2–3?).

[95] A similar alternative is tackled also in *Met.* Z 13, 1039a14–21, where Aristotle claims that if substances are composed neither of universals nor of substances that are present in them in actuality, then substances will be incomposite, so that there will be no definition of any thing.

[96] After claiming that, if animal and biped were accidents of something else, the human being would be an accident as well (186b31–3), at 186b33–5 Aristotle concludes his reasoning with an adversative claim (ἀλλά) that states the two main assumptions of the argument against this hypothesis. The two assumptions are: (1) τὸ ὅπερ ὄν, i.e. the human being, is not the accident of anything; (2) insofar as it is

The difference between these two interpretations concerns the entities that are indivisible: according to the first, only the ultimate constituents of reality are indivisible (in being and account) whereas, according to the second, all things are indivisible (in being and account). The reference to indivisible magnitudes in the following section (185a3) can be taken as providing evidence in favour of the first interpretation, on the assumptions that the indivisible magnitudes are indivisible in being and account and that the two sections follow a unitary train of thought.

Be that as it may, what is clear is that Aristotle is considering a possible alternative to his view, i.e. that reality instead comprises things that are indivisible in being and account. This view will be criticized at 187a8–9, on the grounds that, if being were not something determined (i.e. if it lacked an account of its essence or both essential and accidental properties), it would be impossible to understand.[97]

5. Aristotle's Criticism of Some Post-Eleatic Pluralistic Thinkers (187a1–11)

In the concluding section of the chapter Aristotle tackles the view of some post-Eleatic philosophers (ἔνιοι) who gave in to two Eleatic arguments: the Parmenidean argument resolved at 186a25–32[98] and an argument that is here labelled as the argument from dichotomy.[99] These thinkers gave in to these arguments in that they accepted their conclusiveness. However, they aimed at saving the phenomena. So, they thought that plurality and change could be accounted for only by supposing that absolute not-being exists and that there are indivisible magnitudes (ἄτομα μεγέθη).[100]

composed of, or derives from, animal and biped (τὸ ἐκ τούτων), the 'human being' is predicated of the subject of which 'animal' and 'biped' are predicated.

[97] Cf. also *Met.* Γ, esp. Γ 4 and Γ 7, 1012a21–4, where Aristotle argues that to signify, when applied to a term, means to signify a definition (ὁρισμός).

[98] See section 2.

[99] Both ancient and modern commentators identify the argument from dichotomy with an argument by Zeno against plurality. Unlike Parmenides' argument, the argument from dichotomy is not tackled in chapter 3. Arguably there are implicit references to it at 186b12–14, 35.

[100] The first line of this section suggests that Aristotle is here tackling the view of a single group of thinkers who claimed both that absolute not-being exists and that there are indivisible magnitudes. This hypothesis does not exclude that Aristotle's criticism can be extended to those who gave in to one or other of the two Eleatic arguments mentioned here. As for the identification of these thinkers, irrespective of whether they are a single group or not, various hypotheses have been advanced. Ancient commentators see a reference to Plato and Xenocrates (see Ross 1936, 480). Instead contemporary interpreters

Aristotle's criticism is broken down into two parts (187a3–6, 6–11). In the first part (187a3–6),[101] which relies on the preceding inquiry (186a22–32), Aristotle stresses the *non sequitur* of Parmenides' argument and shows thereby that, to account for plurality, it is not necessary to posit absolute not-being and to violate the principle of non-contradiction: from the premises of Parmenides' argument it does not follow that there is no not-being, for nothing excludes that there is a kind of not-being, which is something that is not something else. In the second part (187a6–11),[102] Aristotle labels as absurd the view that the only way to avoid the conclusion that all things are one is to posit absolute not-being. He labels this view as absurd on the grounds that, if being is only being itself (αὐτὸ τὸ ὄν, ὅπερ ὄν) rather than something determined (τι), it would be impossible to understand (μανθάνει).[103] Arguably, this is the view that being is absolutely unitary and therefore lacks any essential or accidental property. The description of this view fits both the hypothesis that there is only one being of this kind and the hypothesis that there are many (arguably the many indivisible beings and the many indivisible magnitudes mentioned respectively at 186b35 and 187a3). The conclusion is that, if one accepts that being is instead something determined, beings will be many. More precisely: if being is determined by accidental properties, there will be many kinds of being (substance and accidents), and being will be one in number and many in account; if being is determined by its essence and account, being will be internally articulated into parts.[104]

In sum, this concluding section sets out to claim the following:

(1) The view of these philosophers is based on an incorrect argument (Parmenides' argument).[105]

(2) The move that these philosophers make to avoid the Eleatic conclusion that there is only one being (i.e. giving in the principle of non-contradiction) is both unacceptable and unnecessary.[106]

(3) The conclusion of their arguments is absurd, since the indivisible beings (or being) that they posit, insofar as they do not have properties and are not something determined (τι), cannot be understood.

hold that Aristotle is criticizing the atomists (Ross 1936, 480–1; Wagner 1967, 411; Charlton 1970, 63; Horstschäfer 1998, 118–24).

[101] I tackled this part in section 2, 101–2.

[102] I tackled this part in section 3, 112–13.

[103] It is worth noting that in *Met.* Γ 4 the idea that being is something determined and that an ὄνομα signifies 'being or not-being this' (1006a21–5, 29–30) is connected to the validity of the principle of non-contradiction.

[104] In Aristotle's view, it is impossible that everything is accidental (*Met.* Γ 4, 1007a20–b18), which implies that, if being has accidental properties, it also has essential properties.

[105] Aristotle does not provide here an evaluation of the argument from dichotomy.

[106] Cf. also *Physics* I 2, 185b34–186a3.

This passage shows Aristotle's interest in resolving the arguments of his predecessors: first the arguments of the Eleatics and then of some of those who, by giving in to arguments of the Eleatics, did not find the proper solution to the problems addressed in them, and hence did not proceed correctly on the path towards the principles of natural things. Aristotle is here following the steps made by some of his predecessors along the natural path. His purpose is to correct their mistakes and to put natural science back on the right path of research, i.e. the one leading to the true first principles of natural things. Here, his progress along the path followed by his predecessors is not yet complete. Chapter 3 ends with the claim that '[i]t is therefore clear that it is impossible that what is is one in this way', i.e. undifferentiated in being and in account (187a10–11). The inquiry proceeds, from chapter 4 onwards, by considering the way in which what is/what is one is conceived of by those thinkers (labelled by Aristotle as 'physicists') that, unlike the Eleatics, admitted some kind of internal articulation of what is/what is one.

4

One and Many

Physics I 4

GÁBOR BETEGH

After a detailed critical examination of the views of Melissus and Parmenides in chapters 2 and 3, Aristotle in chapter 4 turns back to those who, in his view, practised bona fide natural philosophy. The different groups of these thinkers were already mentioned at the beginning of chapter 2, in the initial categorization of different positions concerning the number and basic characteristics of the principle(s) and being. Yet the painstaking analysis and criticism of Eleatic monism in chapters 2 and 3 had crucial bearings on the course of the discussion: although not properly part of natural philosophy, Aristotle had to come clear with the fundamental problems posed by the unitary being of the Eleatics before moving on to a proper evaluation of the principles of natural philosophers. Aristotle thus does not mechanically continue with discussing the different options concerning the number of principles as the initial diaeresis could have suggested. Rather, he focuses on the various ways in which, by allowing some internal differentiation and change, different groups of natural philosophers, as opposed to the Eleatics, produced the multiplicity out of the one. Aristotle will be especially interested in those who conceive the original one as containing a mixture, and which therefore is already a multiplicity in some way. He then

I am greatly indebted to participants of the Symposium for their helpful discussions and criticisms, and in particular to John Cooper and Sean Kelsey for a set of extremely helpful comments. I had the opportunity to present a previous version of this chapter at the SEEAP workshop in Zagreb, and at the University of São Paolo, where I received helpful feedback from the participants, and especially from Andreas Anagnostopoulos, István Bodnár, Pavel Gregoric, Christopher Noble, and Marco Zingano. Preliminary research for this chapter was supported by the ERC_HU09 BETEGH research grant.

singles out one thinker of this group, Anaxagoras, and consecrates the rest of the chapter to a discussion and refutation of Anaxagoras' theory of universal mixture.

These three chapters are often presented as one of the paradigmatic examples of Aristotle's endoxic method. In particular, the treatment of these thinkers in *Physics* I is often lumped together with the review of the predecessors' position in *Metaphysics* A. Yet, despite some surface similarities, a more careful look reveals that Aristotle's aims and strategy are fundamentally different in these cases. Indeed, it soon turns out that, also in the light of the explicit methodological statements of chapter 1, it is less obvious what Aristotle means to achieve in these chapters within the overall project of the first books of the *Physics*. A connected important question concerns the role of Anaxagoras: what feature or features of Anaxagoras' theory motivate Aristotle's choice of providing a detailed criticism of his theory of matter, whereas he treats the Ionian and neo-Ionian monists very briefly, does barely more than congratulate Empedocles, and provides a mere one-sentence argument to prove that Democritus' theory also takes opposites to be fundamental (188a22–6). Having a better grasp of why Aristotle might think that a proper assessment of the mixture theories, and especially Anaxagoras' theory of matter, should be so central to his project could, I suggest, offer us a key to a better understanding of Aristotle's overall strategy.

1. Another Categorization of Natural Philosophers

[187a10] It is evident, then, that it is impossible for 'what is' to be one in this way [i.e the way the Eleatics conceived it, and as has been analysed in chapters 2 and 3].

Chapter 4. [1] As to the natural philosophers, their accounts take two forms. [2] For some of them made 'what is' one body, sc. the underlying, either one of the three[1] or something else which is denser than fire, but rarer than air,[2] and then generate the others by denseness and rarity,[3] making it thus many. [3] (These are opposites, more

[1] I.e. fire, air, water. None of the monists posited earth as the single underlying body. Cf. also *DA* I 2 405a8–9.

[2] The identity of this intermediate substance theorist has been hotly debated since antiquity (see the extended discussion of Ross 1936). After quite some pondering, I have to confess that I have nothing original or interesting to say on the issue. It is true that at other places Anaximander seems to be taken by Aristotle as the intermediate substance theorist (cf. e.g. KRS 1983, 111–12). Yet Anaximander in the present context is clearly taken to posit not a single homogenous yet intermediate underlying body, but rather a mixture. From the available candidates, I would tentatively opt for Diogenes of Apollonia in view of the importance he attaches to hot air. Diogenes was suggested in antiquity by Nicolaus of Damascus and Porphyry (apud Simplicius *In Physica* 149.13; 151.21), and, as far as I could trace it, the suggestion was revived in modern times by Prantl 1854, 476, n. 16. Another option would be a Heracletianising *anathymiasis* theorist (cf. e.g. *Probl.* 934b34).

[3] Hardie and Gaye in *ROT* (1984) translate 'by condensation and rarefaction'. In view of Aristotle's overall argument in *Phys* I, it is important that he does not mean the *process* of condensation and rarefaction, but rather the *opposites* of denseness and rarity.

> generally excess and defect, as Plato speaks about the great and the small, except that he makes these the matter, and the one the form, whereas these people make the one, the underlying, matter, and the opposites differentiae and forms.)[4] [4] Others, on the other hand, have maintained that the opposites are separated out from the one in which they are already present, as Anaximander says, [5] and those who say that [it] is one and many, like Empedocles and Anaxagoras, for they too separate out the others from the blend. [6] They differ from one another in that Empedocles makes this process occur cyclically, whereas Anaxagoras makes it happen once, and, moreover, Anaxagoras [separates out] an infinite, both the homoiomeries and the opposites, whereas Empedocles only the so-called elements. (187a10–26)

Let me start with a textual question, which might have an effect on the overall assessment of the section. The construal of the second sentence of the chapter is problematic. Ross proposes to emend the received text by excluding ὄν before σῶμα, despite the unanimous evidence of the manuscripts and Philoponus' commentary, as well as the Arabic translation.[5] It seems to me, however, that the received text, although admittedly not very smooth syntactically, makes perfect sense. Indeed, I would maintain, the presence of τὸ ὄν is essential for the correct understanding of the way in which Aristotle connects and contrasts the Eleatics discussed in the previous chapter and the natural philosophers who constitute the topic of this chapter.

Note first of all that the chapter division cuts across a μὲν οὖν–δέ, and hence the first sentence of chapter 4 should be treated as picking up, and responding to, the conclusion of chapter 3. The conclusion of the discussion of the Eleatics states that making 'what is' one *in this way* (οὕτως), i.e. in the way the Eleatics conceived it, is evidently impossible. In the consecutive sentences, constituting the first section of chapter 4, Aristotle continues with the position of the Ionian monists, who also made 'what is' one, but in a different and more tenable way. This articulation between the two groups gets concealed by the elimination of τὸ ὄν. Accordingly, many of the standard translations make the direct object of ποιήσαντες the underlying body (τὸ σῶμα τὸ ὑποκείμενον): 'the first set make the underlying body one' (Hardie and Gaye 1984), 'some make the underlying body one' (Charlton 1984), or 'some regard the underlying body one' (Simplicius 2011: viii and 149).[6] It seems to me, however, that the sentence wants to express

[4] Although the authenticity of this sentence is guaranteed by the unanimous textual tradition, it is a curiously unprepared-for anticipation of the form–matter distinction. The reader is supposed to understand what form and matter mean for Aristotle, although he will only explain it in the next chapters. Indeed, this is the first occurrence of the term ὕλη in the text. This quick side reference to Plato should also be contrasted with the sustained analysis and assessment of his views in chapter 9. At any rate, I don't think Charlton 1970, 63 is right in claiming that Aristotle simply counts Plato as belonging in the first group of natural philosophers.

[5] On this, see Cerami 2018.

[6] This is true even of Cerami who, rightly and with good arguments, defends the transmitted text. Pellegrin 2000, 87 is one of the exceptions: 'Les uns, en effet, ayant fait de l'étant un corps unique qui est le substrat etc'.

philosophers can work with another mechanism, 'separating out'. So Aristotle is now focusing on the different conceptions of such initial unities, and on the different ways in which starting from different types of unities the many can be arrived at.

I therefore suggest the following articulation of the text. At the beginning of chapter 4, Aristotle picks up the discussion of the Eleatic monists by contrasting their inarticulable and immobile solitary being with the monism of the Ionians that contains the possibility of motion and plurality. What is specific to the Ionians is that they conceive 'what is' one by making it a single underlying body, as one of the elements, and allowing motion and change in it, as well as a pair of opposites, the dense and the rare, and thereby a mechanism by which they can generate the plurality out of the one underlying body. There are however other conceptions of the initial one, and correspondingly other mechanisms to account for the emergence of multiplicity out of the one. These are the philosophers, Anaximander, Anaxagoras, and Empedocles, who posit an original mixture, and instead of working with the dense and the rare, use 'separating out' the many of the initial indistinct blend.[10]

If this is correct, Aristotle at this stage does not add a further subdivision to the original diaeresis of the number of beings, but rather starts a new division about the one (Fig. 3):

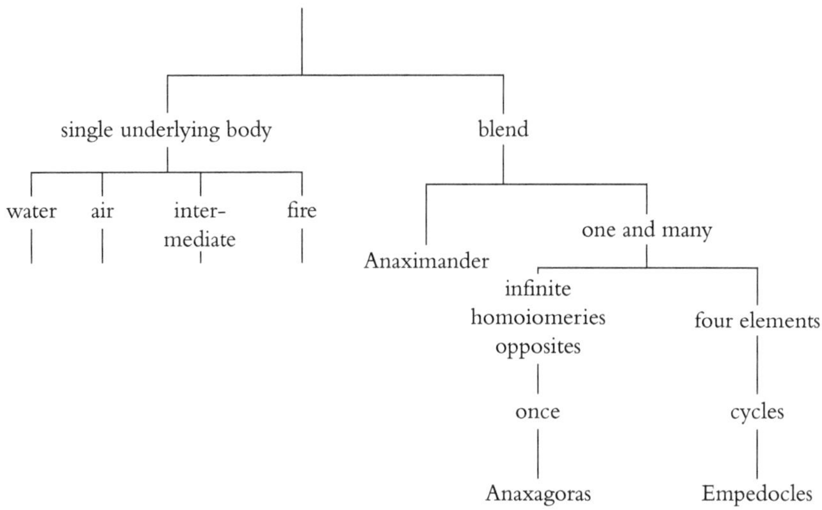

that Aristotle is temporarily bracketing the question of the moving cause, and concentrates on that out of which things are. Cf. Menn and Crubellier in this volume.

[10] It seems that Aristotle does not take the one of this group to be an underlying body. This is also supported by his formulations in chapter 6 (cf. also Anagnostopoulos in this volume. Simplicius still groups Anaximander with the one underlying body group, and separates him from Empedocles and Anaxagoras (154, 1). But this does not seem to respect the natural articulation of the text. He then reports Theophrastus (fr. 228B FHSG) as someone who rather assimilates Anaxagoras' mixture to Anaximanders'.

One important further advantage of this reading is that it does not require a major modification of the two diaereses with which Aristotle started his discussion at the beginning of chapter 2. The first division in chapter 2 can be represented thus (Fig. 4):

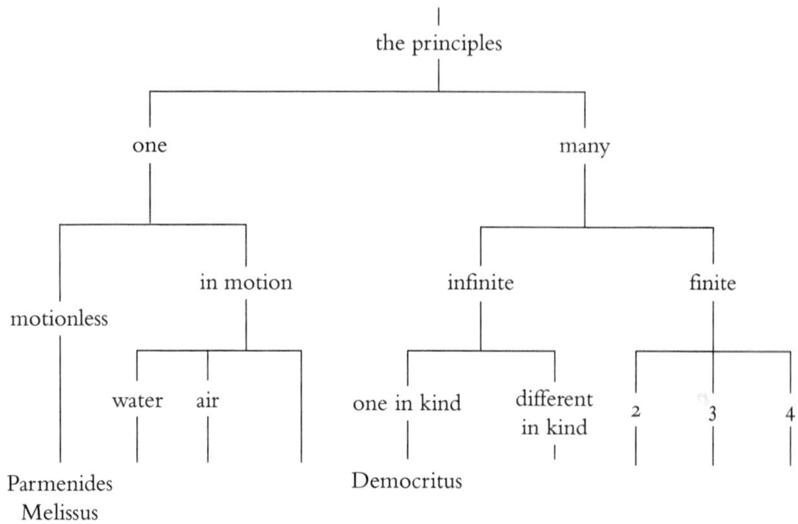

Immediately following upon drawing this scheme, Aristotle charts another division by adding that 'Those who inquire into the number of things that are also inquire along the same lines' (184b22–3) (Fig. 5):

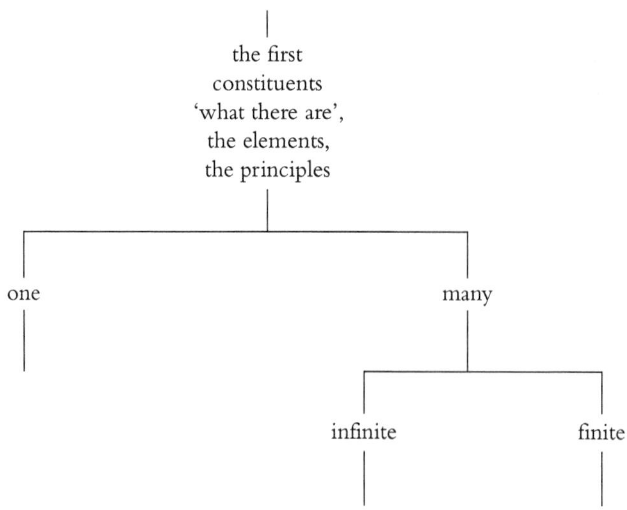

Let us first clarify the relationship between these two diaereses in chapter 2. It seems evident that the division in Fig. 5 is structurally the same as the one in Fig. 4, only more skeletal. Indeed, the question has been raised why at all Aristotle felt the need to add the second one.[11] But there are a number of salient differences between the two. To begin with, the first diaeresis is not about the actual practice of the natural philosophers. Rather, it is a diaeresis based on *a priori* grounds; it is an abstract charting of different theoretical possibilities: 'Necessarily, the principle is either one or more than one, and if one either motionless' (184b15–16).[12] The emphatic introductory ἀνάγκη clearly marks that this is not an empirical or historical statement about the opinions of the predecessors, but rather a grid of the theoretical options by which the actual views can be categorized, without suggesting that all the theoretical options have actually been taken. Aristotle mentions some holders of different options—Parmenides and Melissus for the one plus motionless option, and Democritus for the infinitely many but one in kind slot—not systematically though, but merely *exampli gratia*.

Second, and just as important, the first diaeresis is not about the number of τὸ ὄν, but, in line with the agenda set in chapter 1, about the number of principles. Then, the second diaeresis, the one in Fig. 5, aims to connect the *a priori* diaeresis of the number of principles with the actual practice of the predecessors. First, there is a specification of the topic of inquiry. Aristotle says that these people were ostensibly interested in the number of τὰ ὄντα. Yet, what they were in fact interested in is not how many beings there are around us in the world, but rather the number of primary constituents of the all those beings (ἐξ ὧν τὰ ὄντα ἐστὶ πρώτων, ζητοῦσι ταῦτα πότερον ἓν ἢ πολλά 184b23–4); and to ask this, is in effect to ask the number of principles and elements (ὥστε τὴν ἀρχὴν καὶ τὸ στοιχεῖον ζητοῦσι πότερον ἓν ἢ πολλά 184b24–5).[13] This is precisely why the practice of the natural philosophers *is* relevant to Aristotle's present project. The more historical and empirical, albeit more skeletal, diaeresis in

[11] As Mansfeld 1986, 8 remarks, 'Aristotle appends a note that at first blush is rather puzzling, because it merely appears to repeat what has already been said in the previous paragraph [i.e. what is represented in my Fig. 4]'. Cherniss, as quoted also by Mansfeld 1986, 8, claims that the ones who 'inquire into the number of things that are' cannot be the physicists because 'for in that case the passage would be a mere repetition of the preceding analysis' (Cherniss 1935, 2). He suggests that the reference is rather to Plato's *Soph.* 242c–D. Mansfeld 1986, 9, for his part, effectively criticizes Cherniss's suggestion by pointing out that the diaeresis in Fig. 5 does not correspond to the one in Plato. Mansfeld, however, seems to accept Cherniss's point about why the reference could not be to the natural philosophers. But the point, it seems to me, does not hold water.

[12] So Aubenque 1966², 90; Horstschäfer 1998, 37.

[13] On the relationship between principles and elements, see Menn and Crubellier in this volume. The discussion of the number of beings is obviously connected to the tradition, possibly starting with Hippias, and used also by Plato. Cf. the seminal paper by Mansfeld 1986.

Fig. 5 is a recognition that the abstract, *a priori* categorization of Fig. 4 was in fact observed, at least in outline, by those who were actually dealing with the question.

All this has some important bearing on the interpretation of the classifications at the beginning of chapter 4. What we have just seen about the articulation between the two diaereses in chapter 2 can show that even though the ostensible topic is the number of 'what is', it is to be understood as in fact about the number of first constituents, and thus the number of principles and elements 'from which' things are. Aristotle is now picking up and filling in with details the second, 'historical' diaeresis, which was left skeletal in chapter 2, by giving an in-depth critical assessment of the most relevant theories of his predecessors. But all this should still be understood against the backdrop of the abstract, *a priori* diaeresis of the number of principles and elements.

Seen from this perspective, the division at the start of chapter 4 has three salient features. First, it introduces the concept of τὸ ὑποκείμενον, the underlying body as that which constitutes the unique principle/element of the first group. Second, that it is by concentrating on the possible ways in which the initial unity, be it a singly underlying body or a blend, can give rise to multiplicity, and generation, that we can use the partial results of the natural philosophers most fruitfully. In particular, it will turn out that 'separating out' is the most interesting option. Third, that this survey will also require some readjustment of the initial diaereses. In chapter 2, the distinction between one and many seemed to be absolute, both in the theoretical charting and in the skeletal 'empirical' classification. As opposed to this, the hybrid blend plus one-and-many view, represented by Empedocles and Anaxagoras, now becomes the most intriguing option.[14]

Of course, the one-and-many option, going beyond the rigid monism vs. pluralism dichotomy, is not Aristotle's invention. In the famous diaeresis of the number of 'what is' in Plato's *Sophist* 242C–E, we similarly have two philosophers who maintained that being is both one and many (ὡς τὸ ὂν πολλά τε καὶ ἕν ἐστιν): the more arduous Ionian Muses, i.e. Heraclitus, who claim a constant coming together and separation, and the milder Sicilian Muses, clearly identifiable with Empedocles, who say that the all is at times one, at times many. The difference is once again notable. In Plato's diaeresis Empedocles is ranged in this group, because in different stages of the cosmogonical story he has either one being—his is the Sphairos—or the plurality of elements in the other stages of the cosmogony. In Aristotle, by contrast, the one-and-many must refer to the

[14] On the context-dependency of the classification of Anaxagoras and Empedocles with respect to monism and pluralism, see n. 10.

fact that in the Sphairos—the one—the many is already there.[15] Just as important, Heraclitus is ranged as a straightforward monist by Aristotle, whereas Anaxagoras becomes the most significant representative of the one-and-many group, and most of the rest of the chapter will focus on his views. Although Heraclitus' disappearance from the group remains unexplained, we will soon see why Anaxagoras has become so prominent.

In order to understand this shift from an initially systematic charting to a refocusing on one particular problem, and one particular attempted solution to that problem, it is worthwhile to compare Aristotle's procedure in these first chapters of the *Physics* to what he is doing in *Met.* A. In both texts, the initial unclarity, and the lack of due distinctions at the beginning of the process of understanding, is topical—just as we can expect it on the basis of the methodological preface in the first chapter of the *Physics*. Yet, by the time we reach *Met.* A, Aristotle himself has made enormous progress in making the necessary distinctions in the domain of principles and causes. This is precisely the work done in the first books of the *Physics*. When he comes to discuss the proper objects of wisdom, the primary principles and causes of being, in *Met.* A, he does not need to start the work of making the necessary distinctions from scratch, but can already rely on the results of the first books of the *Physics*. The matter–form distinction, and the fourfold scheme of causes, have already been established and shown their strength and explanatory power in the domain of natural philosophy. The task in *Met.* A is to examine whether the views of the predecessors about the principles and causes would necessitate any modification or expansion of the scheme when transferred to the domain of wisdom. This is all the more adequate as at the start of the discussion in *Met.* A it is not even clear whether wisdom is or is not the same as natural philosophy. This examination needs to be fairly systematic in order to ascertain that the list of causes is complete. For this reason, all those who appear to have contributed something notable to the search for causes and principles have to be taken into account, and Aristotle has to make sure that their respective contributions find their place in the fourfold scheme. This systematic approach is also what justifies the narrative structure and historical aspect of the discussion of *Met.* A. Aristotle is looking for the succession of *πρῶτοι εὑρεταί*, those who contributed with an original insight, in order to see whether that new addition can also be integrated in his scheme.

What makes the interpretative task arduous in *Met.* A is that the predecessors, being at earlier stages of the process of making distinctions, used their principles and causes in indistinct and inchoate ways. For instance, they did not

[15] Cf., once again, the critical discussion of Empedocles in the final part of *Gen. et Corr.* I 1 (314b4–315a25) with the meticulous analysis by Brunschwig 2004.

properly distinguish between the role of opposites, the hot and the cold, as formal causes or efficient causes, or did not clarify the relationship between the good and the efficient cause, while Aristotle himself can do the interpretative work already armed with the differentiations as developed in the *Physics*.

As opposed to this, in *Physics* I Aristotle is still at the beginning of the process of articulating and introducing differentiations into what at first appears as indeterminate and confused. He does not present us already at the start with a worked-out scheme, but leads us through the cognitive process of gradual understanding by making the necessary distinctions and articulations in accordance with the methodological statements in chapter 1. In this process, it seemed promising to start as a first step with a systematic grid of the different theoretical options about the number of principles, supported also by a reference to those who actually investigated into the number of principles/'things that are'/ elements. This is a good, promising start in making distinctions. Moreover, in doing this, Aristotle can already rely on the well-established tradition, used also by Plato in the *Sophist*, of classifying thinkers according to how many *ὄντα* or elements they posit.[16]

It however does not mean that Aristotle is committed to going systematically through each of the options delineated in the initial classification. It may very well turn out—and this is what actually happens by the time we reach chapter 4—that he finds good reasons to shift the focus of inquiry. The discussion of the Eleatic philosophers has precisely this effect. For it draws attention to the problem of the relationship between the one and the attributes, or what can be (truly) predicated of the one. What makes the Eleatic one immobile is the Eleatic philosophers' rigid, indistinct, and monolithic use of 'is',[17] leading to a complete ban on any attributes. No attribute is allowed to belong to the one, no attribute can appear or go away. Similarly, no distinction, internal or external to the one, will enter the scheme. The refocusing in chapter 4 is motivated precisely by this analysis of the Eleatic standstill. Aristotle has realized—and has helped us realize—that progress in the inquiry can be made not by simply enumerating the different options about the number of principles, but rather by examining how differences and change can be introduced into an original one. Henceforth, the focus will not be on the mechanical 'monism or pluralism, if pluralism, finite or infinite, if finite, one, two, or three, etc.', division, but rather on how to introduce distinctions and plurality into the one, starting from the one. This is why, then, Aristotle is now especially interested in those philosophers who have a story about the introduction of distinctions in the one, starting from the one. And this is also why Aristotle, at this point, shows little interest in

[16] Cf. note 13.

[17] Cf. Diana Quarantotto's analysis of Aristotle's diagnosis of the Eleatics in this volume.

those pluralists, like the atomists, who have nothing to say on this issue, because they already start with a full-fledged plurality.[18] And this is also why he considers those like Anaxagoras and Empedocles—who were previously categorized as simple pluralists—from this new angle. After this analysis, Aristotle can return again to the number of principles in chapters 6 and 7, armed also with the results of the present chapter, and, as I will try to show, in particular with the critical assessment of Anaxagoras' theory of mixture.

This is also the point where the opposites become relevant. For, in one way or another, the opposites turn out to be crucial in the transition between one and many. This is, by the way, underlined by the fact that in the context of *Met.* A, Aristotle also emphasizes the role of the opposites, such as the hot and the cold, for the Ionian monists, but there he treats the dynamism between the opposites as the earliest attempt at coming up with an efficient cause, and not something that might lead to the form–privation distinction.[19]

It has often been claimed that the statement about the universal agreement about the principial role of the opposites at the beginning of chapter 5 serves as a general conclusion to the entire discussion of chapter 4:

> All these thinkers in fact make the opposites principles, both those who claim that the universe is one and immobile (for even Parmenides makes the hot and the cold principles, calling these fire and earth), and those who speak about the rare and the dense, and also Democritus with his full and empty, both of which he says exist, the former as what is, the latter as what is not; moreover he makes use of position, shape, and arrangement, and these are genera of opposites, namely position of above and below, and in front and behind; shape of angular and smooth, and straight and curved. It is evident then that all of them in some way or other make the opposites the principles. (188a19–27)

The results of the discussion of chapter 4, however, go well beyond what is stated here.[20] The statement concentrates on the opposites in line with the fact that the opposites will constitute the central topic of chapter 5. But this does not exhaust the import of chapter 4 for Aristotle's subsequent discussion. For once the role of opposites has been consolidated in chapter 5, in chapter 6 he comes back to the underlying one from which generation occurs with the

[18] Incidentally, this might also help to explain the disappearance of Heraclitus from the one-and-many group. The synchronic oscillation and tension between coming together and separation does not lend itself to the analysis of the relationship between one and many which would be useful for Aristotle's purposes.

[19] Cf. Betegh 2012. Horstschäfer 1998, 128 suggests that the dense and the rare are presented as active agents in *Physics* I 4: 'Dieses beiden Bewegungsursachen, die hier durch abstrakte Termini ("Dichte" und "Dünne") bezeichnet werden und die in dem Prozeß der Erzeugung einer Vielheit aus der zugrundeliegenden Einheit die aktive Rolle übernehmen usw'. I don't think the text substantiates such a strong reading.

[20] Cf. also Sylvain Delcomminette's remarks in this volume.

involvement of the opposites. Although he does not make the connection explicit at that point, the underlying unity spoken of by the natural philosophers and announced at the beginning of chapter 4 is obviously relevant to that discussion as well. And it is only this broader context that justifies the detailed assessment of Anaxagoras in chapter 4. For, once again, the extended critical treatment of Anaxagoras goes far beyond substantiating the purported conclusion at the beginning of chapter 5 about the general agreement on the opposites.[21]

Before moving on to Aristotle's assessment of Anaxagoras, let me add two remarks about the first group, the monist natural philosophers, as they are presented in the diaeresis at the beginning of chapter 4, as well as in the statement about the general agreement on the role of opposites at the beginning of chapter 5. What unifies this group for Aristotle is not merely that they posit a single underlying body, but also the way in which they generate the multiplicity, and first of all the other elemental stuffs, from this original unity. They do it 'by denseness and rarity'. Now there is some discrepancy in Aristotle's treatment of this pair of opposites at the beginning of chapter 4 and the beginning of chapter 5. For what distinguishes this group of monists from the Anaximander–Empedocles–Anaxagoras group is precisely that the opposites are *not* already there in the primary underlying body. If so, the dense and the rare of these monists must appear at a later point, and in that sense are not primary. In other words, they are not at the same level at which Parmenides' hot and cold or Democritus' full and empty are, even though they are grouped together with them at the beginning of chapter 5. Although the dense and the rare are not there in the original single underlying body, they are not derivative either in so far as it does not seem to be the case that the original underlying body would somehow produce this primary pair of opposites. In this sense, there is something deeply unsatisfactory in these accounts; a strong reason why Aristotle should concentrate rather on the blend + separating-out model.

Moreover, it is unclear on what basis, or indeed why, Aristotle unifies this group by making the dense and the rare *the* characteristic opposites of these philosophers—especially as he seems to acknowledge other pairs of opposites, the hot and the cold, and the wet and the dry, a little later (188b34). The view is certainly attested for Anaximenes—but, as far as I can see, only for Anaximenes. This is in a way confirmed also by Simplicius, who first remarks that Theophrastus

[21] I thus think that Mansfeld 1986, 10 is not entirely correct in his characterization of the shift when he writes 'Rather remarkably, however, Aristotle no longer adheres, in a pedantic way, to the division of chapter 2. He now argues that, in one way or another, all these previous thinkers had assumed, as (further?) principles, couples of antithetical forces or qualities operative in respect of their corporeal substrate(s), or conversely'. What is so remarkable is precisely that Aristotle does not focus here on the opposites as it would be required by his overall conclusion about the predecessors in the first sentence of chapter 5.

also restricted the rarefaction–condensation model to Anaximenes (149.32). But then Simplicius adds, 'but it is clear that the others too made use of rarity and density, since Aristotle says in general about all that "and then generate the others by denseness and rarity making thus the multiplicity"' (187a15–16). It is a telling sign that Simplicius' only evidence against Theophrastus is that 'because Aristotle says so', despite his notorious and laudable eagerness to provide quotations from the primary sources whenever possible. As a matter of fact, we have indications that other philosophers took other opposites, or quantitative ranges, as more important to indicate qualitative differences among different forms of stuffs. For instance, Heraclitus appears to concentrate primarily on the wet–dry axis. Similarly, Diogenes of Apollonia lists a number of opposites that mark the different forms of air in 64B5 DK—warm–cold, dry–wet, static–in movement, adding also colour and taste. He seems to treat the warm–cold distinction as of particular importance, yet he quite conspicuously leaves out the rare–dense pair. It appears thus that making the dense–rare model the norm is an interpretative move by Aristotle himself. But what could motivate such a move?

I would suggest that there might be an historical reason, coupled with one internal to Aristotle's own physical theory. The historical reason might be connected to Timaeus' description of the way in which the inter-transformation of elements appear to us:

> First, we see (or think we see) the thing that we have just now been calling water condensing and turning to stones and earth. Next, we see this same thing dissolving and dispersing, turning into wind and air, and air when ignited, turning to fire. And when we see fire being condensed and extinguished and turning back to the form of air, and air coalescing and thickening and turning back into cloud and mist. When these are compressed still more we see them turning into flowing water, which we see turning to earth and stones again. In this way, then, they transmit their coming to be one to another in a cycle, or so it seems. (*Tim.* 49B7–C7)

It is clear that the operative qualitative difference in this description is exactly the dense–rare opposition. Inspired by previous commentators, Daniel Graham has argued recently that Plato closely follows Anaximenes here (Graham 2003 and 2006, 75–6, with references). This might well be true—but the view is presented by Timaeus as one which is pre-theoretical and of general currency. Aristotle can very well take this over and treat this as the default view of those who accept the inter-transformation of elements, or elementary stuffs, in the Ionian way.

The internal reason might be the following. In order to make the theory coherent, we need a pair of opposites that is applicable to the whole range of elements. In the framework of Aristotle's own physical theory, neither of the two basic pairs of opposites—wet–dry and hot–cold—will do, because each of them characterizes only two of the four elements. In this sense, the considerable advantage of the Anaximenian opposites is that each of the elements, and their composites, can be assigned a value on the rare–dense axis. If so, Aristotle, from the point of view of his own theory, is being charitable to the Ionian monists when he extends Anaximenes' opposites to the whole group.[22]

2. The Argument for Anaxagoras' Theory

> Anaxagoras, it seems, [A] supposed they are infinite *in this way* [i.e. by positing an infinite number of homoeomeres and the opposites as elements] because [B] he accepted as true the common opinion of the natural philosophers according to which nothing comes to be out of what is not. [C] This is why they speak in this way—that 'all things were together', and that becoming a certain sort of thing consists in qualitative change,[23] while they [made coming into being] aggregation and separation. [D] Moreover, also because contraries come to be from one another; [E] therefore they must have already been there. [F] For if it is necessarily the case that everything that comes to be comes to be either from those that are or from those that are not, and from these two options coming into being from things that are not is impossible (for all who concern themselves with nature agree about this point), they concluded that the remaining option is necessarily the case, namely that becoming is from things that are and are already there, [G] but imperceptible to us because of the minuteness of their bulk. [H] Hence they say that everything has been mixed in everything, because [I] they observed that everything comes to be from everything; but, [J] as they say, things appear different and are called by different names from what predominates on account of its great quantity among the infinite constituents in the mixture. [K] For, [according to them,] there is no such thing that would be purely white or black or sweet or flesh or bone, but that from which each thing contains the most, that is thought to be the nature of that thing. (187a26–187b7)

Aristotle here provides a concise philosophical reconstruction of the argument for Anaxagoras' theory of matter. I shall not try to assess how accurate this

[22] Note also that according to the *Categories*, the dense and the rare (and the rough and the smooth) are not properly speaking qualities, but rather the relative position of the parts (*Cat.* 8. 10a16–22). If so, condensation and rarefaction are not qualitative changes.

[23] The construal of this clause is particularly tricky. After much hesitation I opted for this rendering of καθέστηκεν on the basis of the almost exact parallel in *Gen. et Corr.* I 1 314a14–15: λέγει γοῦν ὡς τὸ γίνεσθαι καὶ ἀπόλλυσθαι ταὐτὸν καθέστηκε τῷ ἀλλοιοῦσθαι. Alexander (*ap.* Simplicius *In Physica* 163. 9–15) suggested that the sentence in our passage is a quotation from Anaxagoras. Zeller, apparently without knowledge of Alexander's proposal, came up with the same idea. Zeller in turn was followed by Ross (Zeller and Mondolfo 1969, 363 n. 19; Ross 1936, 484). This would of course mean that ἀλλοιοῦσθαι would have a more general, non-technical meaning, possibly including all kinds of changes. For a comparison between these passages, and a critical evaluation of Zeller's suggestion, see Brunschwig 2004, 45–8. On the basis of the close parallel with the wording in *Gen. et Corr.* I 1, Brunschwig suggests that τοιόνδε in *Physics* I 4 187a30 is an ancient addition to the text, possibly also inserted in the text of *Gen. et Corr.* used by Alexander, which gave further motivation to him to attribute the phrase to Anaxagoras.

reconstruction is, or how far it can contribute to a better understanding of Anaxagoras' own admittedly highly complex and vexed theory. What I will concentrate on now is merely the structure of the argument as Aristotle reconstructs it, and, connected to it, what aspects of the theory can make it so relevant for the overall project of *Physics* I.[24]

By allowing change and differentiation, the Ionian monists certainly established their place within the field of philosophy of nature. It is also to their credit that they guarantied the regularity of change by identifying a salient pair of opposites—the dense and the rare—that can provide a systematic framework for ordering the elementary stuffs out of which more complex things are constituted. The primary pair of opposites can also function as a general explanatory scheme for the emergence and interchange of elementary stuffs. More generally, the monist natural philosophers gave voice to what will turn out to be the correct solution, namely that change should be explained in terms of the trio of something underlying and a pair of opposites, although they apparently construed all these terms incorrectly. But this is pretty much all we can learn from them, because they have failed to provide any justification for their position, but were 'compelled as it were by the truth itself' (*καίπερ ἄνευ λόγου τιθέντες, ὅμως τἀναντία λέγουσιν, ὥσπερ ὑπ' αὐτῆς τῆς ἀληθείας ἀναγκασθέντες*, *Physics* I 5 188b27–30). Their basic intuition, guided by the truth, will turn out to be correct, but they are not really partners in the philosophical discussion. They might have a story, in form of a cosmogonical narrative, about how the pair of opposites emerges from the original one, but this comes nothing close to an argument or a proper account. A little later, in chapter 6 189a34–b8, Aristotle comes back to pointing out the advantages of having a single starting point: there has to be something of principial status that is not characterized by the opposites. In this respect, the Ionian monists, and in particular the elusive intermediate stuff theorist(s), are even better off than Anaxagoras and Empedocles with their one-and-many starting points. Yet, there is not much else we can say about the monists.[25]

Anaxagoras appears to be a better partner in discussion, first of all, because we can reconstruct the reasons behind his more complex theory. It is not that Anaxagoras provided a full and explicit argumentative support for his views, but he still gave us enough clues so that we can suppose with a fair amount of certainty (*ἔοικε*, 187a26) the arguments on which he based his theory.

What nonetheless renders Aristotle's reconstruction of the Anaxagorean theory vexed is that throughout the passage it is not obvious which claims are

[24] I am grateful to Andreas Anagnostopoulos and Christopher Noble for discussing this difficult passage with me.

[25] Cf. Andreas Anagnostopoulos in this volume.

attributed to exactly which philosophers or group of philosophers.[26] Aristotle's somewhat insouciant use of the third-person plural certainly does not help the task of disentangling this issue. In particular, it is not immediately evident what views and conclusions are supposed to be specific to Anaxagoras and what are those that, according to Aristotle, he shares with others who also posit an initial mixture.[27] On account of these difficulties, I will need to dissect the passage with a very sharp scalpel, and can only hope that I will not butcher the natural tissues of the argument.

What seems certain is that Anaxagoras is singled out in [A] as positing an infinite number of stuffs that are already present in the original mixture and then get manifest by somehow emerging from the mixture. In this, he seems to differ from both the single underlying body theorists, and Anaximander and Empedocles, the other two mixture theorists. The rest of the paragraph should contain the justification of Anaxagoras' specific view, which will be the object of criticism in the next section.

Aristotle seems to zoom in on what is distinctive of Anaxagoras gradually. First, [B] gives the common ground for all natural philosophers: 'nothing comes to be out of not being'. Indeed, we could add, this much is agreed by the Eleatics as well.[28] On the other hand, although it is not stated clearly, the 'they' in [C] seems already to refer exclusively to the mixture theorists, distinguishing the different ways in which they conceive the process through which things are generated from the mixture: 'This is why they speak in this way—that "all things were together", and that becoming a certain sort of thing is the same as qualitative change, while they [made coming into being] aggregation and separation.' It appears then that although starting from the same premise, the mixture theorists arrived at a different conclusion than the monists.[29] The text

[26] As we know it from Simplicius, this question was a matter of debate among ancient commentators as well. For instance (*In Physica* 163. 16–18), Porphyry distributed different parts of my [A], [B] and [C] among Anaxagoras, Anaximenes, and Democritus, and Empedocles.

[27] Just to give one example, the statement in [J] 'as they say, things appear different and are called by different names on account of what quantitatively predominates among the infinite constituents in the mixture', is, by all likelihood, applicable only to Anaxagoras and not to the other members of the mixture group. Yet, Aristotle still uses the plural.

[28] The acceptance of this principle comes back in chapter 8, in a context which is ostensibly a criticism of the Eleatic position. Yet, interestingly, Aristotle starts his discussion by ascribing the principle to those who *ζητοῦντες γὰρ οἱ κατὰ φιλοσοφίαν πρῶτοι τὴν ἀλήθειαν καὶ τὴν φύσιν τῶν ὄντων* (*Physics* I 8 191a24–5). This sounds very much like a reference not to the Eleatic philosophers, but to the Ionian monists. At any rate, at that point Aristotle's emphasis is on the negative outcomes of the unreflective and unqualified use of the nothing-comes-out-of-nothing principle.

[29] The plural of the *λέγουσιν* poses some problem. I don't think Horstschäfer 1998, 135–7 is right in trying to show that the presupposed subjects are still all the natural philosophers, including the monists. Once again, the close parallel with *Gen. and Corr.* I 1 314a14–15 (cf. note 9), together with the Anaxagorean catchphrase *ὁμοῦ πάντα*, is very strong indication. Cf. *Met.* Λ 2 1069b19. Cf. also Brunschwig 2004, 46–8.

at this point does not seem to offer us any further clues as to what motivates the different conclusions of the two groups. For that, we will need to get one step further. We will also need to come back to [C] shortly to explain the distinction between qualitative change and aggregation and separation as two models of coming into being.

[D] then introduces a new premise: 'Moreover (ἔτι δ'), also because contraries come to be from one another'. From this premise the presence of the opposites in the initial mixture can be derived: [E] 'therefore (ἄρα) they must have already been there'. The conclusion does not follow immediately, however. For apparently we also need the additional argument in [F] introduced by a γάρ:

> For if it is necessarily the case that everything that comes to be comes to be either from those that are or from those that are not, and from these two options coming into being from 'what are not' is impossible (for all who concern themselves with nature agree about this point) they concluded that the remaining option is necessarily the case, namely that becoming is from things that are and are already there.

Before trying to explain the way in which this argument is related to the premise in [D] about contraries, let me point out how [F] is connected to the first part of the passage. For the parenthetical remark in [F] reiterates the general agreement about the negation of the second horn of the dilemma that we have seen in [B]: 'for all who concern themselves with nature agree about this point'. Yet the 'they' after the parenthesis in [F] must already refer exclusively to the mixture theorists: all the contraries were already there from the start. So one important difference between the two groups seems to be this. On the basis of the premise that nothing can come into being from 'what is not', the monists concluded that there should be at least something to serve as a starting point of generation, and then picked somewhat arbitrarily one of the elements for that role. However, they apparently used the 'nothing comes from "what is not"' principle as a single self-standing premise. The mixture theorists, by contrast, construed an exclusive disjunction as stated in the first part of [F]: 'For if it is necessarily the case that everything that comes to be comes to be either from those that are or from those that are not.' They then used the shared assumption to negate the second disjunct. The statement in [E], 'therefore they must have already been there', gets repeated at the end of [F], and is presented as the conclusion of the disjunctive syllogism. The (implicit) acceptance of the validity and soundness of this disjunctive syllogism creates a distinctive common ground for the three members of the mixture group. They posited a mixture because of the way they constructed the first disjunct and hence the conclusion of the syllogism: in order that something come to be, it does not suffice that something or other is already there, but that the very thing that emerges as a result of coming into being needs to be already there—and it will be already there as part of an original mixture.

Finally, according to the condition added in [G], members of the mixture group accepted that the mixture contains, or contained, the ingredients in such a way that they are not manifest because of the smallness of their bulk. Note that this is a very loose condition, in so far as it does not necessitate a total mixture of the ingredients.[30]

But how is this related to the premise about the contraries coming to be from one another as formulated in [D]? I would suggest that this is precisely where Anaxagoras' specific solution is introduced. Take the case when some F emerges from an original mixture. According to the shared view of the mixture theorists, F is already there in the mixture. But, as pointed out in [C], it can emerge in (at least) two different ways: by qualitative change or by aggregation and separation. The second option seems simply to construe the mixture as an amalgam which contains also F, or all the ingredients of F. In this case the only thing you need to know about the mixture is precisely this much: it contains F or the ingredients of F. For the rest, the nature of the mixture does not need to be specified. This I think is well supported by the way Aristotle seems to understand Anaximander's ἄπειρον, i.e. as something (also) qualitatively indeterminate. But, *mutatis mutandis*, this can apply to Empedocles' view as well: for the fact that, say, water can emerge from Sphairos, we do not need to know anything about Sphairos' own qualities, only that it is a mixture that contains, in some way or other, also water, so that water can separate out of it (on this see more below).

Now add the premise from [D] that 'contraries come to be from one another'. It means that F has to come to be from something that is, in some sense or other, non-F. So there is a sense in which the nature of the initial mixture will also be part of the account beyond the mere fact that it is a mixture: it has to specify, in some way or other, what that non-F could be, and it has to provide a somewhat fuller account of how F emerges from non-F. I take it that this corresponds to what was ascribed to Anaxagoras in [C], i.e. that he held that 'becoming a certain sort of thing consists in qualitative change'.[31] The coming

[30] On Aristotle's own conception of mixture, and how this loose condition is related to it, see Cooper 2004. On the point that the ingredients of the mixture are not perceptible due to their minute sizes, see *Gen. et Corr.* I 10 328a2–3.

[31] As agreed by the majority of commentators, the most natural reading of the text is that the 'he' who makes coming into being qualitative alteration is Anaxagoras, also because Aristotle makes virtually the same point about Anaxagoras' identification of coming into being and alteration in *Gen. et Corr.* I 1 314a13–15 (cf. notes 9 and 29). For Aristotle there makes the point that the Ionian-type monists are bound to say that coming-to-be is alteration, whereas this is not the case for the pluralists, like Empedocles, Anaxagoras, and Leucippus, because they can also use the separating-out or the composition model. Nonetheless, Anaxagoras 'failed to understand his own utterance'—i.e. that 'all things were together', and thus the plurality is already there—and still reverted to identifying coming into being with alteration. On the connection between separating out and qualitative change in the Anaxagorean context, see the contrast with Empedocles below.

to be of F is a qualitative change in so far as F comes into being from a non-F (although this non-F already contained F), and did not simply separate out from a mixture.

Specifying the difference between an Anaxagorean and an Empedoclean mixture might also help in understanding what is distinctive about Anaxagoras. Take for instance the Empedoclean formula of bone: two parts of water, two parts of earth, and four parts of fire (cf. B96 as quoted by Aristotle in *De An.* 410a1). If this mixture decomposes (through the effect of Strife or otherwise), and water comes out of it, the explanation does not need to refer to the fact that the starting point, the original mixture, was actually bone; it suffices to say that the mixture contained among its constituents two parts of water.[32] Take the same mixture in the Anaxagorean context[33]—ignoring for a moment that there is of course no such mixture in Anaxagoras that would contain only three ingredients. As we will learn in a minute, Anaxagoras accepted also the principle of predominance: [J] 'as they say, things appear different and are called by different names from what predominates on account of its great quantity among the infinite constituents in the mixture'. Indeed, the principle of predominance goes further than appearance and naming, and pertain to the nature of the thing: [K] 'For, [according to them,] there is no such thing that would be entirely white or black or sweet or flesh or bone, but that from which each thing contains the most, that is thought to be the nature of that thing.' If so, the mixture will 'appear' and will be 'called' and will have the nature of fire, in so far as there are four parts of fire to the two parts of earth and two parts of water. Thus, in the Anaxagorean analysis water comes to be not simply from a mixture of the three elements, but from (what is called and is) fire. Moreover, an Anaxagorean mixture is not one in which there would be bits of pure water, quantitatively overpowered by the more numerous or bigger bits of pure fire. In an Anaxagorean mixture of homoeomerous stuffs, all bits are mixtures predominated by what predominates the mixture as a whole. So although fire contains water, fire does not contain bits of pure water, but only smaller bits of fire, each of which contain water. Fire is thus composed of bits of fire all the way through, even though no matter how small a portion of fire you take, there will be water in it. For these reasons, the generation of water will be described not simply as separating out from a mixture, but as a qualitative change from something that is non-water, i.e. fire.

[32] On the point that the constituents of an Empedoclean mixture are like 'bricks in a wall', i.e. unaffected by what they are (temporarily) the constituents of, cf. *Gen. et Corr.* II 7, with further comments below, p. 154.

[33] The interpretation of Anaxagoras' own theory of mixture is of course hotly debated. As indicated at the outset, I am not trying to take sides in this debate in the present context. The following arguments thus simply presuppose what I take to be Aristotle's interpretation of Anaxagoras.

Of course, in the Anaxagorean context if you want to say that water comes out of bone, the starting point will be a mixture that contains bone and in which bone quantitatively predominates, and each portion of which is bone which contains water—as opposed to bone being something that supervenes on, or emerges from, fire, water, and earth mixed in the appropriate ratio as in Empedocles. The important point is that, in the Anaxagorean context, what comes to be—water—is at the same level of analysis as what it comes to be from—bone—whereas bone and water are at two different levels of description for Empedocles. This is why, I would suggest, the coming to be of water is a qualitative change for Anaxagoras, and a separating out from a mixture for Empedocles.

Note also that in the course of the argument we have moved from the cosmogonic mixture, the pre-cosmic *ὁμοῦ πάντα* of Anaxagoras and the Sphairos of Empedocles, to all instances of coming into being, already with the general premise in [F]: 'For if it is necessarily the case that everything that comes to be comes to be'. This is underlined by the fact that in [I] Aristotle refers to an empirical premise—'they observed (*ἑώρων*) that everything comes to be from everything'—which clearly does not apply to the primeval cosmogonic separation, but to the ordinary types of 'comings into being' I was also using as examples in the previous paragraph. If so, by now, the expression 'initial mixture' should refer not (only) to the original cosmogonic mixture, but to all such things which constitute the starting points of any kind of coming into being.

The alleged observational fact in [I][34] is what furnishes the last major element of the Anaxagorean doctrine. For, obviously, if bone contains only fire, water, and earth as for Empedocles, it is not the case that anything can come out of it—indeed nothing can come to be out of bone that contains air as one of its constituents. For instance flesh or blood, being mixtures of all four elements, cannot come to be out of bone for Empedocles. The Anaxagoreans disagree, allegedly on empirical grounds: anything can come out of anything. But if air, or indeed any stuff you like, can come out of bone, this stuff—as we have seen in [E] and [F]—must have already been there in bone, although it was dominated by the quantitatively predominant bone. This is how we finally arrive at the specific Anaxagorean theory of matter: everything is in everything, coupled with the principle of predominance.

It is true, however, that the passage does not seem to contain an explicit argument to the effect that both the pre-cosmic mixture and the individual starting points of comings into being contain not only a very large, but an

[34] Presumably, Aristotle is referring here to the famous argument from nutrition (cf. *GA* I 18 723a9–21 and A45 and A46 DK).

infinite variety of stuffs.[35] This is striking in view of the fact that it is singled out as the hallmark of Anaxagoras' theory in both the initial classification and at the beginning of this section. And this is all the more troublesome as part of Aristotle's criticism in the next section will target precisely this aspect of Anaxagoras' views. On the other hand, in a way this feature of Aristotle's discussion might be a confirmation of what I suggested earlier: what Aristotle is interested in is the way in which different philosophers conceive their original one, and the way in which they produce the plurality out of the one—and not so much the number of stuffs.

Be that as it may, Anaxagoras' theory, reconstructed thus, contains much that Aristotle can find intriguing and appealing in view of his subsequent discussion. First, in chapter 5 188a30–188b3 Aristotle argues that the regularity of change requires that contraries come to be from contraries. This condition, as we have seen, is crucial to the specific version of the mixture theory Anaxagoras has put forward, and is not satisfied by those, like Anaximander and Empedocles, who work with the model of segregation and aggregation. From among the mixture theorists, it is Anaxagoras who apparently came closest to recognizing that the F that comes to be must come to be from some non-F.

Second, and more important, one distinctive feature of Anaxagoras' theory is the conjunction of the theory of universal mixture and the principle of predominance. This means that the character that will come into being as the result of change is already there at the beginning of the change but in a way that at the initial state it is not manifest, i.e. it does not define the character or nature of the thing. It is already there, without however determining the character and nature of the thing. The outcome of the change is and is not there at the starting point. At the end of the day, Aristotle is looking for precisely such an option; his own solution will of course be that the outcome of the change is there in the starting point not as actuality, but as potentiality. It is there, without being manifestly there, without actually determining the character of the starting point. It seems to me that among all the views of the predecessors, Anaxagoras' theory of mixture plus predominance comes the closest to offering a solution to the 'is and is not there at the beginning' puzzle. This, I would suggest, is exactly why Anaxagoras deserves the fullest consideration, a careful and basically sympathetic reconstruction, and a detailed critical assessment. If so, we need to see precisely why his solution proves untenable in order to come up with a viable alternative. This is what Aristotle sets out to do in the next section.

[35] The proposed reconstruction of the argument to this effect in Horstschäfer 1998, 143 is invalid. I think it is preferable to leave the infinity of stuffs not argued for, given also that it is not a stated conclusion of Aristotle's reconstruction.

3. The Criticism of Anaxagoras' Theory

[1] Now, if the infinite as infinite is unknowable, that which is infinite in number or size is an unknowable quantity, and that which is infinite in form is an unknowable quality. If the principles are infinite both in number and in form, it is impossible to have knowledge of the things composed of them. For we assume that we have knowledge of a composite when we have knowledge of what kind of and of how many things it is composed.

[2] Furthermore, if it is necessarily the case that the thing the parts of which can be of whatever size, large or small, can itself be of whatever size (I mean any of such parts that are present in the whole and into which the whole can be divided), and if it is in fact impossible for an animal or a plant to be of whatever size, large or small, it is evident that none of its parts can be of whatever size; for then the whole will be the same. But flesh and bone and similar things are parts of an animal, and fruits are parts of plants. Hence it is evident that flesh or bone or other such things cannot be of whatever magnitude in the direction of either the larger or the smaller.[36]

[3] And again, if all such things are present in one another, and do not come to be, but separate out having already been present, and are spoken of according to the quantitatively predominant, whatever can come to be from whatever (for instance water separating out of flesh and flesh out of water) and if any finite body is exhausted by a finite body, it is clear that it is not possible that everything be in everything. For let some flesh be extracted from water, and again some more flesh separated out from the remaining water, even if what is separated will be always less and less, it will never be less than a certain quantity. So if the separation comes to an end, not everything will be in everything (for there will be no flesh in the remaining water), but if it does not come to an end, but extraction will always be possible, there will be infinitely many equal finite magnitudes in a finite magnitude. But this is impossible.

[4] In addition, if every body must necessarily become smaller if something is extracted from it, and if the quantity of flesh is limited in both largeness and smallness, it is clear that from the smallest flesh no body can be separated; for there would be smaller amounts than the smallest.

[5] Furthermore, the infinite bodies would already contain infinite flesh and blood and brain, separated from one another,[37] no less real, and each of them infinite; but this is absurd.

[6] When he claims that the separation can never be completed, he says the truth, although he does not understand what he is saying. For the affections are not separable; indeed if colours and states are mixed together, and then would get separated out, there would be something white and healthy, which would be nothing else, and would not even belong to a substrate. So, it is a strange mind that seeks after what is impossible, if it intends to separate, while it is something impossible to achieve,

[36] Retaining the δέ in b16 against the emendation to δή as proposed by Bonitz and accepted by Ross.

[37] Retaining the readings of the manuscripts κεχωρισμένα μέντοι ἀπ' ἀλλήλων, without Ross's addition of a negation. For a vindication, see below.

with respect to either quantity or quality—with respect to quantity because there is no smallest magnitude, and with respect to quality because the affections are not separable.

[7] He does not even conceive correctly the generation of bodies of the same form. For in one sense it is the case that clay is divided into clay, but in another sense it is not. And not in the same way in which bricks are from the house and the house from the bricks, but in the way water and air are and are generated from each other.[38]

[8] And so it is better to posit fewer and finite, as Empedocles does. (187b7–188a19)

I shall not analyse each of Aristotle's objections in detail, but concentrate merely on a few salient points.

Let's note first that Aristotle is not attacking certain points of the theory that he in principle could. For instance, as he also makes clear in chapter 5, one of the hallmarks of a tenable account of change is that it is able to explain the regularity of change, in other words that it guarantees that not just anything can come to be from any given thing (if not accidentally) (ληπτέον δὴ πρῶτον, 188a30–6). Anaxagoras' universal mixture, at least in the way Aristotle understands it, fails this test. It is true that Anaxagoras adheres to the principle that F can (only) come to be from non-F. But there is no principled way to say which predominant constituents, determining the nature of the thing at the starting point, will give way to which other constituents, determining the nature of the thing at the end point of the process. Indeed, the alleged observation in [I], 'they observed that everything comes from everything', appears to be a denial of what is a baseline for Aristotle. Yet, remarkably, Aristotle is not castigating Anaxagoras directly for this point.

Second, it is important to note that the epistemological argument in [1] about the unknowability of infinite principles, and hence things composed of qualitatively and quantitatively infinite principles, should be viewed in connection with the methodological and epistemological remarks of the first chapter.[39] Anaxagoras' theory fails to deliver what we expect from the whole enterprise. For we are inquiring into the principles of things in order to attain knowledge and understanding of things. Yet if it turns out that the principles themselves are unknowable, the whole project collapses. For this reason, it is irrelevant whether or not Anaxagoras would have accepted the premise that we know the whole if and only if we know its principles, or constitutive parts. What matters is that Anaxagoras' theory—even if it would have turned out tenable as a physical theory—would be completely futile for the Aristotelian project of a scientific

[38] Retaining the δέ excised by Ross. For a vindication, see below.

[39] On the unknowability of the infinity in quantity and also in kind cf. also *Met.* α 2.

understanding of nature as it is announced in the first chapter of the *Physics*.[40] In view of this, the formulation in [8] (reiterated also in chapter 6 189a15), stating that it is 'better' (*βέλτιον*) to posit a limited number of principles, would seem to be an understatement. But for the same epistemological reason, it is indeed better to posit fewer principles, and not a very large number, even if that number is not infinite.

Arguments [2], [3], and [4] constitute one group. [2] argues that it is not the case that all kinds of stuffs can be infinitely large or small. [3] and [4] then build on this result, and show that if there are minimal quantities of stuffs, Anaxagoras' contention that everything contains everything, the point about universal mixture announced in [E] above, cannot be true.

The argument in [2] has given rise to the famous theory of *minima naturalia*. The theory has received different formulations, varying in their scope. In its most general formulation it states that any natural form can only be instantiated in a definite range of magnitude: there is a natural threshold both in the direction of largeness and in the direction of smallness beyond which the form cannot be realized. This extends not only to living organisms and their parts, but to all natural substances, so that there is a minimum amount of gold, lead, etc. This, at least, is the view that many ancient and medieval, and at least some contemporary, commentators attribute to Aristotle.[41]

I would maintain, however, that Aristotle's argument is more restricted in scope and fundamentally dialectical in nature. Note first of all that Aristotle is careful to point out that he is speaking about parts of wholes in relation to the wholes they are parts of. His point is that the size of the whole fixes the limits of the size of the part. Second, when speaking about wholes, he considers exclusively natural organisms, animals and plants. These considerations already put certain restrictions on the scope of the argument. For there is a sense in which bone and flesh are intrinsically related to natural organisms as wholes they are parts of—this is of course especially true within the Aristotelian framework (on this see more below). Yet, take gold and lead, the favourite examples of the commentators. They can of course be constitutive parts of wholes, but they can also just constitute wholes on their own. At least in the Aristotelian framework, there is no necessity to consider gold as being inherently a part of some whole.

Simplicius' presentation is characteristic of the entire tradition. First he pays due attention to the fact that Aristotle's examples are parts of natural organisms,

[40] Notwithstanding whether *γιγνώσκω* is meant to mean 'to know' or 'to decide', or some mixture of the two in B12: *καὶ τὰ συμμισγόμενά τε καὶ ἀποκρινόμενα καὶ διακρινόμενα πάντα ἔγνω νοῦς*. On this question see, e.g., Lesher 1995.

[41] On this tradition see, e.g., Murdoch 2001; Van Melsen 1960; Cerami 2012.

and that he argues from the natural upper and lower limits of the sizes of these organisms. But then he immediately continues with what seems to be an entirely different consideration, and which is the classical formulation of the theory of *minima naturalia*:

> Every magnitude is divisible to infinity, and that it is therefore possible to have something smaller than whatever you take, let him know that the homoeomeries are not simply magnitudes, but magnitudes of certain kinds, flesh, bone, lead, gold, etc., which cannot be divided to infinity with preserving their kind. As magnitudes, they too can be divided to infinity, but as flesh and bone they cannot. (*In Physica* 167.13–17, trans. C.C.W. Taylor)

In so far as there is no natural limit to the upper size of anything which contains gold as a constitutive part—disregarding for the sake of argument the finitude of the cosmos that Aristotle does not invoke here either—the size of gold does not seem to have any natural upper limit, either in itself or as a constitutive part of another thing. Now what about the lower limit? That the size of an element has such a lower limit is easy to understand in terms of contemporary atomic theory. In so far as a chemical element has a certain number of protons, neutrons, and electrons in their respective shells, the atoms or molecules of that element will have a radius of fixed range of picometres. If so, we simply cannot have anything which is smaller than that size and is still that element, even if we can split the atom of the chemical element into its constituents.

Yet it is utterly unclear to me how such a lower limiting size of non-organic stuffs could be argued for in the framework of Aristotle's own physical theory. If anything, it seems that the argument for such a lower limit could only work from the bottom up, e.g. from the internal structure or composing features of gold, as in the case of the modern theory. But it is unclear what could provide such a limitation when gold, for Aristotle, is a homogeneous stuff composed of a combination of the elements (in which the composing elements are there potentially), and when, to the best of my knowledge, Aristotle never says that there would be such a lower limiting size of the primary elements themselves.[42] Moreover, Aristotle clearly also argues for the lower limiting size from the fact that these are parts of wholes.

The one text that is most often quoted in support of the theory is the famous example of the one drop of wine not mixing with ten thousand pitcherfuls of water, because, Aristotle claims, 'its form dissolves' (λύεται γὰρ τὸ εἶδος καὶ μεταβάλλει εἰς τὸ πᾶν ὕδωρ, *Gen. et Corr.* I 10 328a26–8).[43] But this

[42] On the internal structure of homoeomerous stuffs, see the second part of Cooper 2004. As far as I can see, nothing in that account necessitates a minimal magnitude.

[43] Van Melsen 1960, Cerami 2018.

does not mean that we could not have that drop of wine in itself, before it got dissolved in the huge quantity of water. If so, this is not evidence for the lower limit of the size of a stuff or element.

Note also that Aristotle is careful enough to restrict the argument to real constitutive parts of things: 'I mean any of such parts that are present in the whole and into which the whole can be divided'. There is thus no contradiction with what he says when he turns to affections, such as white or health in [6], and says that separation for Anaxagoras' mind is impossible in terms of quantity because 'there is no smallest magnitude' (ὅτι οὐκ ἔστιν ἐλάχιστον μέγεθος, 188a11–12). Aristotle no longer speaks about real parts here, but about affections. So, even if we assume, as Anaxagoras apparently did, that affections also have magnitude, there is no reason to suppose that *they* would have smallest magnitudes. Obviously, the argument in [2] does not apply to them.

I conclude then that Aristotle's argument is restricted to real parts of natural organisms, animals and plants. Note also that when Aristotle applies the conclusion of [2] in the subsequent two arguments, he sticks to such homoeomeries that are functional, anatomical parts of living organisms, in both cases speaking about the minimum quantity of flesh. In [3] he argues that a limited amount of water can only contain a limited number of minimal portions of flesh, so the extraction of flesh from water cannot go to infinity. In [4] he argues that if there is a minimum portion of flesh, you cannot extract anything out of it, because it would then go below the threshold magnitude, which is impossible.

But doesn't it seriously weaken Aristotle's argument if we restrict it to such organic homoeomeries? Couldn't one object that 'fine, but given that as opposed to bone there is no minimum amount of gold, we could still extract ever tinier amounts of gold *ad infinitum*?' It seems to me that Aristotle's argument can be effective even if he is able to show only that such organic homoeomeries have minima. This is so because it is fundamental to Anaxagoras' theory that *everything* is in everything, and *everything* can be extracted from anything else *ad infinitum*. If Aristotle is able to show that there is a whole range of homoeomeries that cannot behave like this, Anaxagoras' overall theory collapses.

But how does the argument work in the case of organic tissues? First, we can accept from Aristotle that as opposed to forms of stuffs, such as water or gold, the respective forms of living beings can only be realized in a (roughly) fixed range of magnitude. For instance, one could argue that the specific functions of a human being cannot be realized over or under a certain size. But what does this imply for the magnitudes of the homoeomeries that are the constitutive parts of human beings, say, flesh, bone, and blood? Let's see the upper limit first. It is clear that the maximum size of a human being fixes the upper limit of the bone present in any given human being. Fine, but what if we put together the bone qua homoeomerous stuff present in a multitude of human beings? Why couldn't we always keep increasing the magnitude of bone by adding the bone of more and more human beings? One possible answer could go like this: what we would get in this way is an ever bigger pile of bone*s* (in the plural), as opposed to having an ever larger bone. The same would work for flesh (although even more horrific to imagine). What about blood, then? Could not we constitute an ever larger reserve of blood for, say, blood transfusion?

The natural reply from Aristotle would be that all these tissues and fluids, when not in an animal, and not actually fulfilling their physiological functions, are only those tissues and fluids homonymously. Indeed, even within the body they can only be what they are if they are properly connected to the functional whole, the skeletal or vascular system, they are parts of. This is explained for instance in *Parts of Animals* II 9:

> There is a resemblance between the osseous and the vascular systems; for each has a central part in which it begins, and each forms a continuous whole. For no bone in the body exists as a separate thing in itself, but each is either a portion of what may be considered a continuous whole, or at any rate is linked with the rest by contact and by attachments; so that nature may use adjoining bones either as though they were actually continuous and formed a single bone, or, for purposes of flexure, as though they were two and distinct. (654a33–b2, trans. Ogle)

If so, the natural upper limit of the respective sizes of animals do create an upper limit for the magnitude of all the homoeomereous stuffs that are real, functional, anatomical parts of these animals.

The same quote provides us with an answer to the question of the lower limit as well. On the face of it, the lower limit might appear trickier. For why couldn't we take just any tiny, or even tinier, amount of bone of any living organism? Well, the question is what do we mean by 'take'. Of course, we can theoretically consider tinier and tinier parts of bone, or observe it with higher- and higher-resolution microscopes. Actually, this is what it means that it is a homoeomerous stuff. And this is also what seems to be suggested by the quote above: 'but each is either a portion (*μόριον*) of what may be considered a continuous whole etc.' Yet Anaxagoras' theory requires not only such theoretical or 'observational' divisibility, but also that these portions of bone are physically separable. They have to be able to function as components of mixtures in whatever quantities, separated out of mixtures in whatever quantities, and so forth. Here, Aristotle can point out once again that these tinier and tinier bits, separated from their organic functional wholes they are parts of, are merely bone or bones homonymously: 'For no bone in the body exists as a separate thing itself.' So even within the organic body a fraction of a bone, disconnected from the skeletal system, is not actual bone any more. So, if it is the case that the stapes in the middle ear is the smallest bone of a human being, then the size of the stapes is the minimum size of human bone.

I conclude, then, that the text does not warrant the attribution to Aristotle of a more general doctrine of *minima naturalia*. The argument is restricted to organic tissues, and reflects the more general requirements of the Anaxagorean theory of mixture. It is notable, however, that the argument on this reading is heavily dependent on Aristotelian assumptions about the ontology of organic tissues—something that Anaxagoras is not committed to accept. He could maintain that any stuff that shows all the physical characteristics of bone (by bone predominating in it), is bone whether or not it is properly connected to the skeletal system of a living organism. As a matter of fact, Anaxagoras is closer to common intuition and parlance in this. There is nothing implausible in calling the tiny hard whitish chip I found in my carelessly prepared gulyás a piece of bone (and not merely a bone homonymously). Yet, I want to emphasize that this is not an instance of an uncharitable criticism by Aristotle. What matters for Aristotle is whether, or how far, the Anaxagorean theory can get him closer to a solution to the puzzles of change. If so, he is fully entitled to assess it taking into account his own prior theoretical commitments, including those about the ontology of organic tissues, and what is an eye, bone, or blood properly speaking, and what is an eye, bone, or blood only homonymously. Indeed, it is important for him to assess any theory in relation to his own commitments in order to see which aspects of a theory he might keep in building his own theory, and which he should reject.

Let me now briefly turn to [5]. Most editors and translators follow Ross in adding a negation qualifying 'separated' in line 188a4: *κεχωρισμένα μέντοι ἀπ' ἀλλήλων* [οὔ], despite the unanimous evidence of the manuscripts and ancient commentators. It seems to me that the sentence makes perfect sense without the emendation. Imagine a plurality of Anaxagorean-type infinite bodies (note that Aristotle uses the plural). Each of these bodies will contain an infinite amount of each of the homoeomeries. There will thus be more than one infinite magnitude of, say, flesh, in such a way that these individually infinite magnitudes of flesh are separated (*κεχωρισμένα*) in so far as they are contained in the plurality of bodies that they are ingredients of. Each of these separate portions of flesh would thus be infinite individually. But to imagine a plurality of individually infinite magnitudes of flesh, or of any other homoeomerous stuff, is absurd.

The point made in [6] is further evidence regarding what we have seen above: Anaxagoras is being considered and assessed in view of the more general context of *Physics* I. Anaxagoras' point about the ultimate inseparability is a crucial insight, which he however used in a completely wrong way. The fact that none of the opposites or qualities can occur on their own is a fundamental

Aristotelian tenet that has a central role in the argument of chapter 7 as well. It is to the credit of the Anaxagorean theory that it does not allow not only that single qualities and other non-substance items exist separately, but also that bundles of them cannot occur on their own. Any chunk of Anaxagorean matter will have a definite temperature depending on the ratio of the hot and the cold in it, a certain colour depending on what colour predominates in it, and so forth for the other qualities. Yet, there will also always be something predominating in it among the homoeomeries in the mixture—say bone—so that that chunk of stuff will not only be of a certain temperature, of a certain colour, and so forth, but also a certain kind of stuff—say bone. But Aristotle can justifiably point out that this, although Anaxagoras' insight about inseparability is a valuable point, it is not founded on any understanding of the difference between substance and other categories, and on nothing like the acknowledgement of the priority of substance.

Nonetheless, Aristotle's additional remark about the Anaxagorean Mind trying to separate what is inseparable, is highly noteworthy. The ramifications of this brief remark go considerably beyond the physical theory under consideration, and right to the heart of the Anaxagorean conception of mind and its reception—something that I obviously cannot discuss in any detail in the present context. So let me only highlight what I take to be the most intriguing point.[44] Apart from the almost tragic overtones of the picture of this cosmic Danaid, the ἄτοπος nature of mind actually comes in more than one guise. For instead of being satisfied with what is in its capacity to achieve, it seeks what it should know to be impossible. For the kind of separation that would be most appropriate for mind—that is, conceptual separation—would just be possible. This is of course the kind of separation of the non-substantial categories that Aristotle also allows. Strangely enough, Anaxagoras' mind apparently seeks another type of separation—physical separation—which, given the Anaxagorean framework, is ultimately impossible. But, paradoxically, mind's wanting to separate things is precisely what apparently motivates his triggering the cosmic rotation in the primordial motionless One. Without mind aiming at what is impossible, Anaxagoras would be a changeless monist, not much different from the Eleatics.

Yet, is conceptual separation itself possible with such a theory of mixture, where no matter how deeply you go, you will only find that F is that in which F quantitatively predominates? More broadly, this points to a special aspect of the double nature of mind—its cognitive functions as cosmic intellect, and its pragmatic, physical function as a moving cause and cosmogonic agent, and how

[44] For a fascinating recent discussion, to which I am indebted, see Laks 2002.

these two aspects are related to one another. On any account, this is precisely one of the most intriguing parts of Anaxagoras' theory of mind.[45]

[7], the last in the series of objections, targets specifically the way in which Anaxagoras conceives the generation of stuffs of 'the same form' (ὁμοειδῆ).[46] Aristotle distinguishes two ways in which clay, one of such stuffs, can be divided, and claims that Anaxagoras' theory allows only one of these ways, namely division into always smaller bits of clay. As most commentators agree, the other way, that according to Aristotle is excluded by the Anaxagorean theory, is the division of clay into water and earth. I suppose that by 'division' Aristotle cannot mean that water and earth cannot be extracted from clay. Rather, Aristotle must mean that not only smaller chunks of clay, but also two different stuffs, namely water and earth, can also constitute clay. To have more clay we do not need to add another clod of clay, but we can also add some more water together with the appropriate amount of earth.

I assume that the rest of the section explains in what way these constituents, water and earth, are related to the clay and can make up the clay.[47] These constituents do not constitute the clay by simple juxtaposition of particles (as the atomist would have it), nor yet by retaining their own specific earthy and watery natures. So the constituents of the clay are not related to the clay as bricks to the house: 'And not in the same way in which bricks are from the house and the house from the bricks'. The bricks are there all along, with all their characteristics, before the house is built, during the existence of the house, and after the (careful) demolition of the house, so that another house can be built out of them.[48] As opposed to this, constituents of a homoeomerous mixture lose their characteristic properties—are there in the mixture in potentiality—but can then regain the same characteristic properties once the mixture qua mixture is destroyed. If clay is a homoeomerous mixture, water and earth do not have their own natures in actuality while they constitute the clay, but can regain

[45] According to the interpretation of Laks (2002), separation is exactly the link between these two functions: things are discernible in so far as they have their own separate identities. But, of course, this is never achievable in the Anaxagorean world. And thus, in a way, Laks in his conclusion agrees with Aristotle's assessment: 'L'Intellect aura donc au moins en partie travaillé en vain'. But then he raises possibilities in what way the partial results of Mind are, so to speak, still worth the cosmic effort (Laks 2002, §§11 and 12). Ironically, the discerning function that Laks assigns to Mind, making distinctions in an originally indistinct whole, is very close to the methodology advanced in *Physics* I 1. For a criticism of an earlier formulation of Laks' view (Laks 1993) see Lesher 1995, 128–30.

[46] I take it with other commentators that the term ὁμοειδῆ is practically synonymous here with ὁμοιομερῆ.

[47] Ross's deletion of the δέ in line 16 obscures this connection.

[48] Aristotle uses the analogy of bricks and wall in the first part of *Gen. and Corr.* II 7 to describe the relationship between Empedoclean-type elements that, at least on Aristotle's interpretation, are not affected by the composite they are constituents of and that constitute spatially distinct parts of the composite.

them once the clay qua homoeomerous mixture is destroyed. For other purposes, there are of course crucial differences between the constitution of mixtures on the one hand, and the inter-transformation of elements on the other. Yet, it is still significantly closer to the clay case ('but in the way water and air are and are generated from each other'), in so far as water is potentially air, and air can be generated out of water and vice versa (cf. e.g. *Gen. and Corr.* II 7 334b9, 'when either is actual without qualification, the other will exist potentially').

But if it is the case that the generation of such homoeomeries as clay can be explained by reference to their more basic constituents, different from the homoeomeries they constitute, and being present in the homoeomeries in potentiality, then we could just as well reduce the number of stuffs taken as primary in our theory. And this is precisely what (also) speaks in favour of Empedocles: 'And so it is better to posit fewer and finite, as Empedocles does.'

5

Principles and Contraries

Physics I 5

SYLVAIN DELCOMMINETTE

As Titus Maria Horstschäfer has pointed out,[1] two main positions have been held with regard to *Physics* I 5: those who think that Aristotle only begins to develop his own theory of first principles in I 7 tend to treat this chapter as a mere part of the 'doxographical discussion' which would preface it;[2] while others, from Ross onwards, see it as the true beginning of Aristotle's 'positive' or 'constructive account'.[3] Horstschäfer dismisses this dichotomy as too simple: according to him, there is no more a pure 'doxographical account' than a 'doctrinal exposition' in *Physics* I, but a continuous *inquiry* of which chapter 5 is an important step, although it is not Aristotle's last word; hence we should not be puzzled if we find small inconsistencies between what is said in this chapter and what we will find later.[4] More recently, Sean Kelsey moved a step forward and argued that I 5–7 forms a continuous sequence of argument of which I 7 does

This revised version has benefited from remarks and objections offered by many participants of the Symposium as well as from the other papers presented and discussed on this occasion. Special thanks are due to Michel Crubellier, who read and commented on a first draft of the preliminary version, to Gabriela Rossi, who sent me written comments after the Symposium, and especially to Sean Kelsey, who accepted to be my first reviser and provided me with several pages of detailed suggestions and objections, in addition to correcting my English. Two anonymous readers for the Press also provided remarks on the penultimate version. The impact of all these comments on the present version is so important that I have given up trying to keep track of the origin of all the changes. All remaining shortcomings are of course mine.

[1] Horstschäfer 1998, 177–82.

[2] See most notably Mansion 1961, 41: 'Les chapitres 4 à 6 [of *Physics* I] sont une préparation dialectique à l'exposé doctrinal du chapitre 7'.

[3] Ross 1936, 487; among more recent commentaries, see especially Giardina 2002, 82.

[4] See his discussion in Horstschäfer 1998, 167–9, 207, 211.

not constitute the culmination, but rather the resolution of an apparent dichotomy between I 5 and I 6.[5]

Prima facie, the 'doxographical' interpretation might seem to be the most consistent with the first sentence of I 7, where Aristotle makes it clear that he will now speak in his own voice (*ἡμεῖς λέγωμεν*, 189b30), in apparent contrast with the previous chapters (see notably 5, 188b26–7 and 6, 189b16–18). But things become trickier when we take a closer look at I 5 itself. Let us for example compare the first and the last sentences of this chapter. The first begins thus:

> [a] *πάντες δὴ τἀναντία ἀρχὰς ποιοῦσιν* (188a19),

whereas the last reads:

> [b] *ὅτι μὲν οὖν ἐναντίας δεῖ τὰς ἀρχὰς εἶναι, φανερόν* (189a9–10).

There are two main differences between these formulations:

> [1] while in [a] Aristotle only refers to his predecessors, in [b] he seems to speak *in propria persona*;
>
> [2] while [a] treats the contraries as principles, [b] treats the principles as contraries.

[1] suggests that there is indeed something 'constructive' and 'positive' in the chapter, over and above the mere doxographical discussion: apparently, something happens in the course of the chapter that allows Aristotle to draw conclusions he himself endorses. [2] could be explained if the contraries and the principles were to be deemed *identical* in some way, as is suggested by two intermediary formulations:

> [c] *ὅτι μὲν οὖν τἀναντία πως πάντες ποιοῦσι τὰς ἀρχάς, δῆλον* (188a26–7),

and

> [d] *πάντες ... τὰ στοιχεῖα καὶ τὰς ὑπ' αὐτῶν καλουμένας ἀρχάς ... τἀναντία λέγουσιν* (188b27–9)

But in that case, the combination of [1] and [2] is problematic, since Aristotle will apparently not accept in the end that all the principles are contraries, or that the contraries are the only principles, for reasons which will be explained in the following chapter. There is nevertheless no possibility of denying that this *is* the intended conclusion of chapter 5, in view of Aristotle's formulation when

[5] Kelsey 2008, who draws some inspiration from Charlton 1992², 67.

he refers back to it later in chapter 7 (see 191a15–16: πρῶτον μὲν οὖν ἐλέχθη ὅτι ἀρχαὶ τἀναντία μόνον, where μόνον seems clearly to imply that the contraries are the *only* principles).[6] Should we then admit that it is an intermediary conclusion, though a false one? Actually, Aristotle never condemns it as false. At the end of I 6, he famously says that the question whether the elements are two or three is 'full of difficulty' (189b29); and even though I 7 ends with the unambiguous statement that the principles are three (191a20–1), Aristotle takes pains to explain a little earlier that it is in a way also true to say that they are only two, namely the contraries (cf. 190b29–191a3). What makes this conclusion intermediary is thus not that it is false, but that it is only part of the truth, because it is one-sided when taken in isolation from other aspects.

Although I shall not follow all the details of his reconstruction, especially of his reading of I 5, I am therefore sympathetic with Kelsey's view of I 5–7 as forming a continuous argument. As we shall see, this argument is not detached from the discussion of the preceding chapters, however, but is intimately connected to it. I also share Horstschäfer's general stance that what is going on in these chapters is a constructive *inquiry* rather than a doctrinal exposition of any kind. More precisely, I would say that Aristotle develops an elaborate argumentation in order to guide *us*—the readers—to his first principles, an argumentation which stems from, but cannot be reduced to, the discussion of his predecessors' positions, and of which I 5 is only a step.[7] Nevertheless, I will not use this to explain away supposed inconsistencies between I 5 and later chapters—for I think there are none, as long as one accepts that Aristotle does not have to solve all the problems he raises at once. If there is a difference between I 5–6 and I 7, it lies in the fact that in the first two chapters Aristotle is eager to reconstruct arguments that his predecessors *might have used* in order to justify their own position, although they did not use them. Nevertheless, at least as far as I 5 is concerned, these arguments and their conclusions are ones which Aristotle himself endorses too. The further step performed by I 7 will be to integrate these conclusions into a coherent picture, which none of the forerunners saw, thus preparing the way for the solution of the residual difficulties in the last two chapters of the book.

That said, I shall from now on concentrate on chapter 5 itself, although I shall point to later developments when I think they are prepared or called for by it. I have divided the chapter into five sections of variable length, corresponding to the logical articulations I detect in the text.

[6] It is true that one can also construe the word μόνον with ἐλέχθη and understand the passage to say that first it was *only* said that the contraries are principles (see Ross' paraphrase, 1936: 346). But I agree with Hendrik Lorenz below, pp. 282–3, that this construal is less natural given the word order, and in any case does not fit with the explicit conclusion of chapter 5.

[7] For a reading of the *Physics* as a whole which puts the stress on its argumentative progression, see Couloubaritsis 1997.

Section 1 (188a19–27) draws a doxographical conclusion from the discussion of the preceding chapter (and supplementary material it provides), namely that all of Aristotle's predecessors posit contraries as principles.

Section 2 (188a27–30) provides an εὔλογος argument purporting to justify this position, thus making the transition from a doxographical discussion to a constructive inquiry that Aristotle engages in himself *in propria persona.*

Section 3 (188a30–b26) is a long inductive argument purporting to show that the contraries are principles of coming-to-be and passing-away. I have divided it into four subsections: *3.1* (188a30–1) characterizes the whole section as proceeding ἐπὶ τοῦ λόγου; *3.2* (188a31–b8) draws on examples to establish that all coming-to-be and passing-away occurs between contraries or intermediates; *3.3* (188b8–23) extends this thesis to 'composite beings', even if one of the contraries has no name in this case, and points thereby towards the concept of στέρησις (without mentioning it explicitly); *3.4* (188b23–6) reduces the intermediates to the contraries, as being derived from them.

Section 4 (188b26–189a9) turns back to Aristotle's forerunners, whose positions it organizes with the device of συστοιχία.

Section 5 (189a9–10) states the general conclusion that the principles are contraries—[b] above.

Let us examine each section in turn and try to understand their articulation.

1. 188a19–27

> All then (δή) make the contraries principles: those who say that the whole is one and unmoved (for even Parmenides makes hot and cold principles, though he calls them fire and earth), those [who invoke] dense and rare, and Democritus [with] the full and void, of which he says the one is as being and the other as not-being. Again [he makes use of] position, figure and order, and these are genera of contraries: of position, above and below, in front and behind; of figure, angle, the straight and the round.[8] Thus it is clear that all in one way or another make the contraries the principles.

The presence of δή in the first sentence of the chapter suggests that it should not be read in isolation from the preceding one. I think this particle can be given a quasi-inferential sense here. I do not mean that this sentence should

[8] Following a suggestion by Marwan Rashed, at lines 25–6 I keep the text of ms. E, also read by Simplicius, namely σχήματος γωνία, τὸ εὐθὺ περιφερές (the reading of ms. J is almost identical, except that it has an additional τὸ before περιφερές), in place of Ross σχήματος γεγωνιωμένον ἀγώνιον, εὐθὺ περιφερές, which derives from lesser mss. Although somewhat harsh, this reading can be maintained in view of the fact that at *De Caelo* III 8, 307a16–17, Aristotle says that Democritus held that the sphere was a kind of angle (γωνία), which suggests that he held the same about the round. Thus on the present reading, Aristotle would be mentioning two contraries of figure, i.e. the straight angle and the round angle.

be interpreted as stating *the* conclusion of chapter 4,[9] but merely that it states *a* conclusion that can be drawn from it (along with complementary information furnished in this very passage), thereby facilitating the transition to the new point that will be examined in the present chapter. To see this, one has to look, not at the end of chapter 4, but at its opening (187a12–26). There, Aristotle divides the φυσικοί into two main groups: those who posit a single body as substratum and generate all the rest by density and rarity, which are contraries (ἐναντία, 187a16), and those who posit the contrarieties (τὰς ἐναντιότητας, a20) directly in the substratum itself, be it one (Anaximander) or both one and many (Empedocles and Anaxagoras) (187a12–23). It is on this basis that the present chapter claims that *all* ancient thinkers made the contraries principles, supplementing the previous list by mentioning the two main thinkers who were absent from it, namely Parmenides and Democritus—who are certainly those about whom this thesis is the least obvious. Hence the presence of δή, which shows that if this chapter marks a new start, it is nevertheless not detached from the preceding investigation.

Actually, Aristotle will later on qualify this universal claim and speak instead of τῶν ἄλλων οἱ πλεῖστοι (188b27). In the same move, however, he will complete his list by alluding to Xenophanes (ὑγρὸν καὶ ξηρόν, 188b33)[10] and the Pythagoreans (περιττὸν καὶ ἄρτιον, 188b34).[11] Since even Plato was mentioned parenthetically in the previous chapter as invoking contraries, namely the great and the small (187a17), and will be alluded to again at the end of this chapter (189a8), it seems that only Melissus is missing. In any case, what is important is not so much the exhaustiveness of the claim as the very kind of claim it is, which marks a new step in Aristotle's discussion of the ancient thinkers. Up to now, he had focused on the theses of individual philosophers, and especially on two extremes: the Eleatics, who in positing the absolute unity of being leave room for no principle at all (chapters 2 and 3), and Anaxagoras, who admits unlimited principles in all the senses of the term (chapter 4). Here, by contrast, he singles out a thesis which is common to (almost) all thinkers, no matter what option they choose among the possibilities distinguished at the beginning of chapter 2 (184b15–22), and including the extremes just mentioned (for Anaxagoras, see 4, 187a25–6 and 31–2). One could say that Aristotle moves

[9] On this point, I agree with Gábor Betegh's interpretation in his contribution to this volume, pp. 135–6.

[10] From Porphyry onwards, this pair is interpreted as an allusion to Xenophanes' doctrine according to which all things come from water and earth (fr. 29 and 33). See Philoponus, *In Physica*, 125.27–30. (According to the received text, Simplicius, *In Physica*, 188.32–189.1, claims that Porphyry attributes this doctrine to Anaximenes, but this is almost certainly a misquote for Xenophanes.)

[11] The almost complete absence of the Pythagoreans from the discussion might be due to it being obvious that at least some of them (e.g. Philolaus) posited contraries as principles.

from one kind of ἔνδοξον to another: from what appears to some of the most famous and esteemed σοφοί to what appears to all or most of them (see *Top.* I 1, 100b22–3). This move to greater generality is certainly also a move towards greater plausibility, although I would not say that these lines amount to an 'argument from authority' or '*ex consensus sapientium*' to the effect that the contraries are principles, as some suggest.[12] Though such an agreement produces a strong presumption in favour of this thesis, in order to establish it properly, the real work still remains to be done, namely finding reasons why it is correct. In other words, it furnishes the basis for an argument, but it is not an argument *per se*.

The formulation of the thesis Aristotle attributes to the ancient thinkers is a bit surprising, however. Certainly Aristotle does not mean that they made *all* the contraries principles, but only one (or more) particular pair(s). Why use τἀναντία then?[13] I think Aristotle is already pointing at the thesis he wants to extract from the examination of his predecessors, namely that not a particular pair of contraries, but *contrariety itself* is a principle. The whole development of the chapter can be seen as a passage from mere examples of contrary principles to the relation of contrariety itself, which they exhibit more or less clearly.[14] This progression is inductive, in a way which will have to be made more precise later.

Aristotle's testimony about Parmenides raises two problems. First, is he right in identifying the pair of contraries mentioned in Parmenides' poem with hot and cold, on the one hand, and with fire and earth, on the other? Here, he seems to claim that Parmenides himself referred to these contraries under the names 'fire' and 'earth' (cf. προσαγορεύειν, 188a21, and compare *Gen. et Corr.* I 3, 318b6–7 and II 3, 330b14), though in a parallel passage of the *Metaphysics* (A 5, 986b34–5), he implies just the opposite; and as is well known, all we find in Parmenides' fragments is the opposition between φλογὸς αἰθέριον πῦρ and νύκτ' ἀδαῆ, πυκινὸν δέμας ἐμβριθές τε (fr. 8.56 and 59). Second, is Aristotle right in attributing these principles to Parmenides himself, while they occur in the 'way of opinion', which the Goddess describes as 'deceitful' (ἀπατηλόν, fr. 8.52)? These questions need not concern us here. What is clear is that *Aristotle* took Parmenides' physics seriously, and thought that it was actually closer to the truth than the so-called 'way of truth', or at least that it corrected its excesses and should be read in combination with it, since according to him the passage

[12] See e.g. Ross 1936, 489; Charlton 1992^2, 65; Giardina 2002, 82; and contrast Horstschäfer 1998, 169–70.

[13] See Pellegrin 2002^2, 92 and n.1, who translates 'Tous, en fait, posent *des* contraires comme principes'.

[14] Compare Horstschäfer 1998, 167–9, who describes this passage as going from an 'extensional' to an 'intensional' determination of the ἀρχαί, and establishes a parallel with the transition between chapters 2 and 3 of *Metaphysics* Z.

from one to two (no matter what these are) resulted from the 'pressure of phenomena' (cf. *Met.* A 5, 986b31: ἀναγκαζόμενος δ' ἀκολουθεῖν τοῖς φαινομένοις, a phrase which will be echoed in the present chapter at 188b29–30).

As for Democritus, what is especially interesting here—apart from the fact that Aristotle describes the position in his own vocabulary (using θέσις, σχῆμα, and τάξις rather than the Democritean τροπή, ῥυσμός, and διαθιγή), and that he does not seem to give examples of differences in order[15]—is that Aristotle detects two different levels at which contrariety plays a role, namely between void and atoms and between different determinations of the atoms themselves. The same could be said of Empedocles, who according to Aristotle posited contrariety not only between love and strife (188b34), but also between fire and the other three elements (cf. *Gen. et Corr.* II 3, 330b19–21). Actually, what is striking is the diversity of levels at which the contraries Aristotle detects in the thought of his predecessors occur: they can appear in the matter itself (the Ionians, Parmenides, Empedocles, Anaxagoras, Plato), in formal determinations (the Pythagoreans, Democritus), and also in the causes of processes (Empedocles again). This shows that here Aristotle is interested in appeals to contraries quite generally, regardless of the specific role they play in particular theories.[16] Hence the qualification of this intermediate conclusion with a πως (a26).

This relative vagueness also confirms that up to now, we have had no argument to the effect that the contraries are principles, but only to the effect that all of Aristotle's forerunners invoke contraries in their account of coming-to-be. This is only the raw material from which we shall have to extract the rationale, in order to see *in what way* they are (unknowingly) right.

2. 188a27–30

> And this with good reason (εὐλόγως). For the principles must not be derived from one another nor from anything else, but everything from them, and these conditions belong to the primary contraries: because they are primary, they do not derive from anything else, and because they are contraries, they do not derive from one another.

The occurrence of εὐλόγως in this text is classified by Jean-Marie Le Blond among the dialectical uses concerned with theses Aristotle himself accepts.[17] A 'dialectical use' would be one in which εὔλογος does not qualify the thesis at

[15] Unless one accepts Susemihl's emendation [τάξεως] πρόσθεν ὄπισθεν at a25, but this seems unnecessary given Aristotle's usage. See Ross 1936, 488 *ad* a23 and Horstschäfer 1998, 172 n.14.

[16] Cf. Philoponus, *In Physica* 110.14–27.

[17] See Le Blond 1938, 32 and 119. Notice that Philoponus reads εἰκότως: see *In Physica* 110.27, 117.23–4.

stake itself, which might be false, but the justification it can receive, which is deemed logically consistent or satisfying, at least on the basis of some assumptions.[18] I am not convinced that this appraisal is perfectly correct. To be sure, Aristotle is here reconstructing an argument his predecessors might have advanced in order to justify their thesis; but what he is interested in is really the truth (or at least the partial truth) of that thesis; and the argument he offers is actually a step forward in establishing this truth, although it is not sufficient by itself, as we shall see. This is important, for it means that this text marks the transition from the purely doxographical report with which the chapter began to Aristotle's appropriation of the thesis he extracts from it.

The argument begins by stating three criteria that principles must meet:

[Princ_1] Principles must not be derived from one another.

[Princ_2] Principles must not be derived from anything else.

[Princ_3] Everything else must be derived from the principles.

What I have translated by 'be derived from' is expressed in Greek by εἶναι ἐκ + genitive. The ancient commentators insist that this ἐκ should not be understood in the same way as in the subsequent passage (188a30 ss.), where the contraries, which Aristotle will argue are principles, will be said to come from (γίγνεσθαι ἐκ + genitive) their contraries, i.e. from one another: in this latter passage, these commentators claim, ἐκ really means 'after', while in the present passage it denotes either material or productive cause.[19] They are certainly right in detecting a distinction here, in order to avoid Aristotle running into blatant contradiction. But I do not think that it implies a difference in the meaning of the preposition ἐκ: this distinction should much more probably be attributed to the shift of *verb*, from εἶναι to γίγνεσθαι.[20] In both cases, the preposition ἐκ seems to have the same function, namely introducing a principle, or at least a potential candidate to this title, as is attested by the coordination of both expressions in chapter 7 as an elucidation of the phrase αἰτίαι καὶ ἀρχαὶ τῶν φύσει ὄντων.[21] However, this

[18] Le Blond 1938, 29–32.

[19] Themistius, *In Physica* 18.14–20, 20.12–15; Philoponus, *In Physica* 111.19–112.5, 118.3–7, 118.25–119.4; Simplicius, *In Physica* 182.19–26, 186.15–23. Philoponus refers in this context to 'Plato in the *Phaedo*' (111.31), thinking most probably of the anonymous objection and Socrates' reply at *Phaedo* 103a–c—very relevantly, since this dialogue certainly lies at the background of the discussion to come. (I have studied the place of the *Phaedo* in *Physics* I in Delcomminette 2015).

[20] Compare Horstschäfer 1998, 176 and Kelsey 2008, 186 n.8, who nevertheless translates εἶναι ἐκ by 'consist in', which to my eyes has the disadvantage of suggesting material causality.

[21] *Physics* I 7, 190b17–20: *Φανερὸν οὖν ὡς, εἴπερ εἰσὶν αἰτίαι καὶ ἀρχαὶ τῶν φύσει ὄντων, ἐξ ὧν πρώτων εἰσὶ καὶ γεγόνασι μὴ κατὰ συμβεβηκὸς ἀλλ' ἕκαστον ὃ λέγεται κατὰ τὴν οὐσίαν, ὅτι γίγνεται πᾶν ἔκ τε τοῦ ὑποκειμένου καὶ τῆς μορφῆς....*

does not mean that these expressions are synonymous;[22] for whereas a principle (e.g. a contrary) can itself be said to γίγνεσθαι ἐκ another principle (e.g. its contrary), it cannot be said to εἶναι ἐκ another principle. Given that distinction, it is very important to keep in mind that the present lines are not concerned specifically with coming-to-be, nor with the role a principle might have in an account of something coming to be, but only with what it is to *be* a principle, no matter what it is a principle *of*. And they tell us that a required feature of the principles is their *independence* from either one another or anything else, by contrast with whatever they are the principles of, which depends on them *in its very being*. The modalities of this dependence are not described here; but since it is expressed in terms of being rather than becoming, I doubt if the complement of εἶναι ἐκ + genitive refers to material or productive cause: in the present context, it more likely refers to *formal* cause, which is also relevant for things not subject to becoming. This interpretation should become clearer in my comments on 188b23–6 (subsection 3.4). In any case, the translation I have chosen ('derive from') leaves the issue undetermined, as it still is at this point.

$Princ_1$, $Princ_2$, and $Princ_3$ are certainly accepted by Aristotle himself: in fact, they seem to correspond to the minimal requirements anything has to meet in order to qualify as a principle. Then he puts the contenders for the title to the test. It has to be noticed that in doing so, he qualifies these contenders in a decisive way; for he now refers to them as the *primary* contraries (cf. τοῖς ... ἐναντίοις τοῖς πρώτοις, a28–9). This qualification introduces the idea of an ordered series which was entirely absent from the previous report, but which will play an important role later in the chapter (see section 4 below). This shows that Aristotle has distanced himself from the mere evaluation of his predecessors and is already beginning to argue in his own name. The move is certainly not innocent, since it is precisely their primary character which grounds the fact that these contraries meet $Princ_2$. As for $Princ_1$, it is supposed to be met by any pair of contraries *as such*, for a reason which can be extracted from something Aristotle will say in chapter 9, namely that far from being derived from one another, the contraries are destructive (φθαρτικά) of each other (192a21–2), which probably means that they are incompatible, in the sense that the presence of one of them implies the absence of the other.

What is striking is the absence of any attempt to show that these contraries meet $Princ_3$.[23] This does not mean that they do not: as we shall see in subsection 3.4 below, there *is* something which derives from the contraries in its very being, namely what is intermediate between them (see 188b23–4: τὰ δὲ μεταξὺ ἐκ τῶν ἐναντίων ἐστίν); and it could probably be argued (although Aristotle will

[22] On this point compare the discussion by Hendrik Lorenz below, pp. 264–5.

[23] For attempts to explain this lack, see Horstschäfer 1998, 177 and Kelsey 2008, 186–8.

ultimately not agree) that at least within the limits of a given genus, there is nothing outside the contraries and their intermediates, so that the intermediates could be considered as 'everything else' than the contraries. If Aristotle defers this explanation here, it is probably because these intermediates and their role in the account of change have not been introduced yet.

Even if we complete it thus, however, this argument only shows that the contraries meet the minimal requirements for being principles *in general*; it does not explain why they would be especially relevant in an account of *change*. This, I think, is the main reason why it is prefixed by the adverb εὐλόγως: valid as it is, it remains sketchy, because it does not show what role the contraries play in a process of change. That will be the function of the next section of the text.

3. 188a30–b26

[3.1] But we must also examine how this turns out in respect of *logos* (καὶ ἐπὶ τοῦ λόγου...πῶς συμβαίνει). [3.2] One must then first grasp (ληπτέον) that of all beings, none is by nature such as to act on any chance thing or to be acted on by any chance thing (οὐθὲν οὔτε ποιεῖν πέφυκεν οὔτε πάσχειν τὸ τυχὸν ὑπὸ τοῦ τυχόντος),[24] nor does just anything come to be from just anything (οὐδὲ γίγνεται ὁτιοῦν ἐξ ὁτουοῦν), unless one takes it by accident. For how could white come to be from cultivated, unless the cultivated was an accident of the not-white or of the black? Rather, white comes to be from not-white, except not from every [not-white], but from black or some intermediate; and cultivated comes from not-cultivated,[25] except not from every [not-cultivated], but from uncultivated or any intermediate between them there may be. Nor again then do things pass away in the first place into any chance thing, for example the white does not pass away into the cultivated, except sometimes by accident, but into the not-white, and not any chance not-white but the black or the intermediate; the same way, the cultivated passes away into the not-cultivated, and not any chance not-cultivated but the uncultivated or any intermediate there may be. [3.3] And this is similar in the other cases also, since even beings which are not simple but composite (τὰ μὴ ἁπλᾶ τῶν ὄντων ἀλλὰ σύνθετα) follow the same rule (κατὰ τὸν αὐτὸν ἔχει λόγον); but because the opposed dispositions have no name, it remains unnoticed. For it is necessary that all that is tuned come from what is untuned and what is untuned from the tuned, and the tuned passes away into untunedness, and not into any chance untunedness, but the opposed. And it makes no difference whether

[24] Charlton 1992², 11 translates 'nothing whatever is by nature such as to do or undergo any chance thing through the agency of any chance thing'. The problem of such a translation is that it introduces a third term in what are clearly dual relations, as the rest of the text confirms. I therefore construe τὸ τυχόν only with ποιεῖν and ὑπὸ τοῦ τυχόντος only with πάσχειν. Alternatively, one could construe both with both and translate 'none is by nature such as to act on or be acted on, any chance thing by any chance thing'.

[25] As Ross 1936, 489 writes, οὐκ ἐκ μουσικοῦ at b1–2 must idiomatically mean ἐξ οὐ μουσικοῦ, or else one has to follow Pacius and Casaubon and correct the text to obtain that sense (see Pellegrin 2002², 94 n.1).

> we speak of attunement, order or composition: for it is clear that the rule is the same. And in fact a house, a statue and anything else come to be in the same way: for the house comes to be from these things not put together but separated in a certain way, and the statue or anything else which is shaped from shapelessness; and each of these is either a certain order or a certain composition. If then this is true, everything that comes to be and everything that passes away comes to be and passes away from contraries or into contraries and their intermediates. [3.4] But the intermediates derive from the contraries, for example colours from white and black; so that everything that comes to be by nature either is a contrary or derives from contraries (ὥστε πάντ' ἂν εἴη τὰ φύσει γιγνόμενα ἢ ἐναντία ἢ ἐξ ἐναντίων).

Although this bit of text is long and complex, it will be useful to begin by considering it as a whole, since it corresponds to a single argument. I shall turn to each subsection of the text later on.

The first thing to notice is the complete absence of the word ἀρχή in these lines, as well as of the related terms αἴτιον and στοιχεῖον. Does it mean that they do not concern the principles any more? Certainly not: both the sections that precede (1 and 2) and the sections that follow (4 and 5) make it clear that the aim of the present one, which is the heart of the chapter, is to show that the contraries are principles. In order to do that, subsections 3.2 and 3.3 will use the expressions γίγνεσθαι ἐκ + genitive and φθείρεσθαι εἰς + accusative. In both expressions, the subject and the complement of the prepositions ἐκ and εἰς refer to the termini of change. Do these termini correspond to the principles? Not exactly, since they include not only the contraries, but also the intermediates, as the conclusion of the sequence [3.2]–[3.3] makes clear: everything that comes to be and everything that passes away (ἅπαν ... τὸ γιγνόμενον καὶ φθειρόμενον) comes to be and passes away from (ἐξ) contraries or into (εἰς) contraries *and their intermediates* (b21–3). Now these intermediates are *not* themselves principles, as subsection 3.4 will show. There, Aristotle will use the expression εἶναι ἐκ + genitive, not to be confused with γίγνεσθαι ἐκ + genitive, in order to explain that the intermediates *derive from* the contraries; hence according to Princ_2, they cannot be principles themselves. We can thus conclude that the principles we are looking for are the termini of change which cannot be derived from something else, i.e. the contraries.

The way Aristotle formulates the conclusion of the whole section, namely that 'everything that comes to be by nature either is a contrary or derives from contraries' (πάντ' ἂν εἴη τὰ φύσει γιγνόμενα ἢ ἐναντία ἢ ἐξ ἐναντίων, b25–6), remains a bit surprising, however. As we have just seen, what 'derive from' the contraries correspond to the intermediates, which for this reason are not principles; for the sake of clarity, we can leave them out of the picture. There remain the contraries themselves, which *are* supposed to be principles. Now these might seem to play a different role here and in the conclusion formulated four lines above: at b21–3, Aristotle wrote that *τὸ γιγνόμενον comes from* (and *τὸ φθειρόμενον passes away into*) contraries (or intermediates), whereas at b25–6 he says that *τὰ φύσει γιγνόμενα are* contraries (or intermediates). I do not think that this difference is significant, however. For if what something comes to be from is its contrary, it means that it must itself be contrary to it, since contrariety is a symmetrical relation. But the viewpoint is different in the two formulations: whereas the first one, using the verbs *γίγνεσθαι* and *φθείρεσθαι*, focuses on the origin and the result of change, the second one, using the verb *εἶναι*, focuses on the nature of the subject of change itself, which is one of the contraries involved.

Another striking feature of this formulation is the phrase *τὰ φύσει γιγνόμενα* (b25), which might give the impression that Aristotle is focusing on *natural* change. This impression is deceptive. For actually, all the examples he mentions in these lines are examples of *artificial* change. As is well known, it is only in book II that Aristotle will introduce the distinction between natural and artificial change; at this point, he is interested in change quite generally, as he will explicitly state a little later (7, 189b30–2). How to understand *φύσει* in this context? I think it must be interpreted in correlation with *πέφυκεν* at a32–3, which refers to *the nature of change itself*, by contrast with improper descriptions of it which focus on accidental features of such a process (cf. *κατὰ συμβεβηκός*, a34). I will come back to this briefly in subsection 3.2 below.

Let us now turn to the details of each subsection.

3.1. 188a30–1

I single out the first sentence because I think it offers a characterization of the whole argument to come. According to Ross, the expression *ἐπὶ τοῦ λόγου* would mark an appeal to argument, by contrast with the previous appeal to authority.[26] No matter how we understand this so-called 'appeal to authority', the problem with this interpretation is that lines 188a27–30 (section 2) already offered an argument, albeit an incomplete one; if we follow Ross, we can thus make no sense of the *καί* prefixing this phrase, whereas it seems to imply at least a slight contrast. Charlton, on his part, suggests that the phrase means 'considering speech' and makes parallels both with Plato's *Phaedo εἰς τοὺς λόγους καταφυγόντα* (99e5) and with the arguments Aristotle characterizes as *λογικῶς* in the *Metaphysics*, most notably at Z 4, 1029b13.[27] Although these parallels seem relevant to me, I would not say that in any of these passages the focus is on 'things

[26] Ross 1936, 488–9 *ad loc.*

[27] Charlton 1992^2, 66.

said', and especially not here, since Aristotle will show later on that his thesis holds even where one of the contraries has received no name (188b10–11). In response to this objection, one might suggest, as several participants of the Symposium did, that *ἐπὶ τοῦ λόγου* would only apply to lines 188a30–b8, whereas *ὁμοίως* at line 188b8 would extend the conclusion reached by means of an analysis of language to cases where language is of no help. Even thus, however, I do not think one can say that the focus of 188a30–b8 is on language as such, notably because of the complete absence of any verb of speaking in these lines. I would rather suggest that what is going on in this passage, and indeed in the whole section 3, is grounded on *conceptual* analysis, and, from this viewpoint, Charlton's deliberately open translation 'from logical considerations' would do. But only a closer look at the way the argument works will help shed light on this issue.[28]

3.2. 188a31–b8

Let us then focus on the first step of the argument. According to Terence Irwin,[29] these lines contain a deductive argument, appealing to two premises, none of which is justified:

[1] every becoming is an action of one thing on another;

[2] random things do not act on, and are not acted on by, random things.

The conclusion would then be that random becoming is impossible—and Irwin actually counts this passage as evidence for determinist assumptions in Aristotle.[30] To avoid random transitions, Aristotle would then interpret the termini of becoming as contraries rather than contradictories, believing—wrongly—that this would be sufficient to account for the regularity of changes.

I have to say that this interpretation seems to me to completely misrepresent Aristotle's argument.

First, I cannot see how this passage could be taken as evidence for determinism, especially in view of the fact that Aristotle explicitly removes accidents from the picture (cf. a34–6, b4–5)—very sensibly, since according to him, accidents

[28] Horstschäfer's position on this point is ambiguous: on the one hand, he approves Charlton's (1992²) criticism of Ross, but on the other hand, he compares this reasoning *ἐπὶ τοῦ λόγου* to an argument *φυσικῶς*, i.e. concerned with the very thing at stake, whereas the preceding argument (188a27–30) would be 'abstract and formal' and could be compared to an argument *λογικῶς*: Horstschäfer 1998, 182–4, cf. 174–5. This seems to me a little far-fetched.

[29] Irwin 1988, 70–1 and 223–4.

[30] See Irwin 1988, 223 and 510 n.62.

cannot be objects of science.[31] One of the main reasons for refusing to count Aristotle among determinists is that he makes room for accidents; so how could a passage which explicitly leaves accidents aside speak for determinism? True, accidents correspond here to misleading descriptions of changes which are in fact regular under another description; but the possibility of referring to the same thing under different descriptions does actually play a key role, albeit a complex one, in Aristotle's refutation of causal determinism in *Metaphysics* E 2–3 and in his analysis of chance in *Physics* II 4–6.[32] And in any case, even accidental changes such as the products of chance *can* be described as following the general rule formulated in the present passage, and thus as regular in this respect, even though they are not when one considers their cause: when something becomes white by accident (because a tin of white paint falls on it, for instance), it remains true that it must have been not-white before, otherwise this event could not be described as a *change*. What the present passage is interested in is the nature of the process of change *as such*—be it natural, artificial, or accidental—when it (or its subject) is properly described. I think that this is the force of *πέφυκεν* at a32–3 (and thus of *φύσει* at b25), which should be contrasted with the phrase *κατὰ συμβεβηκός* (a34, b4–5), referring to the same process under a misleading description. Now if all change is regular in the relevant sense under a proper description, this has no implication whatsoever on the issue of determinism or indeterminism.

Second, it does not seem to me that the focus of the passage is on action and passion.[33] It is true that it opens with the suggestion that one of the termini of change *ποιεῖ* and the other *πάσχει* (a32–3), but this dynamic description is then dropped in favour of the more neutral terms *γίγνεσθαι* and *φθείρεσθαι*, and it plays no role in the rest of the passage, which as the ancient commentators already saw is merely concerned with the *succession* of contraries in a process of becoming. And this too with good reason. For while it was common among Aristotle's forerunners to attribute an active role to the contraries as causes of change, this will prove problematic in the next chapter and will furnish one of the motivations for the introduction of the substratum as a third principle.[34] One can therefore suspect that Aristotle introduces this formulation in order to pave the way for the following chapter, but then leaves this aspect aside and concentrates for the moment on what makes it possible to see the role of contraries themselves in the process, before turning back to it later when he tries to complete his account by going further than his predecessors.

[31] Compare Horstschäfer 1998, 185.

[32] On this point, see notably Weidemann 1986.

[33] For another, thoroughly different, interpretation putting the stress on action and passion, see Kelsey 2008, 188–92.

[34] See Horstschäfer 1998, 186.

And lastly, I do not think that the thesis of lines 188a31–4 should be interpreted as a *premise* of the argument. In any case, Aristotle cannot hold that this thesis is so obvious as not to need any argument, since it is in direct contradiction with one of the main Anaxagorean theses he has discussed in the previous chapter (see 187b24: γίγνεται δὲ ἐξ ὁτουοῦν ὁτιοῦν).[35] Although Anaxagoras recognized that contraries come to be from contraries (γίγνεσθαι ἐξ ἀλλήλων τἀναντία, 187a31–2), he also 'saw' that everything comes to be from everything (πᾶν ἐκ παντὸς ἑώρων γιγνόμενον, 187b2), which led him to conclude that everything is mixed with everything (πᾶν ἐν παντὶ μεμῖχθαι, 187b1), as necessary to explain how just anything can come to be from just anything. In order to address the observation on which this reasoning depends, Aristotle needs to introduce the distinction between change properly described (πέφυκεν) and change improperly described (κατὰ συμβεβηκός): it is only κατὰ συμβεβηκός that just anything may come to be from just anything, for example white from cultivated, whereas πέφυκεν, everything comes to be from its own contrary. But this is not a premise: this is the very point at issue, which has to be argued for. If that is so, lines 188a31–4 should rather be understood as anticipating in a negative formulation the general conclusion of the whole argument of subsections 3.2–3.3, namely that every process of coming-to-be or passing-away takes place between contraries or intermediates (cf. 188b21–3); and the ληπτέον of a31–2 should be understood as referring to what we shall have to *understand* rather than to what we would have to assume without demonstration—which is a perfectly Aristotelian usage, exemplified elsewhere in *Physics* I.[36]

I therefore concur with Robert Bolton's criticism of Irwin's interpretation, and with his general characterization of Aristotle's argument as inductive rather than deductive.[37] This is consistent with the teaching of *Posterior Analytics* II 19, according to which the principles of a science come to be known through ἐπαγωγή. However, Bolton seems to me to misrepresent the nature of Aristotelian induction, at least as it occurs in this passage, on two counts.

First, he suggests that induction should review a full range of cases.[38] If this were true, one could certainly complain that Aristotle is far from rigorous, for

[35] This back reference is indicated by Horstschäfer 1998, 181 and 190.

[36] See Pellegrin 2002², 93 n.3, and compare εἰληφέναι at 3, 186a11 and ἔλαβον at 8, 191a33. Alternatively, ληπτέον could be understood as positing a premise, although not for the argument which immediately follows from 188a36 onwards, but for the general argument running through I 5–7; in this case, the inductive argument would be used to establish that premise, in keeping with a very common usage in Aristotle (see notably *Top.* II 8, 113b17; II 10, 115a5–6; IV 2, 122a19; VIII 8, 160a36–9).

[37] Bolton 1991, 23–5. This already seems to be the interpretation of Themistius, *In Physica* 20.17–18 and Simplicius, *In Physica* 186.9. I fail to make sense of Horstschäfer's interpretation, according to which we would be confronted to an 'intertwining' (*Verflechtung*) of inductive and deductive argument (Horstschäfer 1998, 189; cf. 185–90).

[38] Bolton 1991, 24–5.

he actually cites very few of them. But if induction consists in moving from the particular to the universal, it does not do so by collecting all the particulars concerned, which would be both impossible and insufficient, but by extracting the *structure* or the *law* which each of them displays (see especially *Met.* A 1, 981a5–12). In order to get this result, enumerating a number of instances is useful, but it is not even required: in the *Posterior Analytics*, Aristotle suggests that in very specific cases, the universal can stem directly from the perception of one single particular (see *APo* I 31, 88a11–17; II 2, 90a24–30). In the present case, induction consists in passing from instances of change between specific contraries to *the relation of contrariety itself* as a general structure displayed in any change at all. This is why the objection of Friedrich Solmsen and David Bostock, who complain that Aristotle has moved to such an 'abstract' level that no comparison is possible any longer with the theories of his forerunners from which he took his impetus,[39] is besides the point: this move from a mere empirical plane to a more 'abstract' or 'conceptual' plane is precisely what is at issue in induction. It has to be added, however, that Aristotle was preceded along this path by Plato, who in the *Phaedo* uses exactly the same method to establish the principle that 'all contrary things come to be in this way, from their contraries' (71a9–10).[40]

Second, Bolton holds that the interpretation of this reasoning as inductive implies that its premises are taken directly from experience, and criticizes Charlton for interpreting Aristotle's remark that white comes to be from black or an intermediate and cultivated from uncultivated or an intermediate as 'a purely logical doctrine'.[41] But such a strict opposition between what is 'empirical' and what is 'conceptual' seems artificial. According to *Metaphysics* A 1 and to *Posterior Analytics* II 19, there are several steps in induction, which starts from sense-perception (αἴσθησις) and goes through memory (μνήμη) and experience (ἐμπειρία) to science (ἐπιστήμη) itself. The level we are dealing with here is the passage from experience to science. How does this passage happen? Aristotle is notoriously stingy with explanations on this point, but what is clear is that our starting points in this case must be the products of experience itself, and not of mere sense-perception. Now these products *are* concepts, which are labelled ἐννοήματα in the *Metaphysics* (see A 1, 981a5–6: ἐκ πολλῶν τῆς ἐμπειρίας ἐννοημάτων) and (first) universals in the *Posterior Analytics* (see II 19, 100a6–7 and 16). These concepts derive from experience, and are thus empirical; but they are also concepts, and thus universal: what distinguishes them from scientific concepts is that they result from a mere collection of particular cases

[39] Solmsen 1960, 79 and 87–8; Bostock 1982, 2006, 3–4; contrast Charlton 1992², 66.

[40] I adopt Sedley's punctuation, removing the comma after γίγνεται at line 10: see Sedley 2012, 149 n.3.

[41] Charlton 1992², 66; see the criticism in Bolton 1991, 25–6, which on this point agrees with Irwin 1988, 510 n.61.

displaying some similarity, but which are not yet unified by a single law or εἶδος that is the *cause* of this similarity (see *Met.* A 1, 981a7–12). In the terms of *Posterior Analytics* I 4, they are only κατὰ παντός, not yet καθ' αὑτά. Aristotle is very clear on the fact that the transition from this first universal to scientific universal does not happen by turning back to sense-perception, but by focusing on the first universal itself (see *APo* II 19, 100b1–3); that is, by investigating its *conceptual* content. And this is what Aristotle does here: for example, against Anaxagoras, who professed to *see* (ἑώρων, 4, 187b2) that everything comes to be from everything, it would be pointless to contest his observation, and this is not what Aristotle does; rather, he introduces a *conceptual* distinction between change considered πέφυκεν and κατὰ συμβεβηκός.

There is therefore no incompatibility in interpreting the present argument as inductive and as based on 'logical', or perhaps preferably 'conceptual', considerations, contrary to what Bolton suggests. The acknowledgement that the white comes to be from the black or from another intermediate colour is not grounded on a review of all the instances of such a coming-to-be, but on a conceptual constraint we observe in order to label something as a coming-to-be—namely, that it takes place between opposites belonging to the same conceptual range.[42] Admittedly, this constraint could not have been discovered without experience, since it belongs to the empirical concept of change whose content ultimately derives from the cases of changes we have previously encountered; but it is manifested by conceptual analysis, which reveals the single *law* that all these cases were following and which is the ground of their being instances of changes.

This interpretation seems to me to be in tune with the methodology exposed in *Physics* I 1. When Aristotle writes there that we should start with universals which are confused (cf. τὰ συγκεχυμένα, 194a21–2) and are wholes better known according to sense-perception (184a24–6), I think he refers to these empirical concepts, which are closer to sense-perception than scientific universals on the path of induction and are merely collected on the ground of their similarity without being properly unified by a single and explanatory law. What remains to be seen is how providing such a law would at the same time correspond to a division of these wholes into particulars (cf. 184a23–4); I shall come back to this point in section 4 below.

Let us return to subsection 3.2 of our text. After announcing the conclusion to be proved (188a31–4), the text breaks into two parts: one concerned with γίγνεσθαι (a35-b3) and the other with φθείρεσθαι (b3–8). In each part, Aristotle

[42] See Graham 1987, 132–3. Interestingly, Philoponus writes that the principle according to which no chance thing comes to be from a chance thing is recognized ἐκ τῶν κοινῶν ἐννοιῶν (*In Physica* 111.8; see also 118.12–15).

uses the same two examples: white coming to be from (or passing away into) black or an intermediate and cultivated coming to be from (or passing away into) uncultivated or an intermediate. The difference between these two examples is that it is unsure whether the second one admits of intermediate (see b2–3: εἴ τι αὐτῶν ἐστι μεταξύ). This indicates that the presence of intermediates is not necessary for establishing the point at issue: actually, it rather raises a further difficulty, which will be tackled in subsection 3.4. We can thus leave it aside until then. Now each example is treated in the same way in both parts. First, Aristotle shows that it necessarily comes to be from, or passes away into, its contradictory negation—white from or into not-white, cultivated from or into not-cultivated. There can be no denying that this claim is purely logical or, one might even say, 'analytic' (in a Kantian sense): the concept of change analytically requires an *opposition* between its two termini, which is at first interpreted in the most general and least determined way, i.e. as a contradiction—in order for something to come to be X, it must have been not-X before. But this is not sufficient, because it would still allow cases of white coming-to-be from cultivated, insofar as cultivated can be described as not-white. This is why the opposition is then specified: white does not come from (or pass away into) any chance not-white, but from (into) black or an intermediate, as does cultivated from (into) uncultivated or an intermediate. At this point, we move from contradictories to contraries. Contraries can be considered as special cases of contradictories, insofar as a contrary is one of the items which belong to the open range corresponding to the negation of its contrary; but contraries differ from contradictories in several respects, three of which are especially relevant here:

[Cont_1] Contraries belong to the same genus (*Cat.* 6, 6a17–18; *Gen. et Corr.* I 7, 324a2; *Met.* I 4, 1055a27–8). Aristotle sometimes seems to deny this and to allow that contraries belong to contrary genera or be themselves genera (see *Cat.* 11, 14a19–25 and *Met.* Δ 10, 1018a25–7), but the examples he provides (justice and injustice, good and evil) are clearly not ultimate genera and can at the very least be subsumed under one and the same category.[43]

[Cont_2] Contraries belong to the same recipient (δεκτικόν), either in species or in genus (*Cat.* 11, 14a15–18; *Met.* Δ 10, 1018a28–9; I 4, 1055a29–30).

[43] See Ross 1924, ii, 291 *ad Met.* I 4, 1055a17–19. The case of *Met.* I 7, 1057b20 is different: it mentions contraries which do not belong to any genus *at all*, i.e. contrary differentiae constituting the contrary species which do, on their part, belong to a genus—namely the same for both. More on this in subsection 3.4.

[Cont_3] Contraries can (though do not have to) admit of intermediates (*Cat.* 11, 11b38–12a25). These intermediates belong to the open range corresponding to the negation of both contraries; thus a change from one of the contraries to one of the intermediates or the other way round meets the analytic requirement of the concept of change formulated in the first step above. The nature of these intermediates will be investigated later (subsection 3.4).

For the present text, I think the crucial point is Cont_1: what is distinctive of contraries by contrast with mere contradictories is that, in addition to being opposed to one another, contraries belong to the same genus; and this homogeneity of its termini too is an essential, necessary condition of change. This is confirmed by a parallel passage of *On Generation and Corruption*, where it is explicitly stated that what makes it possible for contraries to act on one another is the fact that, in addition to being unsimilar (ἀνόμοια) to one another, they belong to the same genus and are thus similar (ὅμοια) in this respect (*Gen. et Corr.* I 7, 323b29–324a3). In the case of contraries that admit of intermediates (Cont_3), these intermediates must belong to the same genus too, so that change can take place along a homogeneous *range*. Up to this point, we still move at a purely conceptual level. But as James Bogen has shown, we have good reason to believe that Cont_1 and Cont_2 are connected, at least in Aristotle's mind.[44] And they certainly should be, because as Aristotle will show in the next chapter, contraries cannot act directly on one another, but only under the assumption of their belonging to a third term, namely the substratum (*Physics* I 6, 189a22–6; see also *Met.* Λ 1, 1069b3–2, 1069b9; 10, 1075a29–32; N 1, 1087a36–7).

How to make sense of this connection? The genus can certainly not be identified with matter understood as an individualizing principle. But this is probably not the way we should take δεκτικόν in the present context, since in the *Categories*, Aristotle makes it clear that the contraries belong to the same subject *either in species or in genus* (ἢ εἴδει ἢ γένει, 11, 14a15). Even thus, however, the genus and the recipient cannot be identified, since as an example of contraries belonging to recipients that are the same in genus, Aristotle cites whiteness and blackness belonging, not to colour, but to body (14a18). However, there may be a sense in which the genus and the recipient can be very closely coordinated, if we take seriously the description of matter as the possibility of being and not being (see *DC I* 12, 283b3–5; *Gen. et Corr.* II 9, 335a32–3; *Met.* Z 7, 1032a20–2; 15, 1039b29–30). In a way, matter is nothing but the mere possibility of receiving contradictory predicates, which include contraries as well as

[44] See Bogen 1992, 9–12.

possible intermediates between them. If this description can be transposed to the recipient of pairs of contraries belonging to any category, then we can say that this recipient, or at least the respect according to which it is subject to the specific change related to such or such a pair, can be identified, not with the genus itself, but with the possibility of receiving the range of predicates belonging to that genus. This would be amply sufficient to show that Cont_1 and Cont_2 are indeed connected, or even that they imply each other.

If I insist on this connection, it is because it suggests that one of the points of the appeal to contraries, by contrast with mere contradictories, is to herald the introduction of the substratum in the next chapter. This is made even clearer in the *Metaphysics*, where Aristotle argues that it is *because* change takes place between opposites of a certain kind, i.e. contraries, or intermediates between those, that it is necessary to posit a substratum.[45] Here (in *Physics* I 5), admittedly, the move is at best implicit; but it gives at least some justification to the claim that the present chapter should not be read as self-contained, but also as providing material for further developments.

Actually, there may be another way in which the substratum is announced in this chapter. For what is referred to by such expressions as (*τὸ*) *λευκόν* or (*τὸ*) *μουσικόν*?[46] The *properties* 'white' or 'cultivated' themselves, or a white or cultivated *thing*? Horstschäfer has reviewed the different interpretations and argued that they must mean the latter.[47] Maybe his strongest argument is that the clause *πλὴν εἰ μὴ συμβεβηκὸς εἴη τῷ μὴ λευκῷ ἢ τῷ μέλανι τὸ μουσικόν* at 188a35–6 can only make sense if *τὸ μὴ λευκόν*, *τὸ μέλαν*, and *τὸ μουσικόν* refer to things rather than properties, since a property cannot be an accident of another property. And in any case, it seems that Aristotle *should* mean this, for according to him what become are not the properties themselves, but only the composites they constitute with their substratum (see most notably *Met.* Z 9, 1034b7–16; H 5, 1044b23–4; Λ 1, 1069b6–7). On the other hand, it is very clear that what are contrary to one another are the properties themselves rather than those composites; and all the argument as I have construed it seems to be concerned solely with these contraries. The problem is that such expressions can have both meanings, as Aristotle expressly recognizes in a different context (see *Met.* Z 6, 1031b22–5), and that at this stage, these meanings have not been distinguished yet.

[45] See *Met.* Λ 1, 1069b3–7: *εἰ δ' ἡ μεταβολὴ ἐκ τῶν ἀντικειμένων ἢ τῶν μεταξύ, ἀντικειμένων δὲ μὴ πάντων (οὐ λευκὸν γὰρ ἡ φωνή) ἀλλ' ἐκ τοῦ ἐναντίου, ἀνάγκη ὑπεῖναι τι τὸ μεταβάλλον εἰς τὴν ἐναντίωσιν· οὐ γὰρ τὰ ἐναντία μεταβάλλει.*

[46] The fact that Aristotle does not use the article in constructions with *γίγνεσθαι* at 188a35–b3 (he does at a35–6, but the construction there is with *συμβεβηκὸς εἶναι*), whereas he uses it in constructions with *φθείρεσθαι* at 188b3–8, does not seem to be significant, since the two cases are supposed to be symmetrical.

[47] Horstschäfer 1998, 191–5.

This does not imply that Aristotle is still confused about this, but rather that he does not yet presuppose a distinction whose necessity will only appear when the substratum is explicitly introduced in chapter 6—although it was already alluded to in chapter 3 (see 186a28–31)[48]—and which will only be clarified in chapter 7. In this respect too, chapter 5 should be read as a step in an argumentative progression rather than as a statement of positive doctrine.

3.3. 188b8–23

Aristotle here pursues his induction by extending it to less easy cases, which at first sight might seem not to follow the general rule. What are these cases? And what is problematic about them?

The difference between the previous cases and the present ones is presented as a difference between two kinds of beings (τῶν ὄντων, b9), namely simple beings (τὰ ἁπλᾶ) and composite beings (τὰ σύνθετα). Despite the fact that the examples of σύνθετα mentioned (a house and a statue) seem to be examples of *ousiai*, I agree with Horstschäfer[49] that this difference cannot be a reference to the distinction between accidental and substantial change, notably in view of the fact that the expression πᾶν τὸ ἡρμοσμένον (b12), which is supposed to describe the σύνθετα, must at least include the result of the tuning of musical instruments, which cannot as such belong to the category of *ousia*. I also agree that it has nothing to do with the distinction between τὰ ἁπλᾶ and τὰ συγκείμενα which will be made in I 7, and that σύνθεσις should rather be understood here as a 'material' composition.[50] I am not sure, though, that the distinction is between items whose parts differ from the whole and items whose parts, in case they have some, have the same nature as the whole (homoiomera being a special case of these); for as Horstschäfer himself acknowledges, it is difficult to apply this last difference to the only examples of ἁπλᾶ mentioned by Aristotle, namely τὸ λευκόν, τὸ μέλαν, τὸ μουσικόν, and τὸ ἄμουσον. Following the suggestion of several participants of the Symposium, I would submit that the σύνθετα are rather those items whose coming-to-be and passing-away involve a complex process of bringing some order into a multiplicity—a process which can be described quite generally as (material) composition and/or decomposition—as is the case for a house and a statue (which according to I 7, 190b5–6 is realized through μετασχημάτισις). The ἁπλᾶ would then be those

[48] This distinction is of course already made in Socrates' reply to the anonymous objector at Plato's *Phaedo* 103b2–c2, a text whose function in the Dialogue might be compared to the transition from chapter 5 to chapter 6 of *Physics* I. See note 19 above.

[49] Horstschäfer 1998, 194–5, criticizing Happ 1971, 290–1.

[50] Horstschäfer 1998, 197–200.

items whose coming-to-be and passing-away involve no such process—no matter what category they belong to.[51]

But why are these σύνθετα apparently problematic? Aristotle's answer is that in their case, 'the opposed dispositions have no name' (μὴ τὰς ἀντικειμένας διαθέσεις ὠνομάσθαι, b10–11). What he here refers to cannot be the dispositions opposed to ἁρμονία, τάξις and σύνθεσις, which *do* have a name (see ἀνάρμοστια, b14 and ἀσχημοσύνη, b20). The fact that this name is formed by prefixing a privative α- is no objection, since this was already the case for ἄμουσον. The dispositions which have no name are rather those opposed to the *specific* compositions which constitute a house or a statue.[52] Aristotle wants to point out that even in these cases, what the coming-to-be comes from and what the passing-away passes into is not the mere absence of composition, but the absence of the *specific* composition corresponding to the specific opposed state (see b14–15: καὶ ταύτην [*scil.* ἀναρμοστίαν] οὐ τὴν τυχοῦσαν ἀλλὰ τὴν ἀντικειμένην; b18–19: ἐκ τοῦ μὴ συγκεῖσθαι ἀλλὰ διῃρῆσθαι ταδὶ ὡδί; compare *Met.* Λ 4, 1070b28–9: ἀταξία τοιαδί).[53] In other words, in these cases too, change can be analysed as occurring between opposite termini *belonging to the same genus*. This does not mean that the change itself is necessarily a continuous and unified *process*—on the contrary, it can involve many different intermediate processes, as is the case of the building of a house (see *EN* X 4, 1174a19–b5)—but only that its termini can be described in a homogeneous way, and thus that it is no exception to the general rule which Aristotle is trying to establish.

For this reason, I cannot agree with Daniel Graham who detects in this passage a shift from contraries to contradictories.[54] It is true that in lines 188b8–21, Aristotle uses the word ἀντικειμένα rather than ἐναντία. But first, in both cases (b11 and b15), he uses it as an adjective rather than a substantive. Second, he will revert to ἐναντία in the conclusion that follows (b21–3) and in that of the whole chapter (189a9–10), which are clearly intended to cover the cases presently examined as well. Third, if we refer to *Categories* 10, the ἀντικειμένα correspond to the genus of which the ἐναντία are one species, so the shift of word need not introduce any substantial difference. The reason why Aristotle

[51] As Stephen Menn has pointed out, Aristotle's reason for focusing on these complex entities and processes might be that they seem to square more adequately with a Democritean than with an Aristotelian account of change, insofar as they result from the ordering of smaller entities. Aristotle therefore has to show that they are no exception to the rule and that the process of change can be described in exactly the same terms in their case too.

[52] Horstschäfer 1998, 201 n.62 compares to *Phaedo* 71b7–8 εἰ μὴ χρώμεθα τοῖς ὀνόμασιν ἐνιαχοῦ, which is certainly relevant given the context. Note, however, that Socrates there talks about the *processes* rather than their termini.

[53] The ancient commentators rightly insist on this point: see Themistius, *In Physica*, 19.25–20.6; Simplicius, *In Physica* 185.31–3; and especially Philoponus, *In Physica* 113.23–115.22 and 120.26–121.3.

[54] Graham 1987, 145–6.

prefers this formulation in the present context, I suggest, is that he wants to insist on the relation of *opposition* between the two termini of change as a one-to-one relation which implies the *specificity* of the ἀνάρμοστια, ἀσχημοσύνη, or ἀταξία involved. Hence the use of the adjective ἀντικειμένας to qualify the anonymous διαθέσεις corresponding to the σύνθετα at issue here, and of the ἀνάρμοστία corresponding to each specific ἡρμοσμένον involved. If that is so, the negative terms of these relations should precisely *not* be interpreted as the mere contradictories of their positive counterpart, for they are already determined in some way.

What are these opposed dispositions, then? It seems clear that they correspond to what Aristotle will later call στέρησις (privation), even though the actual word will only appear in chapter 7 (190b27). Now, privation differs from contradictory negation in two relevant respects:

> [Priv_1] Privation belongs to a determined genus, i.e. the same genus as what it is the privation of.[55]

> [Priv_2] Privation belongs to a determined recipient.[56]

Priv_1 clearly corresponds to Cont_1 above, and Priv_2 to Cont_2. Interestingly, Aristotle suggests that Priv_1 and Priv_2 are correlated,[57] just as Cont_1 and Cont_2 were. It is not easy to decide whether Aristotle also recognizes

> [Priv_3] Possession and privation admit of intermediates;

[55] Cf. *Met.* Γ 6, 1011b19–20: ἡ δὲ στέρησις ἀπόφασίς ἐστιν ἀπό τινος ὡρισμένου γένους. See also *APo* I 4, 73b21–2, ἔστι γὰρ τὸ ἐναντίον ἢ στέρησις ἢ ἀντίφασις ἐν τῷ αὐτῷ γένει. This last text is difficult, and I am inclined to follow Barnes 1993, 118, who brackets the second ἤ. Even if one does not, however, it clearly treats στέρησις as belonging to the same genus as what it is the στέρησις of. It might be objected that privation will later be described as καθ' αὑτὸ μὴ ὄν (I 8, 191b15–16; 9, 192a5) and as οὐδαμῶς οὐσία (9, 192a6), which seems to exclude its belonging to the same genus as what it is the privation of, especially when that is an *ousia*. To this I would reply: (1) that the texts quoted at the beginning of this note and in note 57 clearly locate the difference between στέρησις and ἀντίφασις in the fact that the first belongs, or is attached, to a determined genus, namely the one to which its opposite also belongs; (2) that Aristotle is sometimes less restrictive about privation, for instance at *Physics* II 1, 193b19–20, where he writes that στέρησις is εἶδός πως, or at *Met.* Γ 2, 1003b6–10, where he includes στερήσεις among things which are called ὄντα because of their relation to *ousia*, or again at *Met.* Λ 4, 1070b11–14, where στέρησις is explicitly counted among εἴδη and οὐσίαι; and (3) that in any case, the fact that *all* the contraries can be reduced to possession-privation (see next paragraph and note 65), combined with the fact that contraries *do* belong to the same genus (Cont_1 above), implies that determined pairs of possession and privation must themselves belong to the same genus.

[56] Cf. *Met.* I 4, 1055b7–8: ἡ στέρησις ἀντίφασίς τις ἢ ἀδυναμία διορισθεῖσα ἢ συνειλημμένη τῷ δεκτικῷ.

[57] See the difficult passage of *Met.* Γ 2, 1004a12–16 (ἢ [γὰρ] ἁπλῶς λέγομεν ὅτι οὐχ ὑπάρχει ἐκεῖνο, ἤ τινι γένει· ἔνθα μὲν οὖν †τῷ ἑνὶ ἡ διαφορὰ πρόσεστι παρὰ τὸ ἐν τῇ ἀποφάσει†, ἀπουσία γὰρ ἡ ἀπόφασις ἐκείνου ἐστίν, ἐν δὲ τῇ στερήσει καὶ ὑποκειμένη τις φύσις γίγνεται καθ' ἧς λέγεται ἡ στέρησις), with the commentaries of Ross 1924, i, 259–60 and Kirwan 1993, 84.

he sometimes says so, but what he seems to mean is that these opposites are not universally applicable, precisely because of Priv_2.[58] But it does not matter in the present context: all that matters is that privation thus understood shares the feature of contraries operative in the preceding reasoning, namely the belonging to the same genus as its opposite—which also implies their belonging to a determined recipient, although this further consequence will only be drawn in the next chapter. In other words, what matters here is that the privation corresponding to a determined state is not the mere, undetermined negation of this state, but is already determined by the fact that it belongs to the same genus as what it is the privation of. Hence the privation of a statue is not identical to the privation of a house, even if both can be described as ἀταξίαι of some sort, because they do not belong to the same genus. From this point of view, changes occurring from or into privation are no exception to the rule according to which all changes occur between contraries.

It might be objected that in the *Categories*, the opposition of contraries and the opposition of possession and privation are sharply distinguished from one another (see *Cat.* 10, 12b25–13a36). But the concept of privation in this treatise is clearly different from the one operative in the *Physics*.[59] Most notably, in the *Categories*, privation is restricted to the absence of a property which a subject is naturally capable of having and at the time when it is natural for it to have it (12a29–31). This explicitly implies that privation thus understood can never be a *terminus a quo* of change, but only a *terminus ad quem* (13a31–6): before something possesses the property at issue, there can be no question of privation; and once it possesses it, the only possible change is its loss, which will be privation *proper*. These are the peculiarities that ground the contrast between the two kinds of opposition at 12b25–13a36. Now, it is clear that these peculiarities are incompatible with the concept of privation as it is used in the context of the analysis of change in the *Physics* and the *Metaphysics*, which certainly *is* a *terminus a quo* as well as a *terminus ad quem* of change (see e.g. *Physics* I 8, 191b13–17 and *Met.* Z 7, 1033a8–11). The fact is that privation is said in many ways (see of course *Met.* Δ 22, but also Θ 1, 1046a31–5); and among these, the one that is relevant here, and which Aristotle calls 'perfect' (τελεία, *Met.* I 4, 1055a35), is quite clearly recognized as forming an opposition of contrariety with possession (see *Physics* I 7, 191a14, and maybe also 190b27; V 1, 225b3–4; *Met.* Z 7, 1032b2–4; K 11, 1068a5–6; Λ 2, 1069b33–4). Indeed, Aristotle's fullest

[58] See *Met.* I 4, 1055b8–11, but compare Ross 1924, ii, 292. Contrary to what Ross suggests, *Met.* K 3, 1061a20–2, is no evidence that Aristotle would deem possession-privation as admitting of intermediates, since just and unjust are there treated as contraries, and the fact that they admit of intermediates is seen as a difficulty for the thesis according to which contraries are said κατὰ στέρησιν. As for *Cat.* 10, 13a4–7, see below.

[59] Cf. Bodéüs 2002, 60 n.2.

doctrine in the *Metaphysics* is that this opposition is the *primary* contrariety (πρώτη…ἐναντίωσις, I 4, 1055a33–8), the one that grounds all the others, because in any pair of contraries, one term can be considered as the privation of its opposite (*Met.* Γ 6, 1011b18–20; I 4, 1055b17–29; K 3, 1061a19–20; 6, 1063b17–19; see also *Physics* I 7, 191a6–7). There is therefore nothing anomalous in Aristotle's conclusion that *all* coming-to-be and passing-away, including those which can only be described with the help of the concept of privation, occur between *contraries*.

Another objection has been raised against this conclusion, however. How can Aristotle extend this thesis to substantial change, given his standard and undisputed doctrine according to which *ousia* has no contrary?[60] How can Aristotle maintain that *all* changes occur between contraries in these conditions?[61] Is not the fact that substance has no contrary precisely the reason why Aristotle sometimes makes a difference between change (μεταβολή) in the category of οὐσία, which only occurs κατ' ἀντίφασιν, and in the categories of quality, quantity, and place, which takes place between ἐναντία and can also be called 'motion' (κίνησις) (see *Physics* V 1, 225a34–b5; 2, 225b10–11, b25–6; *Met.* K 11, 1068a1–7; 12, 1068a10–11, a24–5)? However, according to what we have just seen, Aristotle cannot mean that change in *ousia* occurs according to contradiction *stricto sensu*, but only according to privation, which is 'a certain contradiction' (ἀντίφασίς τις, *Met.* I 4, 1055b4), i.e. contradiction in a certain genus. And if the opposition between possession and privation is a certain contrariety, then even substantial change can be said to take place between contraries. In fact, at *Met.* Λ 2, 1069b9–14, Aristotle explicitly states that change takes place between contraries in the *four* categories in which it occurs, including *ousia*.

In this context, the difficulty is rather to understand how Aristotle can maintain that *ousia* has no contrary. Philoponus already faced the problem.[62] Part of his reply consists in suggesting that the principle according to which *ousia* has no contrary only holds for composite (σύνθετος) *ousia*, while the simple forms

[60] The *locus classicus* is *Cat.* 5, 3b24–32. For other references, see Bonitz 1961², 544a60–b2. *Physics* I 6, 189a32–3 may allude to this doctrine, though strictly speaking these lines only say that no *ousia* is contrary to an *ousia*.

[61] See the objection in Solmsen 1960, 80, 90 and Bostock 2006, 10. Actually, the same objection could be raised about quantitative change, since quantities do not have contraries either (see *Cat.* 5, 3b28–30 and 6, 5b11). On this problem, see Bogen 1992, 16–19, who suggests that things of a given natural kind may be assigned maximal and minimal amounts, which can be considered as contraries between which change can occur. He refers to *Physics* V 2, 226a23–32, which explicitly recognizes that change in quantity takes place between contraries and defines increase as change toward complete magnitude (εἰς τὸ τέλειον μέγεθος) and decrease as change away from this. See also *Physics* III 1, 201a6, which cites τὸ τέλειον and τὸ ἀτελές as the 'two ways' in which quantity can belong to something (see 201a3–4), and *Met.* Λ 2, 1069b9–14, where change in quantity is treated as a case of change between contraries.

[62] Philoponus, *In Physica* 112.5–8.

(τά...ἁπλᾶ εἴδη) *are* characterized by contrariety.[63] Although Philoponus is mainly thinking of the forms of the elements (the dry and the wet, the cold and the warm),[64] I suggest that we extend this distinction to other *ousiai* as well: οὐσία understood as εἶδος *has* a contrary, namely στέρησις,[65] whereas οὐσία considered as a compound of matter and form has none. Hence there is no incompatibility between the fact that change in the category of *ousia* occurs between contraries, namely στέρησις and εἶδος, and the fact that *ousia*, understood as compound of matter and form, has no contrary.

This solution, however, presupposes the distinction between the contraries and the substratum, which has not yet been made. Actually, we can find the same kind of ambiguity about the termini of change mentioned in this section as in the previous one: while Aristotle sometimes seems to consider them as *things* tuned (or harmonized, etc.) or untuned (see notably πᾶν τὸ ἡρμοσμένον at 188b12, and ἐκ τοῦ μὴ συγκεῖσθαι ἀλλὰ διῃρῆσθαι ταδὶ ὡδί at b18–19), he quite clearly assimilates them to the *properties* of tunedness or untunedness on other occasions (see especially ἀνάρμοστίαν at b14 and ἀσχημοσύνης at b20). Once again, this unclarity should not be interpreted as implying any confusion on Aristotle's part, but rather as a way of making us feel the need for such a distinction, whose absence will provoke the *aporia* of chapter 6.

Taking all these elements into account, then, I think Aristotle's conclusion *is* warranted at this stage of his discussion: *all* coming-to-be and passing-away takes place between contraries or their intermediates. Let us now turn to the status of these intermediates.

3.4. 188b23–6

As I have noted above, the purpose of this subsection is to exclude the intermediates as contenders to the title of principles, insofar as although they might be termini of change, they can also be derived from something else, i.e. the contraries. To express this, Aristotle shifts back from the phrase γίγνεσθαι ἐκ + genitive to the phrase εἶναι ἐκ + genitive, which as we have seen expresses the relation between something and its principles (cf. section 2 above). It is important to notice this shift: for whereas the intermediates might be what something *comes to be* from (or *passes away* into), they are not what something is *derived* from, but are rather themselves derived from something else, i.e. the contraries.

[63] Philoponus, *In Physica* 112.15–17. See also Ross 1936, 506 *ad* 193b20–1.

[64] See Philoponus, *In Physica* 112.17–18.

[65] See *Met.* Γ 6, 1011b18–19: τῶν μὲν γὰρ ἐναντίων θάτερον στέρησίς ἐστιν οὐχ ἧττον, οὐσίας δὲ στέρησις. Ross 1924, i, 283, *ad loc.* comments: '"privation of the positive, substantial nature"—more commonly in this connexion called εἶδος'.

The way intermediates derive from contraries is explained in *Metaphysics* I 7 (1057b2–32).[66] Aristotle distinguishes there between contraries belonging to a genus and contraries not belonging to a genus, the latter being the differentiae which constitute the first ones by being added to the genus. Similarly, one can distinguish between the intermediate *species* and the intermediate *differentiae* which constitute these species in combination with the genus. Now contraries not belonging to a genus are 'uncomposed of one another, and thus principles' (ἀσύνθετα ἐξ ἀλλήλων, ὥστε ἀρχαί, 1057b22–3); but the intermediate differentiae, which constitute the intermediate species, are 'composed' of these contraries, in the sense that they are 'less' than one of them and 'more' than the other, i.e. they have an intermediate *degree* between the contraries; and 'what is so to speak composed of the more and the less is derived from those than which it is said to be more or less' (τό...τοῦ μέν μᾶλλον τοῦ δ' ἧττον σύνθετόν πως ἐξ ἐκείνων ὧν λέγεται εἶναι τοῦ μὲν μᾶλλον τοῦ δ' ἧττον, 1057b27–9).

Further light can be gathered from earlier passages of the same book, where Aristotle explains that 'the one is the first measure of all things, because we come to know that from which the *ousia* [of something] derives by dividing it either according to quantity or according to form' (πάντων μέτρον τὸ ἕν, ὅτι γνωρίζομεν ἐξ ὧν ἐστὶν ἡ οὐσία διαιροῦντες ἢ κατὰ τὸ ποσὸν ἢ κατὰ τὸ εἶδος, 1, 1053a18–20; cf. 1052b18). This does not amount to a reduction of all beings to numbers nor of all sciences to arithmetic, for Aristotle insists on the fact that the one is different in each category, and even in each genus (2, 1053b25–8). It nevertheless legitimizes an interesting, albeit analogical, use of number as an epistemological device to understand genera other than quantities. Aristotle gives the example of colours: in this genus, the one is white, of which black is the privation; and all the other colours, which derive from white and black, can be considered as 'numbers', i.e. numerical ratios between these contraries (2, 1053b28–34).[67]

These passages show, I think, that when Aristotle says that the intermediates 'derive from' (εἶναι ἐκ) the contraries, he does not mean that the latter 'constitute' or 'produce' the intermediates as material or efficient causes, but rather that *their notion* can be derived from those of the contraries. This seems especially clear in *Met.* I 7, which deals with definitions composed of genus + differentia: if there is any production involved here, it is only at the epistemological level. From this point of view, if the contraries are 'causes' of the intermediates, it must be as *formal* causes: they belong to their definition, since the very nature of the intermediates is to be intermediate (μεταξύ) between the contraries—and the

[66] On this text, see especially Chiaradonna 2005, who insists on its connection with the analysis of change.

[67] On this point, see Bogen 1992, 14–15; and, on the importance of the doctrine of the one as 'measure', see Couloubaritsis 1983, 168–84.

nature of *these* intermediates (e.g. colours) is to be intermediate between *these* contraries (e.g. white and black), namely those belonging to the same genus. This does not prevent, of course, that in some cases, an intermediate can be obtained by a *material* mixing of the contraries, for example an intermediate colour by mixing white and black (see the different explanations of the generation of colours examined in *Sens.* 3, 439b18–440b23); but this possibility should be interpreted as a further consequence of their *eidos* being so, and the two levels should not be confused. For example, it is because grey corresponds to a ratio of (say) 1:1 of white and black that grey paint can be obtained by mixing the same proportion of white paint and of black paint. In any case, while it would be very difficult to transpose such a process to all the pairs of contraries admitting intermediates—cultivated and uncultivated, for instance, or, if it were disputed that they admit intermediates, good and bad or just and unjust, which do (see *Cat.* 10, 12a9–25)—there is no problem in understanding these intermediates as deriving from the contraries considered as formal causes in the way just explained. All this confirms that we remain at the level of conceptual analysis throughout the passage, even though the concepts we are dealing with are of course empirical ones.

The fact that the intermediates derive from the contraries in this way can also explain their role as termini of change. As Aristotle writes elsewhere, 'change takes place from the intermediate, since it serves as contrary to either of two [contraries]; for the intermediate is in a sense the extremes' (*Physics* V 1, 224b30–2; see also 5, 229b14–21). If the intermediates are defined by reference to both contraries, whereas the contraries are purely what they are, then in a way the intermediates can be considered as being contrary to both extremes. Hence one could say that any change takes place between contraries in a way, even those whose termini are intermediates, precisely because these can be reduced to the contraries between which they are located and can thus serve as contraries in a change.

So 'everything that comes to be by nature either is a contrary or derives from contraries' (188b25–6); and among those, since what derives from something else cannot be a principle, only the contraries qualify as principles.

4. 188b26–189a9

Up to this point,[68] most of the other thinkers have practically followed along with us, as we said earlier: all say that the elements and what they call the principles are the

[68] I read *μέχρι...ἐπὶ τοσοῦτον* with Ross, who follows E[1] and refers to Bonitz 1961[2], 464b17–22. Carteron 1926 prints *μέχρι... τούτου* with E[2]*Λ*, but since there is no difference in meaning, it is better to keep the *lectio difficilior*.

> contraries, even though they posit them without argument, as if they were forced by truth itself. But they differ from one another in that some take contraries which are prior and others contraries which are posterior, and some contraries which are better known according to *logos* and others according to sense-perception (for some posit hot and cold, or again wet and dry, as causes of coming-to-be, while others posit odd and even, or again strife and love, and these differ from one another in the way just stated). So they say the same things in a way and different things from one another; different, as it also seems to most, but the same by analogy: for they take them from the same series (συστοιχία), for some contraries envelop, others are enveloped. Thus in this way they speak similarly and differently, and worse and better, some [choosing] what is better known according to *logos*, as was said earlier, and others what is better known according to sense-perception (indeed the universal is known according to *logos*, but the particular according to sense-perception; for *logos* is of the universal, sense-perception of the particular), for example the great and the small according to *logos*, the rare and the dense according to sense-perception.

Here Aristotle turns back to his forerunners and shows how the argument he has proposed in the previous paragraph makes it possible to assess their positions. The most interesting feature of this passage is certainly the use of the concept of συστοιχία (189a1, here translated by 'series') in order to bring order among the different pairs of contraries which have been advanced as principles of coming-to-be.

The concept of συστοιχία has a variety of meanings in Aristotle,[69] but its special use in the present context has received a recent and very thorough treatment by David Lefebvre in relation to the closely connected passage of *Met.* Γ 2, 1004b27–1005a6.[70] Lefebvre shows that although it is possible that this structure has a Pythagorean origin (see *Met.* A 5, 986a23; *EN* I 4, 1096b6)[71] and probable that it was used in the Academy, notably by Speusippus (see *EN* I 4, 1096b6–7), Aristotle uses it as a positive device in key contexts (in addition to the present passage and the one already cited of *Met.* Γ 2, 1004b27, see especially *PA* III 7, 670b21, *Met.* Λ 7, 1072a31, 35 and *EE* VII 12, 1245a2). Its most salient feature is that it can gather terms that, according to Aristotle, may belong to different categories (see especially *Met.* N 6, 1093b18–21, and compare *Physics* III 2, 202b25–7 and *Met.* K 9, 1066a14–17; see also the use of συστοίχως at *Sens.* 7, 448a14). This might be due to its 'archaic', or at least 'external', origin; but in any case this was not considered as a flaw by Aristotle, but on the contrary as an advantage, since it makes the notion serviceable as a truly universal principle of organization, one that can extend over items separated from one another on

[69] See Bonitz 1961², 736b33–60, and 736b61–737a19 on σύστοιχος.

[70] Lefebvre 2008 (on *Physics* I 5, see 311–13). See also Berti 1973, 1975.

[71] *Met.* N 6, 1093b12–13 is more problematic, since it does not specifically refer to Pythagoreans (rather than Academics) and does not imply that the use of συστοιχία is proper to the thinkers discussed (rather than to Aristotle himself).

other grounds.[72] It is especially relevant in the present context, since it makes it possible to unify the account of change in all the categories in which it occurs, at a stage where distinctions between categories have not been introduced yet. Thus just as the *Metaphysics* Γ 2 passage is intended to prove the possibility of a unified science of being qua being despite the fact that being is not said *καθ' ἕν* (see 1005a2–6), the present one can be considered as grounding the possibility of physics as a unified science of change running across different categories.

More concretely, all the pairs of contraries which have been invoked as principles can be organized in two coordinate columns, each of which is called a *συστοιχία*. The terms of each column are connected to those of the other in a one-to-one relationship, namely contrariety. This is the way in which the principles of the ancient thinkers are identical (*ταὐτά*, 188b36, 37; *ὡσαύτως*, 189a3), although the actual pairs they propose are different (*ἕτερα*, 188b36, 37; *ἑτέρως*, 189a3): the identity between these pairs is an identity of the *relation* between the two terms of each pair. Hence each pair can be said to be analogous (*ἀνάλογον*, 189a1) to any other. Here we are very close to what will be said about the analogy of principles in *Metaphysics* Λ 4–5, except of course for the absence of *ὕλη* in the picture so far—a concept which will also be said to be knowable by analogy later (*Physics* I 7, 191a7–12)—as well as of the moving cause.

But there is something more in the concept of *συστοιχία* as it is used here, namely the fact that the terms belonging to each column—and consequently the pairs they form with their contraries—are organized in a *hierarchical order*. The criterion according to which this hierarchy is established is expressed in different ways in our text: some contraries are prior while others are posterior (*πρότερα*...*ὕστερα*, 188b31), some are more knowable according to *logos* while others are more knowable according to sense-perception (*γνωριμώτερα κατὰ τὸν λόγον*...*κατὰ τὴν αἴσθησιν*, 188b31–3, 189a4–5), i.e. some are universal while others are particular (*καθόλου*...*καθ' ἕκαστον*, 189a5–8), some envelop while others are enveloped (*περιέχει*...*περιέχεται*, 189a2), and the account based on some is better (*βέλτιον*) while the one based on others is worse (*χεῖρον*) (189a3–4). Contrary to what Philoponus suggests,[73] it seems clear that these different formulations all refer to one and the same criterion, namely the scale between the most universal and the most particular. This scale is not to be confused with a genus–species

[72] See Lefebvre 2008, 309–11, 319. This does not mean, of course, that this principle of organization is *incompatible* with the distinction between categories: at *Physics* I 6, 189b22–7, Aristotle will say that the first contrariety to which all the others can be reduced (*ἀνάγεσθαι*) is the contrariety in the category of *ousia*, which indeed corresponds to the contrariety between possession and privation that will emerge from the present passage.

[73] Philoponus, *In Physica*, 122.31–123.21 (but cf. 127.3–17). As for Alexander, he seems to have understood that one term of each pair envelops and the other is enveloped, an interpretation rightly criticized by Simplicius who reports it (*In Physica*, 189.19–27).

relation, however, since it organizes terms which can belong to different categories. It rather corresponds to the order of *induction* as I have characterized it above (subsection 3.2). In this context, it is certainly no accident that we find echoes of the methodology of I 1 here—not only the general motif of a path (cf. ὁδός, 186a16) on which we are engaged, but also the assimilation of στοιχεῖα (188b28), ἀρχαί (188b28), and αἰτίαι (188b34–5), and the opposition between two ways of being 'better known'—except that the term καθόλου seems to have regained its more traditional sense in the meantime[74] (if indeed it had ever lost it).

Now, as we have seen, induction does not consist in merely collecting the particular cases under an extensionally defined universal, but rather in finding a single *law* that explains those particular cases. This is exactly what the first pair of contraries is supposed to do in a συστοιχία: not, of course, to *produce* the other pairs, but to furnish their *principle of explanation*.[75] In other words, it belongs to the notion of all the other contraries and can be considered as their *formal cause*.[76] In this sense, one could say that all the other pairs derive from this first pair, in a way similar (albeit not identical) to the way the intermediates derive from the corresponding contraries, i.e. insofar as the very being of the other pairs is specified in terms of this first pair. The being of the intermediates derives from the contraries that define the range to which they belong; the being of the contraries derives from the very first pair of contraries. Hence only this first pair of contraries meets Princ_1, as Aristotle already announced from the outset;[77] and we can now also see that it meets Princ_3, since all the pairs of contraries—and through them all the intermediates between those—derive from them.

What is this first pair of contraries? Although it is not explicitly stated in the present text, it seems clear, on the basis of what we saw earlier, that it is the opposition of possession and privation. This is indeed confirmed by the parallel text of *Met.* Γ 2, 1004b27–1005a5, which says that 'of the contraries, one of the two series is privation' (τῶν ἐναντίων ἡ ἑτέρα συστοιχία στέρησις, 1004b27) and adds that all contraries can be reduced (cf. ἀνάγεται, 1004b28; ἀναγόμενα, b34; ἀναγωγή, 1005a1) to τὸ ὂν καὶ τὸ μὴ ὄν (1004b28) and τὸ ἓν καὶ πλῆθος (1004b28, 34; 1005a4–5) as their principles (ἀρχαί, 1005a1, 4). Since being and not-being should here be understood as belonging to a determined genus and therefore as the opposition of possession and privation,[78] and since the opposition between the one and the many is itself described as an opposition between

[74] As Horstschäfer 1998, 209–10 remarks.

[75] Lefebvre 2008, 309 and 320.

[76] Berti 1973, 1975, 222 and 231.

[77] See 188a28–30: τοῖς δὲ ἐναντίοις τοῖς πρώτοις ὑπάρχει ταῦτα, διὰ μὲν τὸ πρῶτα εἶναι μὴ ἐξ ἄλλων…

[78] See *Met.* Γ 2, 1004a10–16 and the explanations of Berti 1990, 2005.

possession and privation,[79] it is the latter which is really fundamental, as was to be expected given that this opposition is independently described as the 'first contrariety' and the principle of all the others.[80]

This first pair of contraries has a very special status in the *συστοιχία*: on the one hand, it belongs to the series as *one* pair among others, albeit the first one; but on the other hand, as the result of induction, it also corresponds to the identical element present in each of the other pairs, and hence is the reason of the analogy between all these pairs. In other words, it corresponds to the very relation of contrariety considered *in itself*, which may be explicated as taking place between possession and privation understood as concepts devoid of any other content than the kind of relation they have to one another. It has therefore a very different status from all the other pairs, as was only to be expected given what I have said above about the nature of induction; but the device of *συστοιχία* is flexible enough to allow us to integrate it into the same structure as these other pairs, insofar as it is the law which explains them all.[81]

Thanks to this sophisticated device, Aristotle can not only explain how his predecessors were 'practically' (*σχεδόν*, 188b26) in agreement with one another and with himself, but also how far they were right, 'as if they were forced by truth itself' (*ὥσπερ ὑπ' αὐτῆς τῆς ἀληθείας ἀναγκασθέντες*, 188b29–30; compare *Met.* A 3, 984a18, b9–10); and he can even express comparative judgements on the value of their theories (cf. *καὶ χεῖρον καὶ βέλτιον*, 189a3–4), and concede that Plato was closer to the truth (see 189a8)—which is only fair, since as we have seen, Plato was certainly the first one to recognize (in the *Phaedo*) the role of contraries in general in the processes of coming-to-be and passing-away. The path to the principles is induction, and according to Aristotle, all his forerunners were unconsciously engaged on this path, although none pursued it up to the end, some having given up earlier than others.

In fact, their common mistake is comparable to that of young children who call all men 'father' and all women 'mother' (cf. I 1, 184b13–15). These children perceive some similarities between the different men or the different women,

[79] See *Met.* I 3, 1054a23–6, reading Ross' text rather than Jaeger's.

[80] *Contra* Themistius, *In Physica*, 20.28–30, who seems to assume that the first contrariety is that of *σύγκρισις/ὑπεροχή* and *διάκρισις/ἔλλειψις*. Following a similar line, Bonitz 1969, 53–5, wanted to replace *περιέχει... περιέχεται* by *ὑπερέχει... ὑπερέχεται* at 189a2, notably by reference to such passages as *Physics* I 4, 187a16–17; 6, 189b10–11; *Met.* A 9, 992b6–7; H 2, 1042b24–5; but Ross 1936, 489 is clearly right in pointing out that this correction ignores one of the main points of the passage, namely the hierarchical order of the terms belonging to the same *συστοιχία*.

[81] My reading of this passage is therefore different from the one proposed by Stephen Menn in this volume, pp. 49–51. I do not think that this passage can be used as evidence for a distinction between universal principles and principles under universal description, because both coincide in the pair of possession and privation, which is at the same time the first pair and the identical element present in all the other pairs and responsible for the analogy between them.

and because they call one of them 'father' or 'mother', they apply this name to all the other men or women as well. What is mistaken in this process is, of course, that the name 'father' does not designate the respect in which all men are similar to one another. But in order to see this, one has first to realize *both* in what way all men are similar *and* in what way they may differ from one another. This is why, I think, Aristotle says in chapter 1 that going from what is better known to us to what is better known by nature involves the activity of dividing (cf. διαιροῦσι, 184a23; διαιρεῖ, b13) or distinguishing (cf. διορίζει, b14) into the particulars (καθ' ἕκαστα, a24, b13). This does not contradict the fact that the way to the principles is a way towards what is more universal, provided that we understand 'what is more universal' not as 'what has more extension', but as 'what is more *truly* universal', i.e. what is universal *in a scientific way*, by contrast with what is universal in a merely empirical way. As we have seen, the scientific universal differs from the empirical universal by corresponding not to the mere collection of particular cases displaying some similarity, but to the single law or εἶδος which is the cause of this similarity. Now finding such a universal implies recognizing not only the similarity between a set of cases, but also how they differ from one another, in order to identify the relevant *respect* in which they are the same.

This is what Aristotle blames his forerunners for not having seen. They were looking for something all processes of change would share in common and realized that it had something to do with contrariety. But because of the confused way in which they perceived the universal—i.e. change and contrariety as its principle—focusing on similarities without paying enough attention to dissimilarities, each of them picked up *one* pair of contraries (e.g. the hot and the cold) and used it to interpret *all* processes of coming-to-be. This amounts to a confusion between identity and analogy, whereas the latter implies *both* identity *and* difference. On the contrary, Aristotle's method in this chapter has consisted in citing very different processes of becoming—first white coming to be from black and cultivated from uncultivated (subsection 3.2), then 'all that is tuned' from what is untuned, including a house and a statue (subsection 3.3)—and trying to detect what is *truly* common to all these cases, without denying their difference. He ended up with the very relation of contrariety itself, that is to say the relation of possession and privation, which is present in all pairs of contraries as the law explaining them all and thus corresponds to the true—i.e. *scientific*—universal. Moving from the empirical and confused universal to the scientific universal *does* therefore imply an activity of dividing or distinguishing, precisely because the principles we are looking for are not strictly identical in all cases, but only identical by analogy.

5. 189a9–10

That the principles, then, must be contraries is clear.

As I remarked in the introduction, this conclusion reverses the order of the terms: up to now, what Aristotle was arguing for was that the contraries were principles; here, he says that the principles are contraries, suggesting that the contraries are not only *among* the principles, but are *the only* principles. Now, as is well known, this further step will be questioned in the next chapter; and actually, the very first sentence of this next chapter, which is connected to the present one by a δέ answering the μέν of 189a9 and where Aristotle asks whether the principles are two, three, or more in number, immediately suggests that it should not be taken at face value. Moreover, the necessity to add τὸ ὑποκείμενον to the contraries will not emerge out of the blue in chapter 6: it was already announced at different points in chapter 5, albeit only implicitly.

However, given the way chapter 5 proceeded, it can be argued that this stronger conclusion *is* warranted after all. For if the principles we are looking for are simply the termini of change which do not derive from anything else, how could there be any other principle than the contraries? Moreover, as was stressed in the introduction, Aristotle will never say in so many words that this conclusion is *false*; in chapter 7, he will even secure a way in which it is true (190b29–191a3). The reason for this, I think—and here I agree with Kelsey—is that, although change needs three principles in order to be accounted for (form, privation, and the substratum), these are not three *numerically* (ἀριθμῷ) distinct entities: there is only one 'thing' from (ἐκ) which what comes to be comes to be, but this 'thing' can be described either as substratum or as privation, which are *specifically* (εἴδει) different (see I 7, 190a13–21 and b23–4). So there is a sense in which the principles of change are two rather than three, namely the contraries, and the conclusion of the present chapter, even in this stronger form, cannot be discarded altogether.

More generally, I hope to have shown that there is nothing in this chapter that Aristotle could not maintain in the rest of the book. To my eyes, there can be no doubt that it is far more than a mere doxographical discussion (if there is such a thing in Aristotle): it is a decisive part of Aristotle's argument towards the principles. Many points are still in need for further clarification, to be sure; but as it stands, this chapter can certainly claim to have moved the argument a big step forward.

6

A Third and Underlying Principle

Physics I 6

ANDREAS ANAGNOSTOPOULOS

Introduction

Physics I 6 is ostensibly devoted to the question of the number of principles, a question that has been implicitly raised in the opening of *Physics* I 2 and is revisited in I 4. What is this leading question? I take it that it is a question about the principles of substances, and in particular, natural substances.[1] Furthermore, while much of the discussion of I 6 as well as the previous chapter can be understood to focus on a single, perhaps arbitrary, natural substance or kind thereof, the leading question is about the number of principles needed for all natural substance or, as Aristotle will sometimes put it, for the genus.[2] By the

I am grateful to many friends and colleagues for helpful discussions and challenging criticisms: Peter Adamson, Laura Castelli, Tim Clarke, David Ebrey, Pieter Sjoerd Hasper, Mirjam Kotwick, Christopher Noble, Christian Pfeiffer, Christof Rapp, Sirio Trentini, and Joel Yurdin. It is a pleasure to record long-accrued debts to Alan Code and Sean Kelsey, extending to my study of their recent work on this chapter. In addition, I am grateful to the other conference participants for their probing questions.

[1] That it concerns substance is clear within *Physics* I 6, at 189a32–4 and 189b23–7; the restriction to natural substance can be gathered from the discussion of natural beings and natural science elsewhere in book I, and form the overall purpose of the work.

[2] This is evident in the way that I 6 uses the *diaireseis of* I 2 and I 4, such that different accounts of principles amount to different answers to the question of whether, and if so how, being (all being) is one; the connection between the question of principles and the question of being is made explicitly I 2 184b22–5. Thus, the dialectic between Anaxagoras and Empedocles about whether an infinity of contrary principles must be posited (I 6 189a14–17) recasts the debate from I 4 (187a20-6, 188a17-18) about whether finitely or infinitely many basic kinds are present in the 'one' by way of mixture; the portrayal of material monists as positing one underlying principle for all of being, and thus making 'the all' one

end of I 7, Aristotle has explicitly answered the question and announced that it has been settled (190b29–191a3; 191a20–22). Much of the work that prepares this answer is done in I 6. In particular, the chapter is organized around various possible answers to the question: Aristotle first argues that there is a finite plurality of principles, that is, neither just one nor unlimited principles (89a12–20); he then offers reasons for positing more than two (189a20–b1, or more loosely, up to b18); and he finally argues that there can be at most three (189b18–27).

The chapter is not, however, merely about the number of principles; it is more interestingly about *kinds* of principles. Against the backdrop of chapter 5, which argues that the principles are contraries, chapter 6 provides reasons for positing a non-contrary, 'underlying' principle. On the supposition that contrary principles come in pairs, and so number at least two, these reasons are presented as reasons for having at least three principles (189a20–b1). Given the limitation in the last section to at most three principles (one pair of contraries and one underlying principle), we might expect Aristotle simply to conclude that there are three principles. Instead, he presents the question of whether there are two or three principles as a major difficulty at the end of I 6 and this difficulty is resolved in the next chapter. So I 5 and I 6 motivate the account of principles in I 7, which incorporates both contrariety and an underlying principle, not only by providing arguments for both kinds of principles, but also by presenting a difficulty that the account of I 7 resolves.[3]

Though I will comment on the entire chapter, I will focus especially on two issues. The first is an issue that has received significant scholarly attention: how are we to understand the 'two vs. three' *aporia* that remains at the end of the chapter? This issue is bound up with the question of what the upshot of I 5 is supposed to be, as well as with the interpretation of Aristotle's solution to the *aporia* in I 7. The basic source of the *aporia*, I argue, is that I 5's conclusion is that *all* principles are contraries, which contradicts the position argued for in I 6, that there is also a non-contrary, underlying principle, although the dialectic in which these two theses are argued for is decidedly more complex.

Second, I would like to bring out the way that Aristotle often takes on a materialist perspective, in such a way that he ends up downplaying or even

(I 6 189b2–11) draws on the earlier claims that they employ just one changeable principle (I 2 184b15–18) and make all being one in the sense of a single underlying body (I 4 187a12–15). The closing section of I 6, which limits the primary contrarieties to one, also clearly addresses a general question (189b22–7).

[3] Like many of the contributors, I am in strong agreement with Kelsey 2008 that the theory of principles in I 7 does not arise, as it were, *ex nihilo*, but is rather both a development of, and reaction to, the lines of argument found in the previous chapters. Indeed, the resolution of the two versus three *aporia*, I submit, not only shapes the theory presented in I 7, but also contains the crucial insight that Aristotle showcases in his discussions of the Eleatic dilemma (I 8) and the Platonic triad of principles (I 9).

eliding the differences between it and his own approach. In my view (which I cannot fully defend here), Aristotle's way with materialism is careful and deliberate, and is meant to bring along a reader who subscribes to, or is at least more familiar with, a materialist framework. I will refer, in what follows, to two slightly different strands of materialism. The first strand treats material (or in the terms of *Physics* I, underlying) aspects of things as prior to formal (or contrary) ones, so that they alone are their *ousiai* and principles, in contrast with the priority that that form enjoys for Aristotle. It is operative, I think, in one key passage in I 6 (see Section 2.2). But a second strand predominates within *Physics* I, where Aristotle often presents materialist views—specifically monist views—as prefiguring the position he will argue for; in particular, he presents materialists as holding, like he does, that there are both contrary and underlying principles, and indeed at most three principles in all (two contraries and one underlying principle).[4] This similarity, at an abstract level, belies at least two important differences. With respect to *primacy*, as I will call the features by which some ingredient, aspect, or source of a given natural substance qualifies as a genuine principle (whether contrary or underlying) of it, the material monists focus on the most basic or ultimate underlying subject and contrariety.[5] By contrast, Aristotle's own account employs 'higher-level' principles that are determined by the essence of the kind of substance in question.[6] A second difference concerns the *generality* of the leading question. The material monist takes this generality to demand an enumeration of all the principles—fully specified—which account for each and every being;[7] it is a substantive and plausibly empirical question whether there is e.g. just one underlying principle, i.e. whether everything is made of the same, fully specified kind of body, such as one of the four elemental bodies.[8] By contrast, for Aristotle, who is faced with different and higher-level contrary and underlying principles for different kinds of substance, the 'reduction'

[4] However, in chapter 2, the material monists are presented simply as positing one principle, though Aristotle might qualify this when he describes air or water as 'the first principle' on these views (τὴν πρώτην ἀρχήν: 184b15–18).

[5] For example, an underlying nature is primary because it is an ultimate subject; for one underlying nature to be prior to another is for the former to serve as underlying nature for some affection or contrary, or some series of them, which yields the latter. A complementary view about contrariety is that a contrariety is primary if it is the 'first' to characterize a basic underlying nature; one contrariety will be prior to another insofar as the latter is applicable only to—and perhaps comprises affections or modifications of—things already characterized by the former.

[6] See *Met.* Δ 4 1015a7–10 for uses of 'first matter' corresponding to these two approaches. The kind of underlying principle Aristotle is after is sometimes characterized as 'proximate matter'.

[7] A preference to reduce many phenomena to few principles, of the sort Aristotle tends to applaud, might push them towards the most basic matter and contrariety.

[8] A 'fully specified' underlying nature, such that the account does not leave anything out of its nature, need not be—though according to Aristotle (*Gen. et Corr.* 2 5, 328b33–329a3, 329a8–10), it has usually been taken to be—'determinate' in the sense of capable of separate existence.

to just three principles depends on a generalizing move. The three principles he arrives at—substantial form, privation, and matter—are at a level of generality commensurate with the genus at issue, and as such do not serve as the specific underlying principle and primary contrariety for any particular kind of substance. The fact that Aristotle often appeals to material monist approaches to primacy and generality makes his own account of principles to some extent 'incongruous' (Bostock 2006, 3) with much of the discussion in I 5–6 leading up to and indeed motivating it.

1. Neither Just One Nor Unlimited (189a10–20)

The opening line of I 6 is the δέ-clause corresponding to the μέν-clause that closes chapter 5, and which evidently states its main result: that the principles are contraries. It also sets the agenda for the remainder of the chapter: to deal with a question about the number of principles. In particular, this is described as the 'next' (ἐχόμενον) issue, one that is natural to take up at this point.

> That the principles, then (μὲν οὖν), must be contraries is clear;
>
> [chapter 6] The next point (Ἐχόμενον δ') would be to say whether [the principles] are two or three or more. (I 5 189a10–I 6 189a12)

The first difficulty we encounter concerns the upshot of chapter 5. For if all principles are contraries, how can it even be an option, let alone one to be considered 'next', that there are three principles? The tension is exacerbated by the fact that not every conceivable option for the number of principles is privileged to be part of the initial question's agenda; some options (one, infinitely many), appear only in the immediately following justification of their exclusion from the stated agenda of I 6. In this way, Aristotle presents the option of three principles as having a minimal degree of plausibility.[9] One might thus be inclined to understand the upshot of I 5 as the 'weaker' claim that *some* principles are contraries. But this is not the whole story and I think there can be little doubt that Aristotle means for us to consider the 'stronger' thesis that *all* principles are contraries. This is, after all, what the text at the end of I 5 says ('*the* principles') and it corresponds to several other statements of that thesis throughout the chapter.[10] Aristotle does hint at hesitation when he

[9] Moreover, the lack of this minimal plausibility for the option that there is just one principle is grounded in the thought that contraries come in pairs (189a12).

[10] I agree with, and am indebted to, Sylvain Delcomminette (this volume) on this point. In brief, it is only the first lines of I 5 that directly suggest the weaker reading ('all make the contraries

explains why it is reasonable (εὐλόγως 188a27) to make the principles contraries. He points out that primary contraries are neither out of one another nor out of anything else, but does not give any reason for thinking that they satisfy a third stated requirement on principles: that everything else is out of them (I 5 188a27–30); this leaves open the possibility that not everything is out of contraries, in particular, that there is an additional non-contrary principle. But even if the argument Aristotle offers here supports only the weaker thesis, this does not change the fact that the stronger thesis is on the table. Further evidence is found in chapter 7, specifically in Aristotle's resolution of the *aporia* of I 6 (I 7 190b30–191a3) and in the summary of I 5–7 (I 7 191a15–19). As I mentioned, I take the 'stronger' reading to inform the puzzle raised in I 6.

There is, then, a tension in the transition from chapter 5 to chapter 6, which the opening of I 6 downplays. It is, to be sure, only an apparent tension; for Aristotle will explain in I 7 that the principles *are* in a way just two and contraries, despite being in another way (and more precisely) three and not just contraries (I 7 190b29–191a3). But more importantly, the availability of this way of individuating the principles is part of his solution to the puzzle of I 6, so that Aristotle cannot avail himself of it here.[11]

As I mentioned, Aristotle immediately justifies (γάρ 189a12) the restricted agenda of the opening line by explaining why the options it neglects are not (even minimally) plausible. Hence I take Aristotle to mean 'finitely more' with πλείους at 189a11. There is one argument against positing just one principle:

> For they cannot be one, since what is contrary[12] is not one . . . (189a12)

That there is just one principle is ruled out on the grounds that contraries come in pairs and so must be at least two (189a12). This clearly presupposes that there are contrary principles.

We then find four arguments or considerations against positing unlimited principles.

> And [they cannot be] infinite, [A1] since being will not be knowable; [A2] and there is one contrariety in each single genus, while *ousia* is a single genus; [A3] and since it is possible from a limited number and better from a limited number, as Empedocles

principles': 188a19). After that, Aristotle shows, first, that others hold that the principles are contraries (188a19–27); he then argues that this view is 'reasonable' (εὐλόγως) (188a27–30); and finally, in the long argument concerning *logos* (however that is understood: see Delcomminette, Section 3.1), he argues for the thesis by showing that things ultimately come-to-be from (and pass away into) contraries (188a30–189a10). Back references make clear that it is the same thesis at issue throughout.

[11] Again, I have benefitted a great deal from Sylvain Delcomminette's contribution on these issues.

[12] Here I prefer the reading τὸ ἐναντίον to τὰ ἐναντία; the latter makes the text smoother.

> has it, than from unlimited principles. For he supposes that he explains just as much as Anaxagoras explains from the unlimited [principles]. [A4] Further, some contraries are prior to others, and some come-to-be from others,[13] as do sweet and bitter and pale and dark, but the principles must always remain. [A5] That they are neither one nor unlimited is clear from these considerations. (189a12–20)

First, [A1] if there are unlimited principles, being will be unknowable. Aristotle is presumably drawing on the first sentence of book I, in which he claims that knowledge and scientific understanding (τὸ εἰδέναι καὶ τὸ ἐπίστασθαι) is a matter of knowing the relevant causes, principles, and elements—and he makes clear that we are talking about primary causes and principles (189a10–15). A further assumption is needed: that it is not possible to know infinitely many such items. One might be able to know, say, an infinity of numbers insofar as they are constructible from a finite set of axioms. But an infinity of (non-derivative) primary items cannot be grasped in this way.[14] At I 4 187b7–13, Aristotle has made a slightly more nuanced charge against Anaxagoras: that the unlimited is unknowable in that respect in which it is unlimited, whether in quality or number. The quality of the unlimited in quality is unknowable, just like the number (πλῆθος) of the unlimited in number. The unlimited principles of Anaxagoras are unlimited in both respects (if there are ever smaller amounts of everything in each part), so that we cannot know the objects that are made of them.

The second argument [A2] is clearly an abbreviated version of an argument that appears at the end of the chapter, at 189b22–7, and I will therefore postpone discussion of its content for later (see Section 4). However, let me mention certain grounds for suspicion about its placement here. First is the simple fact that the argument is repeated below, and the text shown no awareness of such repetition. Second, while the argument purports to limit contrary principles to *one pair*, the aim of this section is to show that the principles are *not unlimited*; that is in fact all that the remaining three arguments ([A1], [A2], [A3]) are meant to show, and the possibility that there is more than one pair of contrary principles is still on the table in the closing section of the chapter. By contrast, the expanded version at 189b22–7 occurs precisely where it is needed: where Aristotle's explicit aim is to limit the principles to at most three (among them one pair of contraries). Third, there is some linguistic oddity in this line. While [A1] is introduced with ὅτι, [A3] with καὶ ὅτι, and [A4] with ἔτι δέ, the

[13] Here I read ἄλλων (E, Arabic, Simpl.) against Ross, who reads ἀλλήλων (FIJ, Phil.).

[14] In *Met.* α 2, Aristotle makes a similar contrast between grasping an infinity of (primary) formal causes, which is impossible, and grasping the infinity of possible divisions of a line segment, which is possible in a sense, presumably because one grasps the line segment's (general) capacity of being divided (994b16–26).

consideration in question [A2] is introduced with τε, as if completing the first argument [A1], but it does not appear to do so.[15] For these reasons, although present in all the manuscripts—those cited in Ross, anyway—and attested in the commentary tradition, I am inclined to think that this line made its way into the main text from a marginal note based on the later passage, whose similarities are obvious.

We may take the comparison in the next argument [A3] to be between the infinity of kinds—'homoiomers and contraries' (I 4 187a25–6)—already present in the Anaxagorean mixture and Empedocles' four elemental kinds, together with Love and Strife. Whether all these items are construed as exhibiting contrariety in the same way is both questionable and, for the present purpose, unimportant. Given the first argument—[A1] 'being will not be knowable'—it is better to posit a finite number of principles than infinitely many. But Aristotle may rather have in mind a principle of economy that says that fewer is better, when possible, just in virtue of being fewer (Cf. *De Caelo* III 4 302b20–30). The similar remark that closes I 4 claims that both a limited number and fewer (ἐλάττω) are better (I 4 188a17–18).

Aristotle's fourth argument, once again, is as follows:

> [A4] Further, [i] some contraries are prior to others (ἄλλα ἄλλων), and [ii] some come-to-be from others (ἕτερα ἐξ ἄλλων), as do sweet and bitter and pale and dark, but [iii] the principles must always remain. (189a17–20)

This reasoning does not, even if sound, rule out an infinity of principles (whether contraries or not). Rather, it warns us not to treat *all* contraries as principles, since some are posterior. We should not pass too hastily from a large or unlimited number of kinds in the world to the same number of principles. It is thus no accident that this argument comes right after the comparison between Empedocles and Anaxagoras. For while Anaxagoras posits every kind of stuff and every contrary (taking these to be unlimited) as a basic element of the original mixture, Empedocles believes that there are more kinds of stuff and more contraries than the basic ones.

[15] At the conference it was suggested that we can supply οὐκ and ἔσται from [A1], yielding 'and [the principles cannot be] infinite, [A1] since being will not be knowable; [A2] and [there will not be] one contrariety in each single genus' (ἀπείρους δ', ὅτι οὐκ ἐπιστητὸν τὸ ὂν ἔσται, μία τε ἐναντίωσις ἐν παντὶ γένει ἑνί, ἡ δ' οὐσία ἕν τι γένος: 189a12–14). I find this unsatisfying. First, the idea that we should supply οὐκ and ἔσται, and the implied combination of οὐκ and τε is awkward, since we would expect οὔτε. Second, the claim that (if there are infinite principles for substance) there will not be just one contrariety in each single genus presupposes, and so should be supported by, the immediately following claim that substance is a single genus (ἡ δ' οὐσία ἕν τι γένος, 189a14), so that we should expect γάρ in this line rather than δέ. In any case, this suggestion leaves standing my first two reasons for thinking the line has been added here.

I take it that Aristotle is talking about contrary *pairs* being prior to, and/or coming-to-be from, other contrary pairs, so that it is more natural to read ἐξ ἄλλων ('from others') at line 18 rather than ἐξ ἀλλήλων ('from one another'), which would suggest a single contrary being prior to, and/or coming-to-be from, *its* contrary.[16] As for the three (numbered) claims in the argument, it is clear that contraries which are [i] posterior to others cannot be principles. And claims [ii] and [iii] fit well together: those contraries which [ii] come-to-be from others cannot be principles, if [iii] principles must remain. Presumably, if some contraries [ii] come-to-be from others and [iii] do not remain, then those others will be [i] prior to them. But one contrariety might also be [i] prior to another without [ii] the latter coming-to-be from it. And Aristotle's language (ἄλλα ἄλλων ... ἕτερα ἐξ ἄλλων) does not indicate that all the same items are at issue in both cases.

I suggest that Aristotle separates claim [i] from claims [ii]–[iii] in this way so as to address the possible case in which there are posterior contraries which nevertheless do not come-to-be from prior ones. In particular, it seems to me that while [ii]–[iii] best capture a materialist approach, they are not meant to encompass Aristotle's account of principles. It is natural, in light of the preceding text, to look to Empedocles' claim that different colours arise from different mixtures of the elements (DK 31A92; Horstschäfer 222). Ross (490) suggests a reference to Democritus, who mentions colour as well as sweet and bitter as modifications of atoms and void (DK 68B125). This passage is also in line with what I characterized as the materialist approach to primacy, which is shared by the monists (as Aristotle presents them), on which e.g. rarity and density are primary because they are the most basic differentiae of a single underlying body, such that further contraries are in some sense dependent on these. Note that Aristotle employs a similar approach in *Gen. et Corr.* II 2's discussion of the principles of tangible body, which argues for the priority of hot/cold and wet/dry over other contrarieties, hard/soft, viscous/brittle, rough/smooth, and coarse/fine, on the grounds that, for example, the viscous 'derives from [εἶναι + gen.] the moist, for the viscous is a certain affection (πεπονθός) of the moist, for example, oil' (330a5–6). Clauses [ii] and [iii] are, however, difficult to square with the account offered in *Physics* I 7, according to which the primary contrary principles

[16] Recall that the fact that things come-to-be from their contraries was a main reason for positing them as principles in the first place, and this motivation would be completely undermined by the latter reading. There is also no indication in this chapter that one member of a contrary pair is prior to the other, as clause [i] would then have it. Horstschäfer (1998, 222), however, has gone too far in suggesting that the term ἐναντία should here be understood as 'contrary pairs' (he cites what he sees as a similar use of ἐνάντια for contrarieties at I 5 188b29ff.). As Tim Clarke made clear to me, even if Aristotle has in mind the relation between different contrary pairs, there is no need to read ἐναντία as meaning anything other than 'contraries'.

of a natural substance are substantial form and privation; it is not clear what sense to give to the idea of other, posterior contrarieties 'coming-to-be' from this one and not 'remaining'. A model from *Met.* I 7 suggests itself, on which primary *external* differentiae for a genus (lying outside of it) are seen as the primary contrariety for it, prior to and productive of (1057b6: ποιήσουσαι) the contrariety that is primary within the genus. The example there is one of those mentioned in I 6: for the genus colour, the primary contrariety outside the genus is piercing–compressing, and this is prior to the primary contrariety within the genus, pale–dark. But even in this case, there is no indication that the prior contrariety 'remains' even when the posterior one does not. I will argue that when Aristotle returns to the issue of priority among contrarieties at the end of the chapter, he seems to have a quite different conception of it in mind, one that is bound up with the generality of the leading question—priority seems to involve higher generality, up to the level of the genus in question—and which is not naturally understood in terms of what remains vs. what comes-to-be. As for the other thinkers that might be referred to in clauses [ii] and [iii], I see no reason to think that Aristotle has a particular model in mind; he is plausibly casting as wide a net as possible, so as to elicit broad agreement in a stretch of text that is very much in dialogue with his predecessors.

2. A 'Third' and Different Kind of Principle (189a21–34)

In the next section Aristotle offers arguments for positing at least a third principle, i.e. for 'not making them only two' (189a21). These are arguments for (at least) a third principle because they are arguments for a different kind of principle, namely an 'underlying' principle. Along with the arguments in I 5 for the thesis that principles are contraries, these arguments furnish the basis of the *aporia* with which Aristotle closes the chapter:

> So, that there is neither just one element, nor more than two or three, is clear. But which of these, as we said, involves great difficulty (ἀπορίαν ἔχει πολλήν). (189b27–9)

This statement of the *aporia* presupposes that there are *at most* two contrary principles and at most one underlying principle, a thesis only argued for in the final section of I 6 (189b18–27). It is nevertheless appropriate to discuss the puzzle here, because it does not appear to depend on any difficulty in the argumentation at the end of the chapter limiting the principles to at most three: that argumentation is treated as decisive. For the purposes of discussing the source of the puzzle, I will often take that limitation for granted. I begin by distinguishing three proposals about how the two vs. three puzzle arises.[17] These proposals are not necessarily exclusive of one another and all of them, I think, draw on real and important features of Aristotle's argumentation.

To a reader looking for guidance about how the puzzle arises, perhaps the most salient feature of the argumentation of I 6 is that Aristotle expresses tentativeness about the force of the arguments for a third principle.

> Given that [the principles] are limited, there is some reason (*ἔχει τινὰ λόγον*) for not making them only two. (189a21–2)

> To say that the elements are three would seem, if one examines the matter on the basis of these and other such considerations, to be somehow reasonable (*δόξειεν ἂν ἔχειν τινὰ λόγον*), as we said... (189b16–18)

This tentativeness is in sharp contrast to the way he sums up the result of I 5, claiming that it is clear (*φανερόν* 189a10) that the principles are contraries. We will also see that the second argument for a third principle in I 6, which I will call the 'Argument from Substantiality', relies on especially questionable premises. Now, whether this tentativeness gives rise to the two vs. three puzzle or not, Aristotle will view the arguments for a 'third' and non-contrary principle as inconclusive in one sense at least, since the thesis that all principles are contraries (and so just two) will, appropriately understood, be able to withstand them. Aristotle's apparent tentativeness suggests the following account of the puzzle:

> (Naïve) The soundness of certain arguments, in particular those for a third and underlying principle, remains in question.

Here I mean in particular that the arguments for an underlying principle are treated as inherently inconclusive, independently of their standing in conflict with the thesis that all principles are contraries, which I have claimed is argued for in I 5. In fact, this understanding of the puzzle is especially amenable to a view on which the thesis carried over from I 5 is the weaker one, that *some* principles are contraries, since that thesis is compatible with the view argued for in I 6, that there is a (third and) non-contrary principle. Thus, one might hold that while it is clear that some principles are contraries, some less than

[17] I mention, only to leave aside, Horstschäfer's suggestion that the 'two' option can be understood as collapsing a contrariety into a single principle. He proposes that Aristotle has in mind either treating the Platonic dyad as a single principle, or that one might construct an argument for the sufficiency of a single active principle that is parallel to the argument (an assertion really) at 189b18–19 for the sufficiency of a single underlying nature (1998, 246–7; see also Simplicius 208, 8–10). These suggestions seem too distant from the text of I 6 and of Aristotle's solution. Horstschäfer seems to accept nevertheless that the resources are there to generate the puzzle as according to the 'To the Letter' reading (246).

decisive arguments support positing a non-contrary principle. The Naïve reading, as well as the weaker interpretation of I 5, are also suggested by the way Aristotle sums up the situation after arguing for a third principle, which emphasizes that positing three principles is conditional on accepting the account of I 6 alongside that of I 5.

> Therefore, if one holds that the previous account (λόγον) is correct, as well as this one, it is necessary, if one is to preserve them both (διασώσειν ἀμφοτέρους αὐτούς), to posit some third thing. (189a34–b1)

I have claimed, by contrast, that the upshot of I 5 is instead that all principles are contraries, and so only two, which, admittedly, is in some tension with the above quotation at this stage. Nevertheless, I take it that Aristotle's solution not only 'preserves' both accounts but also posits three principles, when individuated most precisely.

In any case, the stronger reading of I 5's conclusion makes for a different understanding of the source of the puzzle:

> (To the Letter) I 5 argues that (all) the principles are contraries (and so, given further assumptions, only two); I 6 argues that there is a non-contrary (i.e. third) principle.

How interesting or serious such a puzzle is will depend on how compelling the arguments on each side are, what is necessary to resolve it, and how, if at all, the soundness of the arguments is to be maintained.

Neither the 'To the Letter' nor the 'Naïve' reading draws on the idea that the status of contraries as principles is put into question. By contrast, Simplicius and, more recently, Kelsey (2008) have pointed to a threat to contraries as relevant to the puzzle. Simplicius nevertheless takes a fairly conciliatory stance. He views this section of the chapter as impugning the status of contraries only when 'taken on their own', so that their status is buttressed by the addition of a third, underlying principle.[18] To my mind he does not adequately distinguish his position from the more mundane one which views I 6 as arguing that contraries are not the only principles. For Kelsey, however, the very reasons offered for positing an underlying principle are at the same time reasons for doubting that contraries can be principles at all (and not merely 'when taken on their own').

[18] Simplicius describes the arguments in this section generally, as well as each separately, in this way. Thus, the contraries do not 'of themselves fulfill the definition of a principle' (αὐτάρκη εἰς ἀρχῆς λόγον: 199,9; cf. 202,4), they do not qualify as principles 'taken on their own' (καθ' αὑτὰ λαμβανόμενα 199,11–12; cf. 199,10, 202,14, 203,30–32) in part because they they do not produce anything if on their own (εἴπερ καθ' αὑτὰ εἴη: 199,13; μόνα ὄντα: 199,14).

In particular, each of the two arguments for an underlying principle in I 6, Kelsey claims, undermines one of the arguments for viewing contraries as principles in I 5. This suggests a different understanding of the puzzle:

(Undermining) The arguments of I 6 for a non-contrary, underlying principle undermine the status of contraries as principles.

One worry about this understanding of the puzzle is that while the options of two (only contrary) and three (not only contrary) principles are certainly among those that might be considered, we might also expect the *result* of undermining the status of contraries—presumably that there is just one, underlying principle—to be one of the alternatives we are puzzled about.[19] But since it is not, the undermining seems to be irrelevant to the two vs. three difficulty. Perhaps the best way to avoid this worry is to hold that the arguments for an underlying principle are inconclusive and puzzling precisely *because* they would undermine the status of contraries, which is treated as unassailable.[20]

The very fact (if it is one) that the status of contraries as principles is undermined, whether this is treated as source of the two vs. three puzzle or not, is also interpretively problematic. If there is such undermining, it seems misleading for Aristotle to say that accepting both λόγοι (i.e. those of I 5 and I 6) involves positing 'some third thing' (189a34–b1), and to present I 6 as arguing against there being 'only two' principles (189a21–2), and in favour of 'say[ing] that the elements are three' (189b16–18).[21] Similarly, Aristotle summarizes the argumentation of I 5–6 by saying, 'First it was said that that the contraries alone are principles, and afterwards that it is also necessary for something else to underlie and that [the principles] are three' (I 7 191a15–17).

There is no indication in these characterizations of the function and upshot of the arguments for a third principle that they might lead us to revoke the status of contraries. We must, it seems, take Aristotle in these passages to refer only to the considerations in favour of an underlying principle, but not the considerations that undermine the status of contraries. But to the extent that there are

[19] I say 'presumably' because it is not obvious that the two arguments have the same result. The first argument seems at best to undermine the status of a 'patient' contrary, that which is thought to be generated into something, whereas the subsequent argument would downgrade any contrary, on the grounds that it is an attribute of a prior underlying substance. Even the result of undermining the status of just the 'patient' contrary would not yield the 'two' in the two vs. three *aporia*; those 'two' are rather two contraries.

[20] Note that the result of undermining is incompatible also with the weaker reading of I 5. On this reading, it is ultimately the arguments for a third principle which would be put into question, as on the Naïve reading.

[21] Perhaps also the introduction of the second argument, which cites a difficulty that arises 'if one does not posit another nature as underlying the contraries' (189a27–9), suggests that it should be an argument for three principles, though this is less clear.

considerations of both kinds, they are (we will see) difficult to separate. I will argue that only the second argument for an underlying principle in I 6 is meant (in part) to undermine the status of contraries, but that Aristotle does not treat this undermining as giving rise to the two vs. three worry. Nevertheless, the second argument brings out fundamental issues on which Aristotle's conception of the natural world differs from that of his materialist predecessors.

2.1. *The poiein-paschein argument (189a21–7)*

Aristotle's first argument is as follows:

> Given that [the principles] are limited, there is some reason for not making them only two. For one might puzzle about how density is such as to make rarity something (ἀπορήσειε γὰρ ἄν τις πῶς ἢ ἡ πυκνότης τὴν μανότητα ποιεῖν τι πέφυκεν), or the latter density. And in the same way also any other contrariety. For Love does not gather together Strife and make something out of it, nor Strife out of it, but rather both some distinct, third thing. And some assume even more [principles] out of which they construct the nature of beings. (189a21–7)

The term ποιεῖν, like the reference to Empedoclean Love and Strife, suggests efficient causation, which does not seem to be at issue in book I: efficient causes are not principles in the sense we are after, i.e. elements, since they are not internal.[22] And so it might be unfair to insist that contraries, if they are such elements and principles, ought to be able to serve as efficient causes and 'act on' one another in this way. Related to this, contraries might not be expected to play this role by many of their proponents. As Code (2018, 166, note 8) points out, however, we can understand the talk of 'making' at 22–3 in a much broader way, so as to cover also the idea that one contrary, by its presence to something, affects that thing and makes it something. This broader construal allows Aristotle to motivate his own theory of principles with an argument that could also apply to the role of Empedoclean Love and Strife.[23] In the same vein, though density and rarity will function as the most basic contraries, while Aristotle's primary contrariety will be at the level of essence and form, the argument is general enough to apply, as Aristotle claims, to any contrariety whatsoever.

The difficulty is not, as Horstschäfer (1998, 225) claims, that if contraries were the only principles, they would have to be out of one another, in violation of the requirement set out in I 5 that principles cannot be out of one another (188a27–8). For one thing, saying that a primary contrary is instead out of some

[22] Aristotle refers to the sought-after principles as elements at I 6 189b16. See *Metaphysics* Λ 4 1070b22–6 for the distinction.

[23] See Hendrik Lorenz's contribution in this volume for a different approach.

other principle(s) would also violate that requirement. Rather, Aristotle is here bringing out a particular shortcoming in the nature of contraries, such that one contrary cannot make another into something, without presupposing that this 'something' is itself a contrary. What kind of shortcoming is this? Usually the argument is understood to turn on the idea that a contrary, specifically the putative 'patient' contrary, because it is an attribute, cannot be affected; only subjects can be affected. Thus, rarity is not the sort of thing that can be made into something (or gathered up), no matter what by, because it is a non-substance attribute rather than a substance and subject.[24] This is certainly true. But at this stage of the argumentation, Aristotle might mean to emphasize a more basic shortcoming stemming from the *relation* between contraries, namely, that they are contrary to one another, without making essential reference to the distinction between subject and attribute.

Consider *Gen. et Corr.* I 7 323b28–9, where Aristotle justifies a point like the one at *Physics* I 5 188a32–3—that ποιεῖν and πάσχειν are between contraries and not between chance things—by pointing out that only contraries or things characterized by contraries are such that one will drive the other out of its nature. Of course, a contrary itself is not driven out of its nature, but rather, we might say, 'driven away' in the sense that it is no longer instantiated; this is a direct result of the relation of contrariety. But for one thing to 'act on' another, in the sense at issue here, and make something out of it, is not for the former to 'drive away' the latter in this way. Perhaps we can assume that for one thing to act on another, the former must be compresent with the latter, or with some part of the latter. But poles of a contrariety, being contrary to one another, cannot be compresent. The upshot of the argument would then be that there must be some third thing, distinct from both the contrary driven out, and the contrary which drives it out.[25] And indeed, the explicit upshot of the argument is that we should not 'make the principles only two' (189a21), and that there is 'some distinct, third thing' (189a26), not that there is specifically something that underlies. On this reading, then, Aristotle might here be arguing for some third thing—some 'non-contrary', we might say—and only in the following lines that it is an underlying subject. Also suggestive of the second reading is the fact that the present passage—which is the fullest version of the argument—emphasizes the role and, as it were, 'nature' of the putative 'agent' contrary (as the subject of πέφυκεν at 189a23); in addition, Aristotle here repeatedly mentions the interaction of contraries with *one another* but does not say that a contrary is not such as to be acted on *by anything*. In the

[24] Among recent authors see Horstschäfer 1998, 225 and Kelsey 2008, 190–1.

[25] In *Gen. et Corr.* I 7, it is taken for granted that contrariety implies generic sameness.

later gloss in I 7 (190b34), though the emphasis is on being a patient, he still claims that contraries cannot be affected *by one another*. At *Metaphysics* Λ 2 1069b6–7 Aristotle merely cites the fact that contraries do not change, which is suggestive of the former reading, but since this passage appears to be part of a summary of *Physics* I, it might also be taken to reflect a (now) clearer understanding of the idea that contraries are always attributes.

In any case, regardless of how the argument might be filled in, if (as I claimed) the result of I 5 is that only contraries are principles and the present argument shows that there is a non-contrary principle, then, on the further assumption that there is at most one pair of contrary principles and one non-contrary principle, a 'two vs. three' puzzle arises. Thus far, we have understood the puzzle only according to the 'To the Letter' reading.

Kelsey (2008) holds that this argument from I 6 undermines the status of contraries as principles in that it conflicts with the final, long argument for positing contraries as principles at I 5 188a31–b26. That argument points out that ποιεῖν and πάσχειν are not between chance things, but rather things come-to-be from, and pass away into, their contraries, or intermediate things, which are themselves out of contraries, so that natural things are or are out of contraries. Kelsey takes this somewhat cryptic conclusion to mean that natural things have contraries as principles. The present argument of I 6, on the other hand, denies that things are 'made from' contraries. Kelsey locates the conflict specifically in the claims that things 'come-to-be from' contraries (I 5) but are not 'made from' contraries (I 6), assuming that what things 'come-to-be' from is what they are 'made from' and that there is just one such thing.

Is Aristotle attempting to bring out such a conflict? Note that the language of ποιεῖν and πάσχειν, central to the present argument, occurs only at the beginning of the I 5 argument; Aristotle then switches over completely to the vocabulary of coming-to-be from and passing away into something. On the other hand, the present argument nowhere denies that things 'come-to-be' from contraries. Furthermore, given that the argument is introduced as an argument for not making the principles only two, it is better to understand it, if possible, in such a way that it does not impugn the status of contraries as principles. One option is to have Aristotle insist that contraries are principles even though things do not come-to-be from them at all, but that would involve a significant departure from the argumentation of I 5. He is better taken simply not to be denying altogether that things come-to-be from contraries. Even his claims in the *poiein-paschein* argument about being made into something are more nuanced: one 'might puzzle about *how*' density can make rarity something, or indeed any contrary can make its contrary something. And when he straightforwardly asserts that 'Love does not gather together Strife and make something out of it' (189a24–5), it is perhaps significant that Love and Strife are both thought of as forces external to the items they act on.

It is helpful to consider a passage from the *Phaedo*, which is a likely source of the argument. Socrates points out that one member of a pair of ἐναντία cannot 'endure and admit' (ὑπομένον δὲ καὶ δεξάμενον: 103b) the other while remaining. An anonymous participant objects that earlier, it was claimed that precisely ἐναντία come-to-be from one another. Socrates responds as follows:

> You do not understand the difference between what is said now and what was said then, which was that a contrary thing (τὸ ἐναντίον πρᾶγμα) comes-to-be from a contrary thing; now we say that the contrary itself (αὐτὸ τὸ ἐναντίον) could never come-to-be contrary to itself, neither that in us nor that in nature. Then, my friend, we were talking of things that have contraries and naming these after them, but now we say that these contraries themselves, from the presence of which in them things get their name, can never admit the coming-to-be of one another (γένεσιν ἀλλήλων δέξασθαι). (*Phaedo* 103b1–c2, tr. Grube 1997, modified)

Physics I parallels this passage's distinction between contraries themselves and contrary things in the shift in language from talking somewhat indiscriminately in I 5 about contraries themselves and contrary things (e.g., the musical, the pale, etc.) to talking of density and rarity in the *poiein-paschein* argument. In the *Phaedo* passage, the claim that contraries themselves do not come-to-be contrary to themselves is not meant to put into question the idea that contrary things come-to-be from contrary things, or even, it seems to me, *in part* from contraries themselves. What is rejected is specifically the idea that some contrary can take on its contrary. Thus, as far as the view Socrates espouses in the *Phaedo* is concerned, nothing precludes the idea that something comes-to-be from the relevant contrary thing, or even that it comes-to-be (in part) from the relevant contrary.[26] I suggest the same is true of Aristotle's argument at 189a21–7.

One difference between the *Phaedo* passage and Aristotle's argument is worth noting: the *Phaedo* passage does not single out what Aristotle calls the 'third thing'. On the other hand, Aristotle speaks of the 'third thing' without (yet) using any phrase like 'contrary thing' that picks out the combination of a contrary itself and the 'third thing'. This might suggest that Aristotle cannot intend to leave the role of the contraries (in combination with the third thing) intact. However, Aristotle's reluctance to talk of 'contrary things' is easily explained by his desire to develop the 'two vs. three puzzle', specifically, the 'three' side: to speak of 'contrary things' would already suggest that the 'three' principles being argued for can also be viewed merely as two and as contraries.

[26] Grube's translation of γένεσιν ἀλλήλων δέξασθαι (103c1–2) as 'tolerate the coming-to-be *from* one another'—as if ἐξ ἀλλήλων—obscures the latter point.

2.2. *The argument from substantiality (189a27–34)*

Aristotle's next argument is as follows:

> [B1] In addition, one might further face this difficulty, if one does not posit a different nature as underlying the contraries. For we do not see that the contraries are the *ousia* of any being. But a principle must not be said of something underlying; for there will be a principle of the principle, for what underlies is a principle, and is thought prior to what is predicated. [B2] Further, we deny that *ousia* is contrary to *ousia*. But how then could *ousia* be out of non-*ousiai*? Or how could a non-*ousia* be prior to *ousia*?

> [B1] πρὸς δὲ τούτοις ἔτι κἂν [28] τόδε τις ἀπορήσειεν, εἰ μή τις ἑτέραν ὑποθήσει τοῖς ἐναν-[29] τίοις φύσιν· οὐθενὸς γὰρ ὁρῶμεν τῶν ὄντων οὐσίαν τἀναντία. [30] τὴν δ' ἀρχὴν οὐ καθ' ὑποκειμένου δεῖ λέγεσθαί τινος. ἔσται [31] γὰρ ἀρχὴ τῆς ἀρχῆς· τὸ γὰρ ὑποκείμενον ἀρχή, καὶ πρό-[32]τερον δοκεῖ τοῦ κατηγορουμένου εἶναι. [B2] ἔτι οὐκ εἶναί φαμεν [33] οὐσίαν ἐναντίαν οὐσίᾳ· πῶς οὖν ἐκ μὴ οὐσιῶν οὐσία ἂν εἴη; ἢ [34] πῶς ἂν πρότερον μὴ οὐσία οὐσίας εἴη. (189a27–34)

In working out an interpretation of this argument, I will consider various options along the way. Let me begin, however, by distinguishing between a 'positive' aspect, which comprises what speaks in favour of a different, and underlying principle, potentially a principle *alongside* contrary principles, as Aristotle's descriptions of the argumentation lead us to expect, and a 'negative' aspect, which comprises those considerations that put into question the status of contraries as principles. Also, note that I read a strong break (and place a period) at the end of line 29 (before 'But a principle': τὴν δ' ἀρχήν), a decision I will explain shortly.

Aristotle's initial claim is that there is a difficulty that arises on the supposition that there is no other *nature* underlying the contraries (εἰ μή τις ἑτέραν ὑποθήσει τοῖς ἐναντίοις φύσιν, 28–9), not specifically no other *principle*; the talk of adding a third principle comes several lines earlier, before the *poiein-paschein* argument ('there is some reason for not making them [the principles] only two': 189a21–2), and Aristotle points out (or perhaps concludes) that the posited underlying nature is a principle explicitly below, once the 'negative' aspect is underway. Given the context, however, I take him to mean from the outset that an additional principle must be posited. Second, the term ὑποθήσει, especially given the dative τοῖς ἐναντίοις, is plausibly understood to mean 'posit as underlying subject', though the text can simply be read as saying that another nature must be posited for the contraries. Aristotle assumes in any case that what is to be posited is something that underlies. His explicit justification for the claim that one faces difficulty if one does not posit a distinct underlying nature is that 'we do not see that the contraries are the *ousia* of any being' (189a29). In positing only contraries, we have left out the *ousia* of things. In line with reading the initial demand as a demand for a principle, we can take Aristotle simply to assume that the *ousia* of things is a principle. This, I take it, is the entirety of the positive aspect.

But this consideration in favour of an underlying principle immediately gives rise to a threat to the contrary principles. It seems that a substance (the *ousia* whose principles we seek, not the underlying *ousia* newly posited) cannot have contrary principles (be 'out of' contraries) at all. As Aristotle claims next, a principle is not said of what underlies (189a30), and this seems to be a general claim about any principle whatsoever (I will consider some alternatives below). The reason given is that what underlies is prior to, and a principle of, what it underlies (31–2), so that, if a principle were said of something, there would be a principle of a principle (30–1). Although unstated, the obvious conclusion that contraries are not principles seems to be intended. The following lines [B2] draw further and related consequences from the claim that '*ousia* is not contrary to *ousia*' (ἔτι οὐκ εἶναί φαμεν οὐσίαν ἐναντίαν οὐσίᾳ 188a32–3), a claim very similar to the initial 'positive' consideration (οὐθενὸς γὰρ ὁρῶμεν τῶν ὄντων οὐσίαν τἀναντία 188a29). From this, on the supposition that *ousia* is out of contraries, it follows that, absurdly, *ousia* will be out of non-*ousia* and also out of something posterior to it (33–4). At bottom, the same fact about contraries, that they are attributes rather than subjects, both requires that we posit an underlying principle and puts into question the status of contraries as principles.

I have placed a strong break after line 29 in order to avoid a reading on which what follows ('and a principle must not be said of something underlying'), is still within the scope of 'for' (γάρ) near the beginning of line 29, giving the reason why one will face difficulty. On this reading Aristotle would justify the need to posit an underlying principle by pointing out that contraries are not principles—more precisely, that while contraries are not substances (i.e. subjects, I take it, on this reading), principles are not said of anything (i.e. they *are* subjects). If so, there is not even a pretence of presenting a consideration in favour of an underlying principle *alongside* contraries, as a third principle. But it is that kind of consideration—in fact only that kind of consideration—that, as I mentioned, Aristotle leads us to expect. Thus, like Horstschäfer (1998, 231), I see an initial, positive step, which leads to negative considerations.[27]

Can the text be read in such a way that it does not in fact put into question the status of contraries as principles? (i) One such reading takes the two instances of the phrase 'the principle' (τὴν δ' ἀρχήν 189a30; τῆς ἀρχῆς 189a31) to refer not to *any* principle, but to just *this* principle, i.e. the *ousia* of things just introduced.[28] Aristotle will then be pointing out that this principle will not be said of a *further* subject (τὴν δ' ἀρχὴν οὐ καθ' ὑποκειμένου δεῖ λέγεσθαί τινος 189a30) (and so is an ultimate subject), a claim that will be supported in the following lines: otherwise there would be a principle of *this* principle. There would then be no general stricture against principles that are attributes rather than subjects.

[27] I thank David Ebrey for pushing me towards greater clarity on this issue.

[28] This is the analysis I presented at the conference.

And afterwards [B2], Aristotle may be taken to point out that this new principle is also not composed out of (further) contraries, rather than pointing out that the (composite) substances for which we seek principles cannot have contraries as principles. The status of the initially posited contrary principles is not directly put into question. (ii) A second such reading takes 'the principle' to mean 'the principle(s) for everything' or 'the only principle(s)'.[29] Thus, Aristotle will be pointing out that contraries cannot be the principles for everything or the *only* principles. And what follows [B2] can, as in the first option, be taken to characterize the newly posited underlying principle, rather than the perceptible (composite) substances whose principles we are seeking.[30] Both these options are problematic. They both require strained readings of the definite article in the phrase 'the principle'. Furthermore, even if they are not specifically targeting the initial contrary principles, certain ideas still provide the basis of an obvious threat to them, especially the idea that no contrary can be *ousia*, which would seem to imply that substances cannot be out of contraries, and the idea that contraries, being attributes, would have their subjects as principles *of them*.[31]

The closest parallel to the argument here is found at *Met.* N 1 1087a29–b4.[32] There, Aristotle argues that since contraries are said of an underlying subject—a point brought out by the observation that nothing is contrary to a substance—and this underlying subject will be a principle prior to the contraries said of it, contraries cannot be 'the principle'. Nevertheless, one key difference is that the argument of *Met.* N.1 explicitly concerns the question of whether contraries can be 'the principle of all things' (1087a31–2) and 'what is strictly a principle of everything' (1087b4), and it is premised on the supposition that *such* a principle cannot, as Aristotle puts it, 'be a principle by being something else' (ἕτερόν τι οὖσαν εἶναι ἀρχήν: 1087a33). In this respect, the argument closely resembles the second non-undermining reading I sketched. However, it is precisely because the argument from substantiality goes further than this argument, that it is not merely questioning the status of contraries as 'strictly principles of everything'. In particular, the argument from substantiality points out that if contraries are principles, then an *ousia* will be out of, and posterior to, what is not *ousia*.

[29] See just below on *Met.* N 1 and Ross 490.

[30] [B2] cannot be read in a way that corresponds more closely to the second option's reading of [B1], that is, so that the problematic scenarios are that a substance is out of *only* non-substance and that it is out of *only* what is posterior.

[31] It is, however, worth noting that the first non-undermining reading (i.e. reading 'this principle'), does not require this second and quite un-Aristotelian idea (that an attribute has its subject as a principle of it). We might take the argument to draw only on the idea that a substance and subject that is itself predicated of some further underlying thing will have that underlying thing as its principle: an underlying thing is a principle of the *thing* for which it is the underlying nature, not of its attributes.

[32] I thank Michel Crubellier and Stephen Menn for encouraging me to bring this passage to bear on *Physics* I 6.

Thus, I favour the reading of lines 189a30–4 initially presented above, on which they do indeed undermine the status of contraries as principles. Of course, it might be part of an appropriate response to the argument to differentiate between 'what is strictly a principle of everything' and a weaker sense of principle, but for now I am concerned with the argument itself. This reading of the argument raises several questions. What is the role of this argument, especially insofar as it purports to undermine the status of contraries, in the development of the two vs. three *aporia*? What is its function within the broader line of argument about the number of principles spanning I 5–7? How, and to what extent, does Aristotle legitimate the status of contraries in the face of the threat it contains, in the remainder of book I?

Before turning to these questions directly (though I will not fully address them), I would like to examine one feature of the argument from substantiality, namely, that it turns on assumptions that run counter to central tenets of Aristotle's natural philosophy, and are instead characteristic of a certain strand of materialism.[33] For the function of the argument in book I will thus reveal something about book I's use of materialism. Consider first the negative aspect. It explicitly assumes that a subject is prior to, and a principle of, what is predicated of it, which is clearly anathema to Aristotle's hylomorphic conception of natural substances, especially insofar as Aristotle grants priority to form. Moreover, it is primarily this assumption, rather than the idea that form is a contrary, which forms the basis of the threat to *Aristotle's* contrary principles, since form is certainly predicated of a subject, even if not a full-fledged contrary. But I would like to emphasize that even the positive aspect turns on what are (in this context) problematic assumptions. As I punctuate the text, the key assumption behind the need for an underlying nature is that contraries are not the *ousia* of things (189a29–30), but the same conclusion can be reached on the basis of the assumption mentioned above (that what underlies is prior to, and a principle of, what is predicated of it). As scholars of Aristotle, we might expect him to maintain the quintessentially Aristotelian assumption that contraries are not the *ousia* of things by denying that substantial form is a genuine or full-fledged contrary. And of course, starting in I 7, Aristotle's construal of the 'contrary' principles as form and privation is already a departure from the kinds of examples that have often figured in the discussion thus far (e.g. hot–cold, pale–dark). Nevertheless, he continues to characterize these principles as contraries (ἐναντία) throughout *Physics* I 7, in particular, when resolving, and referring to the resolution of, the two vs. three *aporia* (190b29–191a3; 191a15–19). Within the framework of

[33] Many of the claims defining what I call a materialist perspective are found in *Categories* 5 and elsewhere (hence 'we deny' at 189b32). But we may view as characteristic of materialism the unqualified acceptance of such claims.

Physics I—i.e. while treating form and privation as contraries—to accept that no contrary is the *ousia* of things is tantamount to denying that form is the *ousia* of things; within this framework, then, the appropriate anti-materialist answer is to reject the assumption that contraries cannot be the *ousia* of things, and this will require rejecting the idea that subjects are always prior to their attributes.

In *Physics* I, there is no other comparably explicit employment of this strand of materialism, yet Aristotle does at times appear sensitive to the needs of an audience sympathetic to it. To see this, note that the assumptions on which the argument from substantiality relies involve assimilating the predicative relation between contraries and the underlying nature within a substance (i.e. between form and matter) to the more familiar predicative relation between non-substance accidents and individual natural substances. And often in *Physics* I, Aristotle relies on the case of non-substantial change as a model through which to understand substantial generation and the principles of natural substance, while appreciating that this might be seen as a difficult and challenging move, and refraining from explaining how to make it. At I 5 188b8–21, for example, he must argue for the application of the 'contrary from contrary' principle to 'composite' (σύνθετα) things (this phrase here does not mean accidental compounds, but appears to designate key cases of substance); this extension is not obvious, thanks to the fact that the 'opposed' (i.e. privative) contrary has no name. At I 7 190b1–10, Aristotle must explicitly justify the extension of the thesis that there is always something underlying to cases of substance; that it applies to substance too, he says, 'will become clear to one who examines things' (ἐπισκοποῦντι γένοιτο ἂν φανερόν, 190b3), whereas this thesis is simply 'clear' (φανερόν, 190a33) for non-substantial coming-to-be. The notion of the underlying nature in relation to substance must be grasped by analogy from cases of non-substance, or perhaps substances in only a looser sense (191a7–12).[34]

Let us now return to the role of the argument in I 6 and in the broader argument about the number of principles in I 5–7. Earlier I mentioned that certain passages in which Aristotle describes the upshot of the argumentation for an underlying principle (I 6 189a21–2, 189a34–b1, 189b16–18; I 7 191a15–17) seem simply to neglect the 'negative' aspect of the argument. This already suggests—and I offer more reasons below—that we should view the undermining as somehow peripheral to the broader line of argument about the number of principles. In fact, the negative aspect makes one of these passages, which immediately follows the argument for substantiality, particularly awkward:

[34] Importantly, Aristotle is not saying that the underlying natures of various (full-fledged) substances are related by analogy; he is rather saying that the underlying nature for *substance* (οὕτως αὕτη [ἡ ὑποκειμένη φύσις] πρὸς οὐσίαν ἔχει καὶ τὸ τόδε τι καὶ τὸ ὄν 190b11–12) can be grasped by analogy with the other cases he mentions: the statue, the bed, and something 'shaped'.

> Therefore (διόπερ), if one holds that the previous account (λόγος) is correct, as well as this one (τοῦτον), it is necessary, if one is to preserve them both, to posit some third thing... (189a34–b1)

I pointed out earlier that it is somewhat misleading at this stage (though not false) to say that the account of I 5 is 'preserved' by positing three principles if the upshot of I 5 is that *all* principles are contraries.[35] A further difficulty is that the emphatic διόπερ suggests that Aristotle is here drawing precisely on what immediately precedes, namely, the most clearly negative part of the argument from substantiality, but I see no way that he can be.

2.3. *Resolving the difficulties*

Aristotle's official resolution of the *aporia* is encapsulated in the following passage, from I 7:

> For this reason, in one sense one should say the principles are two, but in another sense, three. And in one sense contraries, e.g. if one should say the musical and unmusical or the hot and cold or the tuned and untuned; but in another sense not; for it is impossible that contraries be affected by one another. This too is resolved by what underlies being something else (διὰ τὸ ἄλλο εἶναι); for this is not a contrary. Hence, the principles are neither more numerous than the contraries in a way, but two in number, so to speak. Nor are they, on the other hand, altogether (παντελῶς) two, since the being belonging to them is different, but rather three; for the being for man and for the unmusical are different, as well as for the unshaped and for the bronze. (I 7 190b29–a3)

The resolution consists of (i) a distinction between two ways of individuating principles and draws on (ii) the thesis argued for in the first half of I 7, that there is even for substantial generation something underlying, which is 'in being' two, in that (before the generation) it is both privative ('contrary') and non-contrary ('something else'). Accepting that thesis, if we count the principles according to how many different 'beings' there are, we end up with three principles, not all of which are contraries. On the other hand, if we count the privation and underlying nature as a single principle, since they are numerically one thing, we end up with just two principles, which are contraries, but contraries only in a sense. Aristotle here relies on examples such as 'the musical', 'the hot', and 'the tuned', which are contraries in a looser sense than musicality, heat, and tunedness (as well as density and rarity). The move is that of the *Phaedo*: to distinguish

[35] We must, following Kelsey (2008, 185), and as the singular τοῦτον indicates, take λόγος here to refer to the account or position being argued for (in I 6: that there are three principles) rather than to specific arguments.

contraries themselves from 'contrary things'.[36] As he indicates a few lines below (I 7 191a15–17), the two accounts between which his resolution is a compromise are those of I 5 and I 6, as the 'To the Letter' reading maintains.

If the components of this resolution appear too obvious for a puzzle of such difficulty (ἀπορίαν ἔχει πολλήν: I 6 189b27), it is helpful to recall the introductory status of book I. Furthermore, Aristotle has emphasized quite similar distinctions in previous chapters, both against the Eleatics and against Anaxagoras: between non-substance and the substance that underlies it (I 2 185a20ff.); between the 'being' of something and what has such being (I 3 186a26ff.); between contrary affections, which are not separable (not even by the Anaxagorean Mind) and stuffs (I 4 188a5–13.). Furthermore, both components of the resolution remain in the spotlight in I 8–9.

Also, although the *aporia* is explicitly resolved by a distinction between two ways of individuating principles, it arises out of arguments to the effect that all principles are contraries, as well as arguments to the effect that a non-contrary principle is needed. And to the extent that these arguments are plausible, the resolution should also preserve their basic insights. It preserves the account of I 5 by allowing us to say that the principles are just contraries. Aristotle's resolution also clearly mentions the *poiein-paschein* argument ('for it is impossible that contraries be affected by one another': I 7 190b33–4) and claims that it is resolved by the fact that the underlying nature is different from the relevant contrary (privation) (I 7 190b33–5). This preserves the basic insight of the *poiein-paschein* argument, that something different is needed to be affected, so that, in a sense, we do not 'make the principles only two' (189a21). Aristotle, that is, *accommodates* the argument by adding what is in a way a third and distinct principle. He does not—although he could have—emphasize that the contrary is somehow the same as the underlying principle and explain that things therefore can, after all, be said to come-to-be from contraries, or that contraries can, after all, be said to be principles, as we might expect if these ideas were under threat by the positing of an underlying principle. Rather, the upshot of the *poiein-paschein* argument here concerns only the third principle, suggesting that Aristotle does not intend to bring out such threats with the argument in the first place.

Although it does not mention the argument from substantiality, Aristotle's resolution can be seen to preserve its positive aspect to an extent, insofar as it employs an underlying principle. But it does not posit it as 'the *ousia* of things', which is what the argument demanded. Similarly, it can be seen to reject the negative aspect in the sense that it involves contrary principles. But it does not

[36] Aristotle does not similarly collapse together form and *hupokeimenon*, even though form characterizes the *hupokeimenon* once the generation is complete. For an illuminating discussion of this issue, as well as more detail on Aristotle's resolution of the puzzle, see Hendrik Lorenz in this volume.

give us the tools to explain why the negative aspect of the argument does not threaten the status of contraries. In this way, Aristotle's resolution too suggests that the argument from substantiality, and especially its negative aspect, should not be viewed as central to the two vs. three *aporia*.

To fully address the difficulties involved in the argument from substantiality, it would be necessary to explain how substantial form can, in fact, be *ousia*, and in particular how it can be predicated of (i.e. be an attribute of) the underlying nature (i.e. substantial matter) without being posterior to, and having as its principle, that underlying nature—or put in a different way, more familiar to readers of the *Metaphysics*, to explain how this predication is relevantly different from that between a (particular) substance and its accidents. What Aristotle presents as his resolution of the two vs. three *aporia* does not do this. And to the extent that developing and resolving the *aporia* is presented as *the* function of I 5–7, the difficulties raised by the argument from substantiality seem somehow peripheral. Recall again Aristotle's summary of I 5–6, near the end of I 7, which mentions neither a threat to contraries, nor any issue about priority: 'First it was shown that that the contraries alone are principles, and afterwards that it is also necessary for something else to underlie and that [the principles] are three' (191a15–17).

But there are several moves within I 7 and I 9 in the direction of addressing the argument from substantiality, and indeed, in the direction of a hylomorphic approach which grants priority to form as *ousia*. Many of them can be collected under two headings: the use of form–privation as the primary contrariety, which facilitates an identification of form with being, and privation with not-being;[37] and the emphasis on the underlying nature (substantial matter) having (only) a partial claim to being the *ousia* of things, being only to a certain extent 'some this', etc.[38] I would like to close this section with the speculative suggestion that even in the remainder of *Physics* I, we can distinguish between a kind of official line, as it were, which seeks to remain, or at least give the appearance of remaining, more neutral with respect to this materialist strand, and the numerous moves in the direction of a hylomorphic perspective that grants priority to form. By the end of I 7, Aristotle claims to have resolved the *aporia*, and has offered a theory which presents both form and the underlying nature (and in a way, privation) as principles: a hylomorphic theory in a sense. Still, Aristotle quite clearly states at the end of the chapter that it remains unclear whether form or what underlies is *ousia* (I 7 191a19–20)—which would suggest

[37] Still, the use of form–privation might be seen to arise primarily as a way of legitimating the application of the rubric of contrary principles to the case of substantial generation.

[38] Aristotle describes the underlying nature as 'more' or 'rather' *tode ti* than the privation is (I 7 190b25–6), and yet as not quite 'one' and 'being' in the way that a *tode ti* is (I 7 191a12–13), as almost, in a way, *ousia* (I 9 192a5–6) and as desiring the form (I 9 192a16ff.).

that the priority between them also remains unsettled. Even with respect to the claim that form is a principle, perhaps we may say that, though Aristotle accepts that claim, he does not argue that, or explain how, it can withstand the attack on contrary principles in I 6.[39] Finally, though several passages, especially in I 9, are suggestive about the secondary status of matter, we should ask whether the primary, and we might say, 'official', purpose of these passages is to legitimate the status of form, or rather to distinguish matter from privation.[40]

3. Previous Thinkers on the Third Nature (189b2–b18)

In this section, Aristotle invokes the views of his predecessors, both earlier and more recent, in support of the position he takes himself to have motivated. He draws heavily on the classification at the beginning of I 4, focusing first on earlier thinkers' employment of a single underlying nature. To do justice to the accounts in I 5 and I 6, he tells us, we must posit a third principle:

> Just as those say who claim that the all (*τὸ πᾶν*) is some single nature, such as water or fire or the intermediate between them, though it would seem to be rather the intermediate. For fire and earth and air and water are already intertwined with contrarieties (*μετ' ἐναντιοτήτων συμπεπλεγμένα*). This is why those who make the subject something other than these, and those, among the others, who make it air, do not do so unreasonably. For air seems to share in the perceptible differentiae less than the others. And next is water. (189b2–8)

'Those who claim that the all is some single nature' (*οἱ μίαν τινὰ φύσιν εἶναι λέγοντες τὸ πᾶν*: I 6 189b2) are clearly the material monists who make up the first arm of the *diairesis* of I 4: those who make being one by being one underlying body (*οἱ μὲν γὰρ ἓν ποιήσαντες τὸ ὂν σῶμα τὸ ὑποκείμενον*: 187b12–13).[41] This is quite clear from the back-reference at 189b11, but also from the similarity between the two descriptions. A crucial point not found in I 4 is the

[39] As Kelsey (2008, 204) argues, privation is no longer a concern, since natural substances are not 'out of' it (I 7 190b17–20) but this of course does not address the status of form. Alan Code (2018) describes the argument from substantiality as bringing to bear 'logical puzzles' (168), while the *poiein-paschein* argument 'makes use of a specific *physical* requirement for change' (165). While *Physics* I to some degree straddles this divide, it is more clearly respected in the way that *Physics* II and the middle books of the *Metaphysics* respectively treat the kinds of issues raised by the argument from substantiality. *Physics* II crucially argues for the priority of form as nature and *ousia*, but does not take up the kinds of metaphysical considerations we find, e.g., in *Met.* H 6 and Θ 7, which directly address the predicative relation between matter and form.

[40] See the contributions of Hendrik Lorenz and Sarah Broadie in this volume.

[41] 187a12–13, following Betegh, this volume (126), in rejecting Ross emendation.

desideratum that the underlying nature should be less, or ideally not at all, 'intertwined with contrarieties'. He claims that the four simple bodies are all so intertwined, though air is the least intertwined with perceptible differentiae, and next water. In *Gen. et Corr.* I 4, he claims that air is the 'least perceptible' among the elements. Assuming that air is perceptible because its contraries are, this implies that the contraries of air are the least perceptible—perhaps the most plausible reading of the claim that air is least intertwined with perceptible contrariety. However, better still is an intermediate nature between elements.

Given that Aristotle is drawing on the characterization of material monists at I 4 187a12–16, it is plausible that he is referring to the same intermediate-positing theorist in both passages. But there is some support for taking the passage in I 4 to refer to someone other than Anaximander, especially that Anaximander is there categorized with the pluralists.[42] It can be added that in I 6, Aristotle is particularly concerned to emphasize that the intermediate nature is free from contrariety, whereas Anaximander, according to I 4, speaks of separating out contrarieties already present in the one (187b20–1).[43] However, it must be noted that while I 4 (like several other passages noted by Ross, 482–3) mentions an intermediate between air and water, I 6 seems to indicate an intermediate between water and fire (*οἷον ὕδωρ ἢ πῦρ ἢ τὸ μεταξὺ τούτων*: 189b2–3), which would be intermediate with respect to both of Aristotle's basic contrarieties—and so perhaps in a way also intermediate between air and water. Aristotle might be more concerned in I 6 to bring out the ideal version of this view than to accurately describe a historical view.

To understand this desideratum that the underlying nature be free from contrariety, it is plausible to look to the foregoing arguments, which demand a principle of a different kind than the contraries. This consideration might be fleshed out (as suggested earlier in the discussion of the *poiein-paschein* argument) in terms of the need for something to which the active contrary can be present, rather than something 'driven out' by the contrary.[44] The situation will be particularly acute in case the 'intertwined' contrary is opposed to the active one (e.g. if water, treated as basic, is to become fire or fiery). Clearly, the desideratum is meant to distinguish among material monists, and to favour those who posited a single, indeterminate material principle. In this context, the desideratum seems to require that the underlying principle be free from *all* contrariety.

Can Aristotle endorse the desideratum in this form? The closest Aristotle comes to positing a single underlying principle of everything is in his account

[42] See Betegh, this volume (125, note 2).

[43] See also the argument that Anaxagoras' infinite already involves contrariety at *Gen. et Corr.* II 5 332a20–5.

[44] At *Gen. et Corr.* II 5 332a6–17, a related context, Aristotle seems to presuppose persistence when arguing against the view that one of the four elements is matter for all the rest.

of the principles of tangible body in *Gen. et Corr.* II. The underlying principle is 'what is potentially perceptible body' (II 1 329a34). There, however, Aristotle criticizes his predecessors' accounts of a single underlying nature—accounts of the sort mentioned favorably in I 6. He dismisses the possibility of a stuff that can exist without all perceptible contrariety (II 1 329a8–13). Crucially, however, he characterizes such views as positing a *separable body* that is free from contrariety (II 1 328b35, 329a9–10) and his main innovation is to deny that the underlying matter of tangible body is itself a separable body. As potential body, it is inseparable, always existing together with contrariety. Despite appearances, this claim does not signal that Aristotle is giving up on the desideratum from I 6. *Gen. et Corr.* II 1 is best taken to show that the desideratum should not be understood in terms of existential separability, but in terms of essential separability. Thus, the underlying matter of tangible body, and of the elementary bodies, is not *essentially* any of the basic contraries.[45]

Even if the desideratum is understood in terms of essential separability, it still seems that Aristotle can only mean to accommodate a more limited version than the intermediate-theorist of I 4 and I 6 would: the underlying nature for a given kind of substance will be free (essentially separable) from the contrariety (i.e. form and privation) in connection with which it serves as underlying nature, but not necessarily from *all* contrariety. *Gen. et Corr.* II 1 is again helpful. Since the elemental contrarieties are basic, an underlying nature that is essentially separable from them will also be essentially separable from all other contrarieties; the more limited version of the desideratum plausibly implies the stronger one (free from all contrariety), on the supposition that the underlying nature in question is a single and ultimate underlying nature. Thus, though the kind of intermediate underlying nature Aristotle presents as best satisfying the desideratum is free from all contrariety (and indeed thought to exist separately) this need not reflect a desideratum stronger than Aristotle can endorse. Admittedly, Aristotle does little to dispel the most natural reading of these lines in isolation, as demanding that any underlying nature be essentially (or indeed also existentially) separable from all contrariety.

Earlier I pointed out that there is good reason for thinking that Anaximander is not the intermediate-theorist described in I 6, and that he is, in certain respects, better viewed as a pluralist in *Physics* I, alongside Empedocles and Anaxagoras. In I 4 and I 6, none of these thinkers are presented as having posited an underlying principle. By contrast, Empedocles and Anaxagoras are presented as

[45] One could claim that existential separability from *each* pole of a contrariety taken separately is at issue in I 6, while *Gen. et Corr.* II 1 denies existential separability from the contrariety as a pair, but the language of I 6 suggests that separability from the contrariety as such is at issue (*μετ' ἐναντιοτήτων συμπεπλεγμένα*: 189b4–5).

employing the underlying principle and matter in *Metaphysics* A 3.[46] Of course, these thinkers' basic stuffs are, according to Aristotle, intertwined with contrariety; thus Horstschäfer understands the desideratum as a critique of Empedocles (238). But this critique would apply to most of the material monists as well. Moreover, its force against pluralist thinkers is far from clear. For example, Empedocles might well have no use for a distinction between contrary and 'underlying' aspects of his elements, given that he does not allow for transformations between them, or even (according to Aristotle anyway) that the elements are subject to genuine affection.

It is rather this failure to countenance genuine affection, which we may put by saying that the pluralists' basic bodies do not underlie contraries in the way Aristotle here envisages, that explains why they are left out of consideration here. Within *Physics* I, this is implied by the characterization of Anaximander, Empedocles, and Anaxagoras as generating from a 'one' through separation, and for the latter two thinkers, from a mixture (I 4 187a20–3)—generation *from* the primary bodies is not considered in *Physics* I.[47] But such 'generation' *of* the basic bodies is not the only possibly relevant phenomenon. A sustained and more general attack on the ability of pluralist theories to allow for genuine affection, and thus to account for a wide range of changes, is undertaken in *De Generatione et Corruptione*. Aristotle's contention in the opening chapter that the pluralists 'must' distinguish generation and destruction from alteration (*Gen. et Corr.* I 1 314a8–12, 314b1–8) is clearly grounded in their unacceptable identification of generation and destruction—including generation *from* the elements—with aggregation and segregation. And while it is obvious that the atomists (who, in *Physics* I, disappear from view after chapter 2) do not accommodate genuine affection at all, instead identifying and distinguishing a range of processes (alteration, mixture, generation) as various sorts of atomic arrangement and rearrangement, Aristotle holds that Empedocles cannot accommodate genuine affection either, so that he ends up employing, or is committed to, essentially atomistic accounts of various processes.[48] Empedocles' apparent rejection of elemental transformation is claimed

[46] Ross 1936, 490 and Horstschäfer 1998, 227, take Aristotle to be referring to Empedocles' elements with the observation that 'some assume even more (καὶ πλείω) [principles] out of which they construct the nature of beings' (189a26–7) at the end of the *poiein-paschein* argument. That may well be, though the anonymous ἔνιοι after mentioning Love and Strife suggests different thinkers, while καὶ πλείω seems to pick up on the enumeration of all the principles ('not only two' (189a21); 'some third thing' (189a26)), thus possibly referring to additional *kinds* of principles.

[47] This Empedoclean separation involves being acted on by a contrary, but not taking on that contrary.

[48] Here I focus, like Aristotle does, on Empedocles. Regarding Anaxagoras, recall that Aristotle not only finds his 'everything in everything' theory incoherent, but also claims that for him, the character of a blend is determined by that of the most prominent ingredient (I 4 187b2–4), so that 'alteration' would be brought about by the same kind of separation that 'generation' consists in; there seems to be no role for genuine affection.

to imply a rejection of any alteration on the part the elements (314b18–28, cf. also *Met.* A 8 989a27–30). Both Empedocles and the atomists are criticized for their use of pores (*Gen. et Corr.* I 8), to which Aristotle contrasts his view that the potential to be affected is the basic explanatory cause of affection, while pores are a merely facilitating, but not strictly necessary, factor. Similarly, Aristotle claims that mixtures like flesh and bone, on the Empedoclean theory, will come-to-be by 'composition' (σύνθεσις), so that the ingredients will retain their identity, lying next to one another in small portions (II 7 334a27–30). Earlier, Aristotle presents the same basic conception of mixture in atomistic terms and contrasted it with one that involves the ingredients being affected by one another (I 10 327b31–328a31). Elsewhere, I believe that, drawing on the unalterability of the elements, Aristotle ties the above conception of mixtures as *materially* non-homoiomerous (non-uniform in body) to a conception of them as *formally* non-homoiomerous (non-uniform in formal properties, such as colour).[49]

Returning to the text of I 6, Aristotle next reiterates the point from I 4 that the material monists apply a basic contrariety to their single underlying nature, and then explains that the resulting triad of principles is similar to a more recent, Platonist view.

> In any case, all shape this *one* (τὸ ἓν τοῦτο) with the contrarieties, density and rarity and the more and less. And these are in general clearly excess and deficiency, as was said earlier. And even this view seems to be an old one, namely, that the one and excess and deficiency are principles of beings, except not in the same way; rather, the early thinkers say that the two act, while the one is affected, whereas some of the later thinkers say rather the opposite: that the one acts while the two are affected. (189b8–b16)

Aristotle associates density and rarity, the preferred basic contrariety of the material monists, with the Platonic dual principles 'the more and less' and 'excess and deficiency'. Also at I 4 187a16–20, he has described the Platonic dual passive principle as 'excess and deficiency', treating it as a generalization of 'the great and the small', which seems to be the more perspicuous description.[50]

[49] See *De Sensu* 439b20–440a30. Aristotle criticizes the view that a new colour arises from white and black bits being set beside one another (παρ' ἄλληλα τιθέμενα): 'But is clear that, when [the bodies are] mixed, the colors too must be mixed, and this is the chief reason for their being many colors, not by being on the surface or by placement beside one another (τὴν παρ' ἄλληλα θέσιν). For it is not the case that one color of mixed things appears from far away but not from close up, but rather [one color] appears from everywhere' (440b13–18). Empedocles must choose between genuine properties of the elemental bits, which (like the elemental contrarieties) do not change, and properties that do somehow arise and pass away, but not through any genuine affection of the elemental bits. I have benefitted a great deal from Ierodiakonou 2005 on these issues.

[50] ταῦτα δ' ἐστὶν ἐναντία, καθόλου δ' ὑπεροχὴ καὶ ἔλλειψις, ὥσπερ τὸ μέγα φησὶ Πλάτων καὶ τὸ μικρόν, πλὴν ὅτι ὁ μὲν ταῦτα ποιεῖ ὕλην τὸ δὲ ἓν τὸ εἶδος, οἱ δὲ τὸ μὲν ἓν τὸ ὑποκείμενον ὕλην, τὰ δ' ἐναντία διαφορὰς καὶ εἴδη (187a16–20). Note that the reference to matter here is unexpected and indeed suspicious, as Betegh, this volume, points out (126, note 4).

Even the seemingly more recent triad of 'the one and excess and deficiency' is really just an inversion of the older view.

Let me make two further remarks on the mention of a Platonic underlying principle here. First, Aristotle hints at the kind of generalizing move that is presupposed in his limitation of the principles to three, though 'excess and deficiency' seems to be first introduced as a generalization over different *candidates* for a primary contrariety. Second, Aristotle is clearly drawing on the Platonic provenance of the phrases 'excess and deficiency' (*Statesman* 283c–285c, mentioned also at I 4 187a16–17), 'the more and the less' (*Philebus* 24a–25a, 41d), and 'the great and small' (in the unwritten doctrines, mentioned also at I 4 187a17).[51] In *Physics* I 9, which contains a preponderance of references to the Receptacle of the *Timaeus* (see Sarah Broadie in this volume), Aristotle will again refer to the Platonic underlying principle as 'the great and the small' and present the (now contrasting) triad of a formal contrariety and single underlying nature as his own (192a6–13). Within *Physics* I, then, Aristotle seems to treat these phrases as more or less interchangeable characterizations of the same Platonic principle. This is not a harmless move. For it suggests, first, that all of these phrases are in principle applicable to natural substances, but also that Aristotle understands the above dialogues and the unwritten doctrines as developing the same basic theory (at least in application to natural substance), perhaps in different versions. It demands a unified interpretation of Aristotle's Platonist target across book I.

Aristotle describes the interim conclusion of the discussion in the next lines, which I have already mentioned: 'To say that the elements are three would seem, if one examines the matter on the basis of these and other such considerations, to be somehow reasonable, as we said' (189b16–18).

4. No More Than Three (189b18–29)

He continues:

> But [to say that they are] more than three, [would] no longer (οὐκέτι) [seem reasonable]. (189b18)

'No longer' (οὐκέτι 189b18) does not indicate that Aristotle has already made the case for a limitation to at most three principles. Rather, I suggest that Aristotle is indicating that he will draw on the foregoing determinations about the functions of contrary and underlying principles, and that the hard work is

[51] See *Physics* IV 2 209b11–16 and 209b33–210a2 with the discussion in Steel 2012, 194–5.

done; the remaining limitation is comparably easy. But even if Aristotle feels he is on more solid ground here, the interpreter may not. It is often unclear what kind of concrete case or theory Aristotle has in mind, in part because the argumentation is so abstract. Moreover, in contrast to much of the rest of the chapter, in this part Aristotle seems to presuppose distinctive features of his own approach to the chapter's leading question, as opposed to the approach of his favoured material monists. Here I mean what I earlier called the Aristotelian approaches to (i) *primacy* and (ii) *generality*, respectively: (i) taking the primary contrariety and underlying nature in an individual case to be those determined by, and in a sense at the level of, the Aristotelian essence of the thing; and (ii) resorting to generalization in response to the generality of the leading question. Insofar as these are indeed features of the account of principles Aristotle is trying to motivate, his argument should, if possible, be consistent with them.

I will first present an interpretation of the last section [C3], an independent argument for limiting contrary principles to two, which I suggested above (195-6) is the source of an earlier passage (189a13–14, i.e. [A2]). I argue that this section provides Aristotle's main argument for limiting contrary principles to two (i.e. for a single primary contrariety), and that it presupposes his approaches to primacy and generality. Regarding the previous two sections [C1–2], whose function is far less clear, I explore, without adjudicating between, different interpretive options, some of which, perhaps, have not been adequately appreciated. Here is the whole passage:

> [C1] For with regard to being affected (πρὸς μὲν γὰρ τὸ πάσχειν) the one is sufficient (ἱκανόν). [C2] And if, [the elements] being four, there will be two contrarieties, [then] there will have to be some different intermediate nature for each taken separately, though if they can generate from one another, being two, one of the contrarieties will be superfluous. [C3] At the same time (ἅμα δέ), it is also impossible for the primary contrarieties to be more [than one]. For *ousia* is a certain single genus of being, so that (ὥστε) the principles will differ from one another only by being prior and posterior, but not in genus. For in one genus there is always a single contrariety, and all the contrarieties are thought to be brought back to one. (189b18–27)

I will present my interpretation of [C3] from the start in terms of Aristotle's own primary contrariety in I 7. As I see it, the argument is addressing the relations between putative contrary principles at different levels of generality, and treats the priority of certain such principles over others as a matter of higher generality, up to the level of the genus. Here is a reconstruction:

1. *Ousia* is a single genus of being.
2. In each genus of being (a) there is one primary contrariety and (b) all the contrarieties can be brought back to one.

3. The principles (i.e. different contrarieties) will differ only in priority, and not in genus (1, 2b).
4. There is one primary contrariety for *ousia* (1, 2a).

Steps 1 and 2a make for a neat argument for step 4, the stated aim of the argumentation. Step 3 is clearly meant to follow from step 1 (with which it is connected by ὥστε), and seems also to depend on step 2 (γάρ at 189b25). I suggest, in particular, that the two claims making up step 2 serve different functions, 2a supporting the main conclusion and 2b supporting step 3.

What 'principles' are meant in step 3? And why does Aristotle claim that some principles will be prior to others? Given that he aims here to limit the primary contrarieties to one, he is not positing a priority relation among the three principles, e.g. between form and matter or between form and privation, but rather between different contrarieties. But how then can they all be described as 'principles'? The model from *Met.* I 7, of 'commensurate' contrarieties, one prior to the other, suggests itself, insofar as, e.g., piercing–compressing and pale–dark, though at the same level of generality, could both be called 'principles' for the genus of colour, in line with Aristotle's phrasing here. However, even on this view it is somewhat misleading to say that these principles do not differ in genus (189b24–5), since one is in a sense outside of the relevant genus. But more importantly, it is hard to see what the analogues of *both* contrarieties might be for the genus *ousia*. We must allow, I think, that Aristotle is speaking here also of contrarieties that are more specific than the primary contrarieties for the genus *ousia*, for example, he is considering the relationship between the primary contrariety form–privation and a posterior one such as *canine* form–privation. The latter contrariety comprises 'principles' (as Aristotle calls them here) for the more specific genus *canine*. And it is only in a sense true that these two contrarieties do not differ in genus: they do not differ according to whether they fall under the genus *ousia*, though they will differ according to the sub-genus *canine*. On this reading, the priority between different contrary principles can be understood as the priority of a more general contrariety over a more specific one.[52] Aristotle is making a kind of generalizing move, and views it as tantamount to a move to what is prior for the relevant genus.

It seems to me that [C3] is not only consistent with, but positively presupposes, the Aristotelian approaches to primacy and generality.[53] I say this

[52] On this reading, he will not be insisting that *any* two contrarieties differ in priority. The important claim is that there is no difference with respect to falling inside or outside of the genus *ousia*.

[53] This claim, if correct, provides one more reason for thinking that 189a13–14 was originally a marginal note based on this passage. For Aristotle's presupposition of his own approach to primacy and generality might be out of place earlier in the chapter, where he is very willing to rely on materialist approaches.

not because there could not be a single primary contrariety for a genus in any other way. Rather, it is hard to see why one should otherwise be moved to accept, without justification, that there is a single primary contrariety in a genus. In other words: why accept that there is just *one* contrariety for a given substance if we are not thinking of the contrariety composed of *what something is* and the corresponding privation? And why think the number of contrarieties for the genus is the same as the number of contrarieties for a given substance unless, in specifying the contrarieties for the genus, we are generalizing over individual (kinds of) substances? Moreover, even though there could be a single contrariety for natural substance (e.g. density–rarity) on a materialist approach, when Aristotle employs such an approach, as I have mentioned, he insists that there are irreducibly two primary contrarieties, namely, the two elemental contrarieties as principles of tangible body in *Gen. et Corr.* II 2. This discrepancy might seem to indicate that when he limits the primary contrarieties to just one, in *Physics* I 6, Aristotle is no longer 'engaging in physical enquiry' (Bostock 2006, 4). But it is surely better taken to indicate that different questions and contexts within natural science require different approaches. In fact, even with regard to the elemental bodies, it is possible to insist that each has a single form, which is nevertheless composed of two contraries.[54]

This analysis of [C3] raises a fundamental question about the remainder of the passage: insofar as [C3] presupposes Aristotelian approaches to primacy and generality, and indeed views generality as a mark of priority, does it cohere with [C1–2]? Let me make some preliminary observations. First, [C2] seems to be addressing the leading question in its generality, rather than an individual case, insofar as it addresses the possibility of two contrarieties generating from one another, and perhaps also of each acting on a distinct underlying nature. If so, it is preferable to read [C1] (which gives no indication either way) as concerned with the general question as well. Second, as for the distinctive Aristotelian approaches to primacy and generality, again, it is preferable if [C1–2] are consistent with them. But it is unclear whether [C1–2] presuppose them, or need to. Consider that [C3], though it does not explicitly assume an Aristotelian approach, at least contains a suggestion of logical considerations about the genus *ousia*, but there is no such indication in [C1–2]. Moreover, the fact that [C1–2], as I argue below, make less categorical claims than [C3] could in part reflect an openness to other approaches at this stage. But to say more about this issue requires looking more closely at the content of [C1–2], to which I now turn.

I will first present a version of a common, and I think basically sound strategy, championed recently by Horstschäfer and Ross, which is, with one minor

[54] For a different approach to the form of the elementary bodies, on which it consists of (primarily) just one contrary, so that the two texts might be brought into better alignment, see Krizan 2013.

modification, consistent with my reading of [C3]. This strategy takes [C1–2] to function as a self-standing argument for the claim that there is at most one underlying principle [C1] and, on that basis, one primary contrariety [C2]. Clearly [C1] ('For with regard to being affected the one is sufficient' 189b18–19) concerns the underlying principle, picking up on the phrase *τὸ ἓν τοῦτο* at 189b8 as well as the *poiein-paschein* argument's contention that there must be something to be affected. [C2] then shows, on this strategy, that if there is more than one primary contrariety, there will have to be more than one underlying principle. On this reading, then, [C3] will be arguing for a conclusion that has already, to some extent at least, been established. Let us look in more detail at the two rather cryptic clauses of [C2]. The first is as follows:

> And if, [the elements] being four, there are two contrarieties, there will have to be some different intermediate nature for each taken separately (*εἰ δὲ τεττάρων ὄντων δύο* [20] *ἔσονται ἐναντιώσεις, δεήσει χωρὶς ἑκατέρᾳ ὑπάρχειν ἑτέραν* [21] *τινὰ μεταξὺ φύσιν·*). (189b19–21)

The reference to an 'intermediate nature' picks up on the material monist views just canvassed, in particular, the preferred view on which something between the elements is the underlying nature. According to the common strategy, the consequent says there will be two additional and underlying principles, so that we now end up with six, rather than the 'four' (20) contraries with which we began.[55] Here is the second clause:

> Though if they [the contrarieties] can generate from one another, being two, one of the contrarieties will be superfluous (*εἰ δ' ἐξ ἀλλήλων δύνανται γεννᾶν δύο* [22] *οὖσαι, περίεργος ἂν ἡ ἑτέρα τῶν ἐναντιώσεων εἴη*). (189b21–2)

Note that Aristotle is saying that the two contrarieties interact, not that two opposed contraries interact (if that is the scenario, it must be inferred); he is saying that the two contrarieties generate something, rather than that they are generated;[56] and he is saying in particular that they generate from one another reciprocally (*ἐξ ἀλλήλων*), rather than just the one from the other. Now, according to the common strategy, the notion of two contrarieties acting on one another signals that they operate on a *single* underlying nature, unlike in the first clause. The second clause describes the case in which at least one member of contrariety A–B acts on an underlying nature characterized by at least one of C–D and vice versa. In this case, it turns out that one of the contrarieties is not, in fact,

[55] The phrase '[the elements] being four' (*τεττάρων ὄντων*: 189b19) can be understood as saying that we make an initial move to at least four principles, without ruling out that there are more than four.

[56] Porphyry (Simplicius reports) comes close to an interpretation of this kind, by referring to heat and cold as generated from the rare and the dense (*In Physica* 206, 3–10).

a principle, but 'superfluous'. The second clause thus states the contrapositive, so to speak, of the first: a single underlying principle implies a single primary contrariety. Why should one of the contrarieties turn out to be superfluous? Ross (491) suggests that A can generate from D in the relevant way only if A and D are themselves contrary to one another (perhaps for the reasons spelled out at I 5 188a31–b21), so that the two putative contrarieties A–B and C–D collapse into one. On a less trivial interpretation the two contrarieties need not be the same contrariety in order to 'generate from one another'; rather, as the interpretation of [C3] suggested, one might correspond to a sub-genus of that to which the other corresponds. I will return shortly to the question of whether the claim of [C2], on the common strategy, is also plausible on a more materialist approach to primacy and generality.

One potential worry about the common strategy as I have presented it is that the phrase ἅμα δέ ('at the same time': 189b22), with which [C3] begins, suggests that what follows is an argument for a different but related claim, or a more decisive argument for the same claim. And indeed, we will see that the relative clarity and strength of [C3] contrast sharply with the character of [C1–2]. Thus, I think the most plausible version of the common strategy is to take [C1–2] to show that it is implausible or improbable that there is more than one contrariety, leaving the decisive argument for [C3], which shows that more than one is impossible (καὶ ἀδύνατον: 189b22–3). This can be achieved simply by taking the word ἱκανόν in [C1] to leave open the possibility of more than one principle.[57]

I will now explore alternative readings of [C1] and [C2] in turn, which are as far as I can tell independent of, but consistent with, one another, and which give rise to additional readings of the whole passage. The claim that a single underlying principle is sufficient for the whole genus may sound too strong simply to be asserted, especially since it seems to dismiss material pluralism out of hand. One possibility, found in Philoponus (*In Physica* 148, 28–30), is that Aristotle is saying that a single underlying principle is sufficient for a single contrariety. This makes it slightly easier to read ἱκανόν strongly: more than one would be excessive or even implausible. Moreover, the claim is no less plausible when understood as dealing in generalizations, i.e. as saying that a contrariety that is just one by generalization demands an underlying principle that is similarly just one by generalization. An advantage of this reading is that [C1] and [C3] together provide a clean argument for the conclusion that there is at most

[57] Another possibility is to focus, like Horstschäfer, on the move from showing that a second contrariety would be 'superfluous' (189b21–2), on the grounds (as far as I can tell) that a single contrariety could account for all things, to showing that more than one is 'impossible' (189b23).

one primary contrariety and one underlying principle; accordingly, the μέν at the beginning of [C1] (189b17) is best read as answered by the δέ at the beginning of [C3] (189b22). Further, this reading could explain why [C2] treats the possibility of four elements as one in which there are two primary contrarieties, rather than one contrariety and two underlying principles. Yet another advantage of this reading—for the interpreter anyway—is that it takes much of the weight off the puzzling passage [C2], which is no longer essential to the argument.[58] Finally, even if this is not the way to understand [C1]—it is certainly not what [C1] literally says—Aristotle may still intend for [C3] to have this function. For example, if we read [C1–2] as on the (modified) common strategy, Aristotle might still intend that we supply the argument just sketched, so that [C3] provides the decisive ground for limiting not only primary contrarieties, but also underlying principles to one.

Turning now to [C2], I begin with a concern about the common strategy: that it puts some strain on the notion of contrarieties generating from one another in the second clause. On the one hand, as noted above, this notion should be narrow enough to ensure that one or the other contrariety is superfluous. On the other hand, the notion should be wide enough to encompass any relevant way in which there might be a single underlying nature for two contrarieties, if the function of this clause is to (help) rule out that scenario. This means, first of all, that certain approaches are not relevant, i.e. not under consideration at all. For example, the kind of materialist approach Aristotle has relied on to motivate his view, and which he himself employs in *Gen. et Corr.* II, makes it difficult to see why both constraints should be satisfied. Hot–cold and dry–wet have the same underlying nature. Do they qualify as 'generating from one another'? If so, they provide a counterexample to the inference that one contrariety is superfluous, but if not, the second clause will not rule out there being, in this way, just one underlying principle for two primary contrarieties. In fact, Aristotle offers no reason to think that if there is a single underlying nature, the two contrarieties need to interact in any sense whatsoever. Taking the second clause to rule out the possibility of a single underlying principle for two primary contrarieties is most plausible if Aristotle's own approach to primacy and generality is to be assumed, and in particular, that a kind of underlying nature is individuated according to the contrariety for which it serves as underlying nature, e.g. if to be the underlying nature for a contrariety A–B is to be potentially A and potentially B. Note that in this case, it is not sufficient to

[58] On roughly the same lines as in the following two paragraphs, one might take [C2] to explore certain constraints on positing more than one primary contrariety, without taking it to rule out that possibility.

understand the case of two contrarieties generating from one another as the case in which they operate on a (numerically) single stuff; rather, it must be understood as the quite specific case in which they share *essentially* the same matter. The required assumptions become more complex if we take the case of two contrarieties generating from one another to allow for non-commensurate contrarieties.[59] Is this problematic? On the one hand, Aristotle presupposes aspects of his own distinctive approach to principles anyway in [C3]. But on the other hand, this interpretation requires that we read into the text quite specific conceptions of the underlying principle, and of what sense is to be given to the idea that there is a *single* underlying principle, which other thinkers would not only not adhere to; they might not even understand them. And this becomes more problematic given that there is no indication of a focus on logical considerations concerning relations between genera (as we find in [C3]), but also because there is no mention whatsoever, in this clause, of the underlying nature. These worries are avoided if we do not take the second clause [C2] to address the case of a single underlying principle for two primary contrarieties at all.

Consider again the consequent of the first clause: 'there will have to be some different intermediate nature for each [contrariety] taken separately' (δεήσει χωρὶς ἑκατέρᾳ ὑπάρχειν ἑτέραν [21] τινὰ μεταξὺ φύσιν· 189a20–1). The common strategy supposes that it implies two additional (i.e. fifth and sixth) principles, in addition to the four contraries, presumably on the basis of the emphasis on 'each taken separately'. But while perhaps most natural, this is by no means required. The function of this phrase could be to specify that some nature that is intermediate with respect to the first contrariety is needed, and also that some nature intermediate with respect to the second contrariety is needed—it must be intermediate with respect to each taken separately—without implying that these are two distinct intermediate natures. And once this possibility is made clear, there is nothing in the construction of the sentence to rule out the more radical possibility that no additional principles are required, i.e. that each contrariety serves as the intermediate nature for the other. Though one might reasonably think that at this point in the chapter, treating one contrariety as the underlying nature for another cannot seriously be considered, recall that the Platonic dual underlying principle has just been mentioned, and that we have not yet clearly distinguished between contraries themselves and 'contrary things'. The second clause, as I mentioned, though usually taken to concern the number of underlying principles, does not explicitly refer to an underlying principle. What the text says is simply that each contrariety generates from the other (ἐξ ἀλλήλων), not that each generates from some underlying nature characterized by the other. This is especially striking given that Aristotle

[59] Thus, we must allow that they operate on a single matter, in the relevant sense.

has just argued that a contrary makes something out of some third, non-contrary thing. If the first clause is read as making no commitment to additional (i.e. fifth and/or sixth) principles, the phrase 'generate from one another' is most plausibly taken to refer to the case in which each contrariety serves as patient for the other.[60] What about the consequent (περίεργος ἂν ἡ ἑτέρα τῶν ἐναντιώσεων εἴη: 189b22)? It might still be understood as claiming that one or the other contrariety is superfluous, namely, superfluous *as a contrariety*: the duality of the Platonic underlying principle has no relevant function, as Aristotle will point out later in I 9 (192b7–12). For this reading, though, the antecedent needs to be understood as claiming that one contrariety can generate *everything* from the other;[61] otherwise both will be needed as active principles. It seems better to understand the consequent as claiming that 'what is different from the contrarieties' (now taken together, not separately), i.e. the distinct underlying 'one' of [C1], is not needed. On the somewhat radical alternative I have been sketching, [C2] will point out that, if there are two primary contrarieties, then, first, there must be a different intermediate nature for each taken separately; second, if that condition is satisfied by virtue of each contrariety serving as the intermediate nature for the other, then no *additional* underlying nature is needed (whether we then say that there is no underlying principle, or that there are two, or even four, is unclear; we should in any case not say that there is exactly one). But we need not worry about that possibility because, as [C3] informs us, there is just one primary contrariety.[62]

To sum up the remarks thus far, [C3] provides the decisive, and perhaps the only, argument for limiting contrary principles for the genus to two. [C1–2] could then, in a version of the common strategy, (i) be taken to support the same conclusion, but more weakly. Alternatively, (ii) [C1] might be read as saying that a single contrariety implies (or is best supplemented with) a single underlying nature, which, along with the conclusion of [C3] could furnish the premises for limiting underlying principles to just one, so that [C2] becomes inessential. On the final alternative, closer to the text of one crucial phrase (ἐξ ἀλλήλων) but admittedly less natural in other respects, (iii) [C1] and [C2] jointly explore,

[60] Not, as Charlton 1970, 69 suggests, that one contrary generates from its contrary, without any underlying principle, which is surely (his claim to the contrary notwithstanding) much more of a departure from the text that Ross account of these lines, and, as Horstschäfer 1998, 242 notes, clearly neglects the need for a third principle.

[61] I am unconvinced by Simplicius' suggestion that Aristotle has in mind that even if each contrariety generates just some things, at least one will be superfluous because it is not a principle of *all* things (*In Physica* 205, 14–23). Philoponus also supposes that we must understand each contrariety generating all things here (*In Physica* 149, 20–150, 2).

[62] [C1–2] can then be taken to primarily concern the number of underlying principles—one is possible, though not the only option—while [C3] (ἅμα δέ at 189b22 again answering πρὸς μὲν γὰρ τὸ πάσχειν at 189b18–19) switches the focus to the number of primary contrarieties.

but do not settle, the question of the number of underlying principles, while [C3] takes up, and settles, the question of the number of contrary principles. Only the second alternative (ii) involves a fully explicit and decisive argument for a single underlying principle, on the basis of [C3] and the claim (on this alternative, in [C1]) that a single contrariety requires (or is best supplemented by) a single underlying principle. But as I mentioned, [C3] may still be intended to provide the basic ground for limiting underlying principles to one as well (as one might think, for Aristotle, it ought to). It is worth noting that the possibility which would provide a counterexample to this claim, namely, the possibility of one primary contrariety with multiple underlying principles, is not explicitly considered.

Aristotle closes the chapter as follows:

> So, that there is neither a single element, nor more than two or three, is clear. Which one of these, as we said, involves great difficulty (ἀπορίαν ἔχει πολλήν). (189b27–9)

Aristotle has not, evidently, mentioned precisely that this question involves great difficulty. He is referring back to the opening statements of each of the two arguments: 'For one might puzzle about (ἀπορήσειε) how density is such as to make rarity something' (189a22–3); 'In addition, one might further face this difficulty (ἀπορήσειεν), if one does not posit a different nature as underlying the contraries' (189a27–9). Both of these uses of ἀπορεῖν seem to refer to difficulty or puzzlement that arises if one does *not* posit a non-contrary, underlying principle. For this reason, they indicate difficulty on just one side of the two vs. three *aporia*, and so do not quite ground the claim at the end of the chapter on their own.

7

The Complexity of the Subject in a Change

Physics I 7, PART 1

BENJAMIN MORISON

In the first part of this famous chapter, Aristotle gives us his official theory concerning the nature of the ὑποκείμενον in a change. He does so with an eye on the question which he turns to in the second half (190b17ff.), namely what the principles (ἀρχαί) of natural things which come to be (τῶν περὶ γένεσιν φυσικῶν, 191a3–4) are. The background to the enquiry is this: since coming to be is a crucial feature of those substances whose principles he is trying to discern, he thinks that by investigating the conditions which obtain when those natural substances come to be (first half of *Physics* I 7), he can shed light on what the principles are of those substances *themselves* (a task undertaken in the second half of *Physics* I 7). It is easy to forget, especially when reading the scholarly literature on *Physics* I 7, that the aim of *Physics* I is not to give an account of the principles of *change*, but rather to give an account of the principles of the *things which change*—natural substances.[1] Aristotle builds up to an account of

I am grateful to the audience at Delphi for their spirited and incisive comments on the chapter as delivered there, and especially to Tim Clarke, Sean Kelsey, Hendrik Lorenz, Stephen Menn, Carlo Natali, Oliver Primavesi, Christof Rapp, and David Sedley for their detailed comments. I have also benefitted greatly from discussion with Arnold Brooks, Adam Crager, John Cooper, Hendrik Lorenz, Scott O'Connor, Jacob Rosen, and Christian Wildberg, and from the meetings of the Princeton Classical Philosophy Reading Group and the Princeton-São Paulo workshop on *Physics* I.

[1] Bostock (2006) explicitly asserts that Aristotle is not interested in the principles of natural objects, despite what Aristotle says in I 1: 'it soon becomes clear that the principles he is interested in are not so much the principles of natural *objects* (τὰ φύσει ὄντα) but rather of natural *processes* or *changes*, and in particular *generations*' (1).

the principles of natural substances through an enquiry into change; the account of the principles of natural substances draws on the investigation into the conditions which prevail when a natural substance comes to be.

The account he offers in the first half of I 7 of what the ὑποκείμενον in a change is emerges from his philosophical discussion in chapters 5 and 6, and solves the outstanding problems about how to do justice to (i) the role of contraries in change, and (ii) how to account for the fact that there must be something underlying the contraries. I shall be arguing for the view that Aristotle discerns, in change, a novel sort of item which he identifies as the ὑποκείμενον, the 'underlying thing', or 'subject'[2] in a change. This item is a metaphysical complex, consisting of two features or aspects, something which *persists* through the change (in cases of substantial coming to be, the matter) and the *privation* of the feature which characterizes the item which stands at the end of the change (in cases of substantial coming to be, privation of form).

The argumentative thrust of the first half of chapter 7 is to identify this complex ὑποκείμενον in the case of non-substantial changes, for instance the paradigmatic case where an unmusical person becomes musical, and then extend the resulting analysis (at 190b1) to the more difficult case of a substantial change, where a substance comes into being. This will allow Aristotle in the second half of the chapter to use his insights about the components of this complex ὑποκείμενον in a substantial change to help him single out the privileged role of matter as being an aspect of the being of the resulting natural substance. But the direction of the enquiry is this: we first discern the role of this complex ὑποκείμενον in change, and then use the insights we gain from seeing its (dual) role in coming to be to discern the principles of natural substances as such.

As we shall see, Aristotle will characterize this metaphysically complex ὑποκείμενον in change as something which is 'one in number' but 'two in form': a unified item to be counted in the ontology, but one which is irreducibly metaphysically complex. What are examples of such subjects, especially in the case of substantial changes? The prime example Aristotle will give in this first half of the chapter is σπέρμα: seed. It is a readily intelligible thought that things like *seed* are, as we might put it, subject *to* change. They are, quite obviously, things which are ripe for engaging in change. They are items whose role in the science of nature cannot be understood except as being items which will *become* natural substances. Any good student of nature will want to make room for such items in her ontology, and will want to theorize about their behaviour and their contribution to the coming to be of a plant or an animal. (Other examples

[2] When I use the term 'subject' here, I do not mean the notion of a subject *of predication* which we find in the *Categories* (and which is referred to in I 7; see below on 190a31–b1). I mean the notion of a subject for or in change.

Aristotle gives include a lump of bronze which has not yet been fashioned into a statue, or a lump of stone which has not yet been chiselled into a herm, although these items presumably will not be of any interest to the student of *nature* as she engages in her study of the natural world.)

Now is the time to turn to the text of the first half of *Physics* I 7, to see how Aristotle manages to clarify the nature of the ὑποκείμενον in non-substantial and substantial changes.

> [189b30] Let us then give an account in this way, by first reviewing every kind of coming-to-be. For it is in the nature of things to speak first of what's common and on this basis investigate what's specific to each thing.

I translate λέγωμεν as 'let us give an account', i.e. of what the principles of natural things which come to be are, as announced in the second half of this chapter (190b17–191a22). Aristotle approaches that question, as he says, by way of clarifying the nature of change quite generally, leading up to an analysis of substantial change, all of which occupies the first half of the chapter. One way of seeing that there is this divide between the two halves of the chapter is that in the first half there is no mention of principles or natural bodies—at least, not under those descriptions. There is plenty of discussion of things which come to be, but that list also prominently includes non-natural objects such as statues and houses, and there is also discussion of types of change *other* than coming-to-be, such as alteration, locomotion, and growth. Thus the first half of the chapter discusses *change* and *coming to be* as a lead-up to the remarks in the second half on the principles of being of natural bodies which are generated, i.e. beings which naturally come to be.

As I said earlier, it is only the second half of I 7 which finally answers the question posed in I 1, of what the principles of natural substances are. The first half of I 7 undertakes an intensive study of change, and Aristotle leaves a few hints that you could think of this study as being an enquiry into what the principles of change are.[3] (We just need to remember that establishing what the principles of *change* are is not the same thing as establishing what the principles of *natural substances* are.) Although Aristotle eschews the language of principles in the first half, presumably so as not to encourage confusion with the project of the second half, there are a few textual pointers to encourage us to think of the first half of the chapter as, *inter alia*, an investigation into the principles of change. Notably, right at the start, Aristotle announces a programme: first he is going to review 'every kind of coming-to-be' (πάσης γενέσεως) by speaking of

[3] Cf. Ross's summary of 189b30–2 in his 'Analysis' (1936): 'Let us state the *principles* involved in all coming-to-be, before we pass to the several kinds' (345; my emphasis).

what's common in all instances of coming-to-be, and then move on to investigate what's different in those cases. Aristotle is here referring us back to the programme for enquiry he laid out in *Physics* I 1, enquiry into 'the principles and elements and causes' of things, and, as Kelsey remarks (2008, 181), Aristotle seems particularly to be recalling I 1, 184a23–4, where he said that we must proceed from the general to the particular. Aristotle also claims here in I 7 that this mode of enquiry—proceeding from discerning the common elements in the different manifestations of something to discerning the differences among those manifestations—is κατὰ φύσιν (189b31). This should remind us of when Aristotle said back in I 1 that 'the journey from things better known and clearer to us is *by nature* (πέφυκε) a journey towards things clearer and better known by nature' (184a16–18): κατὰ φύσιν at 189b31 is the counterpart of πέφυκε at 184a16. So we start with change, the general phenomenon, which, in an unarticulated way, is clear to us: we see it going on all around us. As we bring its elements or constituents into focus, we thereby (i) obtain the means for offering a proper scientific definition of it, and (ii) obtain the means to make principled distinctions between the different types of change. Accordingly, in the first half of this chapter, we (i) discern the features common to all of the kinds of change *and* (ii) see how these features differ across the two major types of change, non-substantial change and substantial change. Indeed, Aristotle makes it clear in the first part of the chapter that the discernment of the elements *common* to all changes—i.e. the discernment of the triple structure of a (doubly) complex subject plus a form—is far from easy, because of the case of substantial change: at 190b1–3, he tells us that that structure will become clear in substantial change once we think through its analogy with non-substantial change; the language of something *becoming* clear to us on investigation also recalls the idea in chapter 1 that at the end of our investigation, things which are clear by nature, and which were not clear to us at the start, should have *become* clear to us (184a22–3). Thus, Aristotle's discussion in the first half of I 7 of the common features of change enables us to draw relevant distinctions between the types of change, following the programme of enquiry laid out in I 1. And that programme of enquiry was supposed to be a programme for discerning the principles of something. Hence, it seems reasonable to take Aristotle as engaging, in the first half of I 7, in the project of discerning principles of change. But, to repeat, the enquiry of the first half of I 7, discerning the principles of change, is not the same as the *overarching* enquiry of *Physics* I, namely discerning the principles of the natural substances themselves.[4]

Suppose that it is true that the first half of the chapter does give us some principles concerning change, and suppose that the three principles of change

[4] I am particularly grateful to Sean Kelsey for discussion of the issues in this paragraph.

we identify are form, matter, and privation. Notoriously, Aristotle slides between identifying principles as items and identifying them as propositions. But presumably if an item is a principle, then some privileged proposition about that item is a principle. (So if points and lines are principles of geometry, then some propositions about points and lines will feature among the axioms of geometry.) Let us assume then that if an item like form is itself a principle, there will be a propositional principle in the science of nature which mentions it. So perhaps Aristotle envisages that there is an overarching propositional principle in natural science which goes something like this: *Changes involve some* ***matter*** *characterized by a* ***privation*** *taking on the* ***form*** *of which the privation is the lack.* This principle will serve to guide the scientist to find, for any change that she encounters, three items playing those roles in that change and serves to link in an explanatory way the four notions of change, matter, privation, and form. Part of having the *epistēmē* of the science of nature will therefore involve knowing that principle. I take it that what we are offered in the first half of *Physics* I 7 is a 'dialectical' justification for that principle. By 'dialectical' here I simply mean that Aristotle doesn't attempt to derive this principle demonstratively from other principles (since it is itself a principle, this would be impossible); rather, we are given very good reasons for believing it: we see against the background of the previous chapters how it emerges from careful consideration of the views of Aristotle's predecessors, taking the good parts of those views and rejecting the bad parts, and we see how it comes about from discerning the various constituents of change.

> [189b32] For we say that one thing comes to be from another and something from something different, speaking either of simple things or composed things. I mean this in the follow way. For it is possible for a man to come to be musical, [189b35] and it is possible for what is not musical to become musical or the man who is not musical to become a musical man. [190a1] So then I call 'simple' that which does the coming to be, namely the *man* and *that which is not musical*, and [I call] 'simple' the thing which it comes to be, namely *musical*. Whereas [I call] 'composed' both what it comes to be and what does the coming to be, whenever we say that the man who is not musical comes to be a musical man.

Here Aristotle starts his investigation of coming-to-be in general. He first of all observes that both when 'one thing comes to be from another' (*ἐξ ἄλλου ἄλλο*) and when 'something comes to be from something different' (*ἐξ ἑτέρου ἕτερον*), there are both simple items and complex items involved.

'We say that one thing comes to be from another and something from something different': nobody is quite sure what distinction Aristotle means to point to by using two different locutions here, the first employing *ἄλλο* and the second employing *ἕτερον*. The 'and' is not epexegetic (why would *ἕτερον* be used to elucidate *ἄλλο*?), so we are left trying to find a difference in meaning between *allo* and *heteron*. Ross refers *ad loc.* to *Metaphysics* N 1, 1087b29 and remarks that 'ἄλλο points to numerical difference, ἕτερον to difference of quality'. If one went that route, one might think that Aristotle meant the following:

(i) 'one thing comes to be from another': e.g. *this statue* coming to be from *this bronze*, or *this musical man* coming to be from *that unmusical man*.

(ii) 'something comes to be from something different': e.g. *musical things* coming to be from *unmusical things*.

But it is far from clear that Ross's *Metaphysics* passage supports such a differentiation between ἄλλο and ἕτερον, and some scholars deny that one can find any distinction in meaning between the two expressions at all (André Laks vigorously denied the distinction at the conference, and in this he is supported by Jonathan Barnes in his commentary on Porphyry's *Isagoge*, 164–6 and 345–8).[5] Traditionally, grammarians claim that ἕτερον denotes one of a pair (LSJ: 'one or the other of two'), whereas ἄλλο denotes something other than the thing in question (LSJ: 'another, i.e. one besides what has been mentioned') where in principle there could be many things which are other than that thing (Stephen Menn and David Sedley both suggested this understanding of the distinction at the conference). If one went that route, one might think that Aristotle meant the following:

(i) 'one thing comes to be from another (ἄλλο)': e.g. this statue coming to be from bronze (bronze is other than a statue, but there are many things which are other than a statue).

(ii) 'something comes to be from something different (ἕτερον)': e.g. a musical thing coming to be from an unmusical thing ('musical' and 'unmusical' form a pair).

The great advantage of this interpretation is that it fits nicely the two types of example Aristotle will go on to give on the γίγνεσθαι ἐκ construction: (i) ἐκ χαλκοῦ ἀνδριάντα γίγνεσθαί φαμεν (190a25) and (ii) ἐκ μὴ μουσικοῦ μουσικός (190a6–7; cf. ἐξ ἀμούσου μουσικὸν γίγνεσθαι at 190a23). This provides us with a reason *why* Aristotle should have bothered to have said something so repetitious as 'we say that one thing comes to be from another and something from something different' in the first place: he wanted to be sure that we had both such cases in mind.[6]

[5] Barnes 2003. I'm grateful to Stephen Menn for this reference.

[6] In offering this interpretation, I do not mean to commit myself to the thesis that ἕτερον and ἄλλο always have these differing senses or uses. But we need to come up with some difference *here in this text* between the two phrases Aristotle employs—otherwise why did he phrase things the way that he did?

Now, it is very telling that Aristotle begins his investigation into coming to be quite generally by announcing that we employ the locution 'coming to be *from* such-and-such'. This locution dominates the first half of this chapter, and for good reason. Aristotle will identify the subject in a change (the ὑποκείμενον) with that from which the thing which comes to be comes to be. And of course, the ὑποκείμενον will be identified in the second half of the chapter as one of the *principles* (ἀρχαί; see Lorenz in this volume, p. 270) and Aristotle clearly means to exploit the idea that the *principle* of something is that *from* which it is or comes to be.[7] So when Aristotle introduces the idea here that we say that one thing comes to be from another, he immediately goes on to clarify what he has in mind ('I mean this in the following way', 189b34). What follows in the rest of the first half of the chapter is precisely an extended discussion of what exactly the item *is* from which something comes to be (in the case of substantial coming to be), including prominently extensive discussion of when one can employ the locution 'coming to be from such-and-such'.

In order to settle these questions about coming to be from something, Aristotle first has to make some important distinctions, and this is what he immediately turns to at 189b34–190a1, by giving the example of a man becoming musical: 'For it is possible for a man to come to be musical, and it is possible for what is not musical to become musical or the man who is not musical to become a musical man'. I have translated ἔστι both times as 'it is possible'. An alternative translation would be 'it is a fact that...' to yield something like this: 'It is a fact that a man becomes musical and it is a fact that what is not musical becomes musical or the man who is not musical becomes a musical man'. On either translation, Aristotle is appealing to facts (out there in the world, so to speak) which will serve, once suitably analysed in 190a1–5, to justify (γάρ, 189b34) why we say that one thing comes to be from another, referring to both simple and complex things (189b32–4).[8]

[7] See *Physics* I 5, 188a26–8, and *Metaphysics* A 3, 983b24–5. I'm grateful to Sarah Broadie and Arnold Brooks for help here.

[8] Aristotle is often thought in this chapter of the *Physics* to appeal to the way we use language to ground claims about the world. See for instance Bostock 2006, 4: 'there is no denying that most of the discussion in chapter 7 seems to be conducted on a conceptual level, indeed one that pays much attention to the niceties of linguistic usage'; 'It is perfectly clear throughout these paragraphs [189b32–190a31] that Aristotle is concerned to comment simply on the way we speak, and to show that it conforms to this general scheme ... The discussion so far, then, is of a conceptual or linguistic nature' (5). This seems not to do justice to Aristotle's actual procedure. No doubt, he makes a claim in 189b32–4 about something we say ('we say one thing comes to be from something else and one thing from another'), but it is a claim which is *backed up* or *justified* or *explained* (γάρ) by an appeal at 189b34–190a1 to the way things *are* (a man comes to be musical, what is not musical becomes musical, etc.), the further analysis of which takes place in 190a1–5. In 190a5–8, Aristotle does document a difference between two natural language locutions as to the scope of their application; he then returns to those locutions later in the chapter (190a21–31) to show how his distinction between the thing which comes to be that remains and the thing which comes

At 190a1–5, he discerns three 'simple' components or elements in the change where a man becomes musical:

(1) The man (one of the γιγνόμενα; one of the things which does the coming-to-be, 190a2);

(2) The unmusical (the other γιγνόμενον, 190a2);

(3) The musical (ὃ γίγνεται, 190a3).

How are we to understand these items? The man (1) is easy enough: a familiar everyday substance. What about the other items (2) and (3)? Here are some options for item (3):

(a) 'The musical' denotes the man (i.e. the bare substance);

(b) 'The musical' denotes the quality of musicality, a repeatable universal, to be found in other people too;

(c) 'The musical' denotes this man's quality of musicality, a 'trope', not to be found anywhere else except in this man;

(d) 'The musical' denotes the musical man, i.e. the man along with the condition of being musical (a complex item);

(e) 'The musical' denotes the musical item, i.e. the man, but under the description 'musical'.

(e) is the most popular candidate in the literature for what phrases such as 'the musical' refer to in Aristotle texts: 'translators and commentators seem agreed that the factors distinguished when a man learns music are not the man, ignorance of music, and knowledge of music, but the man, the thing which is ignorant of music, and a thing which knows music' (Charlton 1970, 70). What is the difference between (a), (d), and (e)? This is not an easy philosophical matter, but here is a sketch: (a) is the man, the bare substance; (d) is the conjunction or hybrid of the man and the condition of being musical—an entity whose being or definition includes the condition of being musical (i.e. the condition of musicality); (e) is the man, but under the description of 'musical', where no extra hybrid entity such as a *musical man* (the entity described under option (d)) needs to be brought into the ontology, because to say that *the musical* is there at the end of

to be that doesn't remain can *explain* this difference in scope of application, but this is in the spirit of showing the explanatory value of the metaphysical distinction (perhaps that's what Bostock meant by showing that our way of speaking 'conforms' to a scheme?). Aristotle is not arguing for a distinction among beings *on the basis of* linguistic turns of phrase.

the change but not at the beginning is just to say that 'musical' is a true description of the man at the end of the change and not a true description of him at the beginning.[9] For a canonical enunciation of this view (which was also the most prevalent view at the conference), here is Bostock again (2006, 5): 'Before the change we have an object which can be described as a man (as an underlying thing) or as a thing that is not musical (as having a privation); it is the same thing that is described in these two ways. *Qua* underlying thing it persists throughout the change, in the sense that we have the same man at the end as we had at the beginning, but it can now be described rather as a musical thing (i.e. as having a certain form)'. I shall come in due course to the question about what the underlying thing is in the change, but this should suffice for the moment to distinguish options (a), (d), and (e).

Which of options (a) through (e) does Aristotle intend here? In other words, which simple item does Aristotle mean to pick out when he says that the musical is one of the three simple items involved in the change? (a) is to be rejected because the man was the first simple γιγνόμενον to be discerned (the one labelled (1) above), and so if the musical is identified straightforwardly with the man, we cannot distinguish between two of the simple items. (d) can't be right because these are supposed to be the *simple* γιγνόμενα, and Aristotle will be turning to the *complex* ones shortly. That leaves options (b) and (c) (which are akin), and (e).

It seems to me that on balance the evidence does not favour option (e) at this point in the text. The issue will recur later on when Aristotle identifies the ὑποκείμενον in a change as the γιγνόμενον, and I shall return to it then, but for now, here are some initial reasons why I reject the usual interpretation:

(1) At this point, we are considering metaphysically simple items (we turn to metaphysically complex items at 190a3–5), and the metaphysically simple items in play here can only be the man, unmusicality, and musicality.[10]

(2) Perhaps in answer to (1), commentators want to argue that the (un)musical thing is *one in number* with the man himself. But Aristotle never says this, here or elsewhere (see p. 243 for discussion of the locution 'one in number' as it occurs in our text at 190a15–16).

(3) When enumerating the principles later in I 7, Aristotle twice refers to *privation* (στέρησις) as being a principle in addition to form and matter (190b27 and 191a14). I take all three of these items, privation, form, and matter, to correspond to the simples Aristotle identifies here, respectively,

[9] Option (e) is the 'intensionalist reading' discussed in detail by O'Connor in his 2015 paper, and identified there as the most common interpretation of I 7.

[10] Interpreters often assume that Aristotle is not here talking about *metaphysical* simples but *linguistic* simples. But Aristotle is calling certain *things* simple, not expressions.

> 'the unmusical', 'the musical', and 'the man'. So 'the unmusical' refers not to the *deprived thing* (the man who is unmusical, the thing which *has* the privation) but rather to the *privation* itself, lack of musicality. Similarly, 'the musical' refers not to the *thing which has the form* (the man who is musical, the thing which *has* the form) but rather to the *form* itself, musicality.

On the basis of these considerations I want to explore the possibility that Aristotle means to identify as simple the items mentioned under options (b) and (c), namely musicality itself, or unmusicality itself, whether these be particular tropes or universals. (It may prove impossible to choose here between the two options.) This, at least, is the interpretation that this chapter will explore and advocate. On this interpretation our three *simple* items are Man, Musicality, and Unmusicality, and the two *composed* or *complex* items are the unmusical man (τὸ γιγνόμενον) and the musical man (ὃ γίγνεται).

We need a way of describing the simple items with a notation that can be applied to changes quite generally (Aristotle takes himself, after all, to be making a point about *every* coming-to-be). There are two simple γιγνόμενα, and one simple thing which the γιγνόμενον becomes (ὃ γίγνεται). We will learn in lines a9–13 that of the two simple γιγνόμενα one remains, and the other doesn't. So let us designate the three simple items involved in change as:

(i) SGR (**S**imple *Gignomenon* which **R**emains);

(ii) SG~R (**S**imple *Gignomenon* which does **not R**emain);

(iii) SHG (**S**imple *Ho* ***G**ignetai*).

Then our three simple items are: SGR, SG~R, and SHG. The two complex items are SGR+SG~R, and SGR+SHG.

My interpretation identifies the SG~R, in the case we are considering, as the privation of musicality (i.e. unmusicality), and the SHG as musicality. One obvious difficulty it therefore faces is that it seems to commit Aristotle to saying that unmusicality (the SG~R) *changes*;[11] after all, it has been identified as being the simple γιγνόμενον, a thing which does some changing. Do I really want to interpret Aristotle as saying that unmusicality becomes musicality (τὸ μὴ μουσικὸν γίγνεσθαι μουσικόν, 189b35)? The answer is that I do think Aristotle is saying this, or something similar: privation becomes form. Of course, I don't think that this is how he thinks the change should be understood in all of its details, but no one should think that the characterization of the change in terms of the

[11] This difficulty was pressed at the conference by Andreas Anagnostopoulos, and separately by Jacob Rosen.

simples alone is the preferred description for Aristotle.[12] For Aristotle, the preferred description of the change is the description in terms of the complex (συγκείμενον) items, the unmusical man and the musical man.[13]

> [190a5] In the case of one of these it is not only said that *that* comes to be, but also that something comes to be *from* that, e.g. something comes to be musical from not-musical, but in the case of the other this is not said in all cases. For it is not that something came to be musical from man, but rather man came to be musical.

'In the case of one of these' can be interpreted as one of the two simples which are γιγνόμενα, namely SGR and SG~R. In the case of one of them, SG~R, we say that it comes to be, i.e. that it undergoes coming-to-be. This is not surprising, since it is a γιγνόμενον, which we agreed meant 'engage in the coming-to-be'. If we were to spell this out fully, specifying *what* the SG~R comes to be, i.e. what predicate is taken on (a specification which Aristotle does not offer here), we would say 'SG~R comes to be SHG', or in the language of the example: 'The unmusical comes to be the musical'. Aristotle tells us that we *also* say that from SG~R something comes to be. So, in the example, we say 'From unmusical, something comes to be'. If we spell out fully what that thing comes to be, obviously it would be *musical*, and this Aristotle makes explicit (ἐκ μὴ μουσικοῦ μουσικός, 190a6–7). So, in summary, for the item SG~R, we have two locutions we can use:

(L1) SG~R comes to be, and what it comes to be is SHG;

(L2) From SG~R, something comes to be, and what it comes to be is SHG.

Notice that here, in the case of the locution with ἐκ, I am adding in a dummy subject 'something' for the γίγνεσθαι-verb. This is because in those cases,

[12] For an analogy, consider that if we describe a change as one where some bronze becomes a statue, we don't think that the bronze has literally turned into a statue; rather, we suppose that the bronze persists and the form of statue is imposed or added.

[13] It may strike many readers as bizarre that Aristotle would think or say that unmusicality becomes musicality. But in many places Aristotle says strange things about qualities such as musicality and unmusicality. For instance, in *Metaphysics* Δ 9 Aristotle says that when a man is musical and white, white and musical are accidentally (κατὰ συμβεβηκός) the same (τὸ αὐτό) in virtue of belonging to the same thing (1017b27–8), and that man and musical are accidentally the same because one is an accident of the other (1017b29); in both these cases, words such as 'white' and 'musical' clearly refer to the *quality* white and the *quality* musicality, since it is the qualities which belong to or are predicated of the man. (Aristotle distinguishes these cases of sameness from the one where the musical man and the man are accidentally the same (1017b31–3).) If a philosopher allows himself to identify two qualities which belong to the same subject as the same (in a way), it does not seem much of a stretch that one quality (in a way) becomes another, if that second quality comes to qualify the bearer of the first quality, replacing it in the process.

Aristotle's sentences run something like this: 'from SG~R' + γίγνεσθαι-verb + 'SHG'. One might be tempted to make the SHG the *subject* of the verb, but the SHG must be the complement after the γίγνεσθαι-verb—that much we know from the grammar of ὃ γίγνεται, where the ὃ is complement after γίγνεται. So, for instance, in the sentence ἐκ μὴ μουσικοῦ μουσικός (190a6–7, with an understood γίγνεσθαι), μουσικός, which has already been identified as the SHG at 190a3, must be the complement after the understood γίγνεσθαι. But then in English we need to add a subject to get sense. Hence why I translate 'something comes to be musical from not-musical'. (Similar considerations apply to Aristotle's further discussion of these locutions at 190a26–9.)

'In the case of the other': now we are talking about the other γιγνόμενον, namely SGR. Of course, we say of it that it comes to be—this is not surprising, since it is a γιγνόμενον. And what it comes to be is SHG, hence the example that Aristotle gives us: 'Man came to be musical': ἄνθρωπος ἐγένετο μουσικός (190a8). But Aristotle tells us that the other locution, '*from it* something came to be', cannot be used in *all* cases of SGR. In other words, you cannot say in all cases 'from SGR something came to be'. A case where we cannot use that locution is right in front of us: we do not say that from man (the SGR) something came to be musical (SHG), οὐ γὰρ ἐξ ἀνθρώπου ἐγένετο μουσικός (190a7–8). (Notice again a fact in the world is offered as a reason for why we speak a certain way.)[14] So when it comes to SGR, the other simple γιγνόμενον besides SG~R, the counterpart of L1 from above is available:

(L1) SGR comes to be SHG;

But we *cannot* always use the counterpart of L2 from above:

(L2) From SGR something comes to be SHG.

So far Aristotle has pointed to a difference we mark in speech between the roles of the two simple γιγνόμενα, SG~R and SGR. That is, he has pointed out that in the case of the change from an unmusical man to a musical man, there is a difference, marked in speech, between the role of one simple γιγνόμενον, the unmusical, and the other simple γιγνόμενον, man. This difference is the following. In the case of the SG~R (the unmusical), we can always use locution 2: 'From SG~R something comes to be SHG'. However, in the case of SGR (the man), we do not in *all* cases use locution 2: 'From SGR something comes to be SHG'. To make good on this claim, Aristotle has produced a case where this is not so: in the change where a man changes from being unmusical to being musical, we do not say 'From man (the SGR) something comes to be musical (SHG)'.

[14] I'm grateful to Sean Kelsey for spotting this.

Aristotle states his claim about locution 2 in this way: it cannot be used in all cases of one of the simple *γιγνόμενα*. He gives us an example of where it can't be used ('From man something comes to be musical'), thus making good on that claim. However, by saying that it can't be used in all cases, he gives us to understand that there are at least *some* cases where we *can* say 'From SGR something comes to be SHG'. He does nothing in this part of the chapter to justify this claim by giving an example, but he does give an example in the *next* part of the chapter. For it turns out that we *can* say that from bronze (an SGR, i.e. a simple *γιγνόμενον* which remains) something comes to be a statue (the SHG) (190a25). So when in 190a7 Aristotle says that locution 2 is not used in all cases of the simple *γιγνόμενα*, he is nodding ahead to his claim later that in *some* cases one *can* say 'From SGR something comes to be SHG'.

> [190a9] Of the things which come to be in the way we say simples come to be, one comes to be while remaining, and the other while not remaining: for while he is becoming a musical man, the man remains and has being, while the not musical, i.e. the unmusical, neither remains simply nor in composition.

Aristotle continues with his distinction between the simple *γιγνόμενα*, and divides them now into one which remains and one which doesn't: SGR and SG~R, in our terminology. Notice again that he appeals to the way things are ('for while he is becoming...') to justify his assertion about which simple remains and which does not. He does not appeal to ordinary language, or linguistic practice, or anything like that, to back up his theoretical claims. (See note 8 above.)

Later, the distinction among the *γιγνόμενα* between the one which remains and the one which doesn't will be put to use to *explain* why the two locutions L1 and L2 are not both universally available for talking about change. In other words, the distinction made in 190a9–13 will be employed by Aristotle to account for the differences he observed in the immediately preceding lines 190a5–8. More on that below (on 190a17–31).

How are we to understand the expression 'the man remains and has being'? The point, I assume, of 'and has being' is that there is not merely a *trace* of the man at the end; the item with which we finish, the musical man, is a fully fledged man. The man remains and has being as a man even at the end.[15] The words could be thought of as forming a hendiadys: the man *remains in his being*. Aristotle is

[15] Might ἄνθρωπος at 190a11 be the complement after 'remains and is'? If so, then Aristotle is emphasizing the point I am making in the text, that not only does the man remain during the change, but he is not changing insofar as he is a man. Alternatively, might μουσικὸς ἄνθρωπος be the complement after 'is' (as Michel Crubellier and Sean Kelsey suggested to me)? If so, then Aristotle would be making a rather different point from the one I claim he is making, namely that the man at the end of the change is a musical man (whereas during the change he was *becoming* a musical man).

emphasizing that there is an element involved in the change which survives unscathed as far as its own nature goes. This fact is important because any candidate for being the *ὑποκείμενον* of a change needs to have *some* aspect or part which survives the change, otherwise that candidate could not qualify as the subject in the change. (There would be no basis for thinking that a change had taken place as opposed to one item being replaced wholesale by another.)

> [190a13] With these distinctions in place, it is possible to grasp this from all things which come to be—assuming one looks at it in the way we have been explaining—that there should [190a15] always be something underlying, namely that which does the coming to be, and that thing, even if it is one in number, is nevertheless in form not one. For I mean by 'in form' and 'in account' the same thing. For being for a man and for an unmusical thing are not the same.

We must grasp that the underlying thing (the *ὑποκείμενον*) is the *γιγνόμενον*, and we must grasp that even if it is one in number, it is two in its being or account.[16] Strictly speaking, Aristotle does not *assert* here that the *ὑποκείμενον* is always one in number yet two in form, but rather just says that even if it were to be one in number, it would be two in form. However, he will show that the *ὑποκείμενον* is indeed one in number and two in form (and will assert it again later at 190b23–4).

[16] The thing that we must grasp on the basis of all the distinctions that Aristotle has made is this: *δεῖ τι ἀεὶ ὑποκεῖσθαι τὸ γιγνόμενον, καὶ τοῦτο εἰ καὶ ἀριθμῷ ἐστιν ἕν, ἀλλ' εἴδει γε οὐχ ἕν*. I assume that the first part means 'There must be something underlying, i.e. that which does the coming to be', taking *τὸ γιγνόμενον* to be in apposition to *τι* which is the subject of the verb. The grammar of the second half of this sentence (from *καὶ τοῦτο εἰ*...) is worth spelling out: *καὶ τοῦτο εἰ καὶ ἀριθμῷ ἐστιν ἕν, ἀλλ' εἴδει γε οὐχ ἕν* is to be understood as a whole conditional sentence conjoined to the previous thought by the first 'and', and *ἀλλ'...γε* is *in apodosei* meaning 'nevertheless': 'and that thing, even if it is one in number, is nevertheless in form not one'. This construction (*ἀλλὰ ...γε in apodosei* after either *καὶ εἰ* or *εἰ καί*, in their meaning of 'even if'), is extremely common in Aristotle. See for instance: *Cat.* 5, 4a28–9 (*εἰ δέ τις καὶ τοῦτο παραδέχοιτο, ἀλλὰ τῷ γε τρόπῳ διαφέρει*); *APo* I 31, 87b28–30 (*εἰ γὰρ καὶ ἔστιν ἡ αἴσθησις τοῦ τοιοῦδε καὶ μὴ τοῦδέ τινος, ἀλλ' αἰσθάνεσθαί γε ἀναγκαῖον τόδε τι καὶ πού καὶ νῦν*); *Top.* IV 5, 127a10–13 (*εἰ δ' ἄρα καὶ ἐπὶ τούτου δεῖ συγχωρῆσαι ὅτι ἀήρ ἐστι κινούμενος τὸ πνεῦμα, ἀλλ' οὔτε γε ἐπὶ πάντων τὸ τοιοῦτον ἀποδεκτέον καθ' ὧν μὴ ἀληθεύεται τὸ γένος, ἀλλ' ἐφ' ὅσων ἀληθῶς κατηγορεῖται τὸ ἀποδοθὲν γένος*); *Physics* VIII 3, 254a25–8 (*εἰ γὰρ καὶ κατ' ἀλήθειαν οὕτως ἔχει καθάπερ φασί τινες, εἶναι τὸ ὂν ἄπειρον καὶ ἀκίνητον, ἀλλ' οὔτι φαίνεταί γε κατὰ τὴν αἴσθησιν, ἀλλὰ κινεῖσθαι πολλὰ τῶν ὄντων*); *DC* I 4, 271a17–19 (*Εἰ δὲ καὶ αὗται ἐναντίαι, ἀλλ' οὔτι γε αἱ ἐπὶ τοῦ ὅλου κύκλου φοραὶ ἀλλήλαις διὰ τοῦτο ἐναντίαι*); *Meteor.* I 6, 343b34–5 (*οἱ γὰρ ἀστέρες κἂν εἰ μείζους καὶ ἐλάττους φαίνονται, ἀλλ' ὅμως ἀδιαίρετοί γε καθ' ἑαυτοὺς εἶναι δοκοῦσιν*); *Met.* B 6, 1002b27–30 (*καὶ γὰρ εἰ μὴ καλῶς διαρθροῦσιν οἱ λέγοντες, ἀλλ' ἔστι γε τοῦθ' ὃ βούλονται, καὶ ἀνάγκη ταῦτα λέγειν αὐτοῖς, ὅτι τῶν εἰδῶν οὐσία τις ἕκαστόν ἐστι καὶ οὐθὲν κατὰ συμβεβηκός*); Γ 6, 1011b8–9 (*καὶ εἰ τὸ αὐτὸ καὶ ἥμισυ καὶ ἴσον, ἀλλ' οὐ πρὸς τὸ διπλάσιόν γε τὸ ἴσον*); Z 16, 1040b36–7 (*ὥστε καὶ νῦν εἰ μὴ ἔχομεν τίνες εἰσίν, ἀλλ' εἶναί γέ τινας ἴσως ἀναγκαῖον*); *EE* II 10, 1227a1–2 (*εἰ γὰρ καὶ μὴ διακριβοῦσιν, ἀλλ' ἅπτονταί γέ πῃ τῆς ἀληθείας*), and no doubt many more.

Aristotle thus states that if one looks at things in the way he has explained—articulating cases of coming-to-be into three simple items, SGR, SG~R, and SHG, and the two complex items SGR+SG~R and SGR+SHG—then we will see in all cases (and not just in the example case of an unmusical man becoming musical) that *two* theses hold:

(1) There is always something underlying (ὑποκεῖσθαι), namely that which does the coming-to-be (τὸ γιγνόμενον).

(2) Even if that thing is one in *number*, in *form* it is not one.

Thesis (2) tells us that the γιγνόμενον that Aristotle has in mind as underlying, or being the subject in, the change, is a complex item ('in form it is not one'). This means that the underlying thing he has in mind here is the complex γιγνόμενον he identified at 190a4, SGR+SG~R, the complex consisting of the simple γιγνόμενον which remains and the simple γιγνόμενον which does not remain.[17]

There are two controversial issues here which I shall take one by one. First, what items are said to be 'one in number' but 'two in form', and what do these locutions mean? Second, what does Aristotle mean anyway by talking about the 'underlying thing' in a change?

(1) What does it mean to say here that this complex item is 'one in number' but 'not one in form'? This complex is one in number in the sense that its components—SGR and SG~R—come together to make one metaphysical item. In our example, that is the unmusical man; in another of Aristotle's favoured examples, it will be the as-yet-unshaped bronze. These are identifiable, countable items in the world. They are not one in *form*, because even though the two items together constitute a unity of a sort, the two components are different in being, as Aristotle says in 190a17: 'being for a man and being for an unmusical thing are not the same'. As with all other occurrences of μουσικόν and ἄμουσον where they appear on their own, I am assuming that Aristotle is referring to the metaphysical *simples*, musicality and unmusicality, so that he is saying—reasonably enough—that being for a man and being for the privation of musicality are

[17] Kelsey takes a different view. He glosses the view Aristotle states in these lines in the following way: '"things that come to be" (τὰ γιγνόμενα) are always "subjects" (ὑποκείμενα)' (192), implying that *every* γιγνόμενον is a subject or ὑποκείμενον. But we have already been introduced to a γιγνόμενον which is definitely not a subject, namely the simple γιγνόμενον which does not remain (SG~R), e.g. the unmusical, which I have interpreted to be a reference to the privation. This difference in interpretation can be traced to the fact that Kelsey interprets the phrase τὸ ἄμουσον as a reference to *the man who is unmusical*, and not to the privation.

different. The two *simple* items, the man and unmusicality, differ in form, even though the *complex* (συγκείμενον) item they come together to form is one in number. Aristotle is making a point about how this complex γιγνόμενον is a combination or coalescence of two distinct things (man and unmusicality) to make one thing (an unmusical man).

Now, in explicating this part of the sentence I have deviated in two vital respects from the orthodox interpretation. The orthodox interpretation has it that Aristotle here says that (i) the *man* and the *unmusical thing* are one in number, and that (ii) to be an unmusical thing is to be an unmusical *subject*, rather than the metaphysically simple item unmusicality—in other words, the orthodox interpretation is interpretation (e) sketched above. I suspect that commentators have been led to think (i) on the basis of already assuming (ii). But against (i), the fact is that, grammatically speaking, 'one in number' is here a one-place predicate, applying to τὸ γιγνόμενον, that which does the coming to be. In saying that the γιγνόμενον is one in number, Aristotle is making a point about a certain kind of metaphysical unity that the γιγνόμενον has. Similarly, when he says that 'in form [τὸ γιγνόμενον is] not one', Aristotle is denying that the one-place predicate 'one in form' applies to the τὸ γιγνόμενον. So Aristotle is saying that the complex τὸ γιγνόμενον—the unmusical man, for instance—is something which is one in number but two in form. So in summary, against (i), Aristotle *just doesn't say* that the man and the unmusical are one in number. He only says that the τὸ γιγνόμενον is one in number (and he then points to the two components of that τὸ γιγνόμενον, namely the unmusical and the man).

Now, Aristotle does use the locution 'one in number' in a variety of ways. He certainly can use it to apply to something which has a dual nature where the two components *coincide* (rather than combine as two aspects of one complex thing). Take, for instance, *Physics* VIII 8, 262a21: the mid-point of a line is both a beginning (of one line) and an ending (of another), and so it is two in *logos*, but in number one: it's just a single particular *point*. He makes the same claim, using the same language, about the instant in time which divides the before and the after (*Physics* VIII 8, 263b13). Note carefully, however, that in such cases as these, the two things which are one in number but two in form (i) are both Fs and (ii) together amount to just one F in number. In the case of the two points, they are both points (one is the beginning point of one line, the other is the end point of the other) and together they amount to just one point in number. In the case of the two instants, they are both instants (one is the beginning instant of one stretch of time, the other is the end point of another stretch of time) and together they amount to just one instant in number. Notice that this use of 'one in number' is making a point about the *coincidence* of two things of comparable metaphysical type (point, instant) to amount to one thing of that metaphysical type (point, instant).

Could Aristotle have been meaning in *Physics* I 7 to use this way of being one in number, the 'coincidence' kind according to which two things of the same metaphysical type coincide to make just one thing of that type? I don't think so. According to the orthodox interpretation of I 7, Aristotle is saying that the man (the ἄνθρωπος) and the unmusical thing (the ἄμουσον) are one in number. Waive for the moment the obvious problem that Aristotle doesn't predicate oneness in number of those two items here in the text, but rather of the γιγνόμενον (for whatever reason, interpreters don't seem terribly perturbed by this). Could Aristotle philosophically employ the coincidence meaning of 'one in number' when talking about the man (the ἄνθρωπος) and the unmusical thing (the ἄμουσον)? I doubt it. This claim is different in kind from the ones we have just seen in *Physics* VIII 8, since there we had two things of equal metaphysical rank (two points, two instants) being said to coincide to become one. And we do not have in *Physics* I 7, for instance, two things described as a man (one meeting condition X, the other meeting condition Y) together amounting to just one man. We just have two items, one described as a man and the other described as unmusical. If Aristotle meant to refer to an unmusical *man* by using the description 'the unmusical', this would be the only text in the corpus where a substance and a qualified substance are said to be one in number.[18]

(2) What is the notion of ὑποκείμενον that Aristotle is employing here, such that it is a *discovery* that it is this complex γιγνόμενον (e.g. the unmusical man)? What does it mean to say that this complex γιγνόμενον *underlies* or *is the subject* in a change? Aristotle sometimes uses the word ὑποκείμενον in these chapters to refer to a subject of predication, and this is a use which is familiar from (for instance) the *Categories*. This use occurs in 190a34–b1, where Aristotle wheels in the familiar doctrine that substances are the only items which are not predicated of some other subject but all other items are predicated of them, to explain why it is that the non-substantial items he has just mentioned, *poson* and *poion* etc., must be predicated of some subject. But it is one thing to be a subject of *predication*, it is another to be the ὑποκείμενον in or of a *change*, and this latter is the notion that Aristotle is employing here.

Aristotle's notion of the ὑποκείμενον is, at a first pass, this: the ὑποκείμενον is the thing which undergoes the change, i.e. the thing which does the changing.[19]

[18] The closest text is the famous one in *Metaphysics* Δ 6, where musical Coriscus and Coriscus are said to be one in a way (1015b23–6). But Aristotle is there explicating another kind of unity, namely unity κατὰ συμβεβηκός (1015b16, 17, 21, etc.). Note that he also says there that musical and Coriscus are one (accidentally), because one is an accident of the other (1015b22–3), referring with the word 'musical' to the quality of musicality (since that is the thing which is predicated of Coriscus).

[19] Cf. Code, 359: 'the substratum of a change is to be identified with that which comes to be the product'.

More precisely, since Aristotle has already claimed that there are at least three items which might fit the description 'that which undergoes the change' (τὸ γιγνόμενον), the ὑποκείμενον is *the item which undergoes change, whose nature makes the change intelligible as a change.* So understood, the ὑποκείμενον will provide insight into the principles of natural things, for natural things are those things which have a privileged relationship with a certain kind of change, namely generation, since they are the products of generation (this is the topic of the second half of the chapter). So when Aristotle says that what underlies in a change is the γιγνόμενον, he is telling us that the item which undergoes change, whose nature makes the change intelligible, is the γιγνόμενον.

How does this actually work? Well, the γιγνόμενον Aristotle here has in mind is one in number and two in form, so the γιγνόμενον he is referring to must be the *complex* one he mentioned at 190a3–5, i.e. SGR+SG~R, the combination of the simple γιγνόμενον which remains together with the simple γιγνόμενον which does not remain. Aristotle's idea is this: the thing which undergoes the change, whose nature makes the change intelligible as a change, is SGR+SG~R. By identifying this item, we at once discern both the aspect of the changing thing which *remains* during the change (SGR), and the aspect of the changing thing which is *destroyed* during the change (SG~R), thereby capturing the peculiar nature of change, that in it something must be different at the end (otherwise nothing has *changed*) and something must be the same at the end (otherwise no *thing* has changed). These two aspects of the ὑποκείμενον are rightly touted by Aristotle to hold the key to understanding where his predecessors went wrong. So for instance in *Physics* I 9, Aristotle praises the Platonists—albeit in a backhanded way—for having seen that the ὑποκείμενον is complex, although they botched it by not discerning *distinct* explanatory roles for the two items constituting the complex: 'they got this far, that they said there must be some underlying nature, but they made it one: for even if someone makes it a dyad, saying that it is the big and the small, nonetheless he makes them the same, for he overlooked the other nature [i.e. privation]' (192a9–12; for the identification of 'the other nature' with privation, see Broadie's contribution in this volume). Aristotle explicitly criticizes these Platonist predecessors for making their ὑποκείμενον 'one in power' (δυνάμει μία, 192a2): in other words, they failed to discern the different roles each component plays in change. Matter (SGR) and privation (SG~R) have those different roles, and so, on the correct account, the ὑποκείμενον is a complex thing, both of whose aspects will play an explanatory role in accounting for change.

These complex γιγνόμενα are features of the natural or physical world. They are instances of matter, in a state of lacking a form or feature of a certain kind. We are going to get an example of this kind of item later, namely *seed*. This kind of item is a special kind of item in the world. It is in its nature to *lack* a certain form. So the full specification of what an acorn is, for example, should make reference to the fact that it is not now, but is of a nature to become, an oak tree. Notice that acorns are beings which are characterized by a privation of form (the form of oak), and which also have a material component. This material component is what persists during an acorn's maturation into an oak tree. The privation component or aspect of the acorn does not persist; at the end of the process of maturation there is a (young) oak tree and hence no privation. Thus the acorn, since it is a complex of matter plus privation of form, also does not persist; it is no longer there at the end of the process. So much, I take it, concurs with facts which, as Aristotle would put it, are familiar to us.

In I 9, 192a10, the 'underlying nature' is clearly conceived of as complex, as it is here in I 7, 190a15, and again at 190b24. But at other points in Aristotle's presentation, he seems to use the expression *τὸ ὑποκείμενον* to pick out only the *persisting* part of the complex *ὑποκείμενον*, i.e. the matter (or the item such as man which persists during a non-substantial change), *without* the privation. Candidate occurrences of this usage would include 190b13–17, where Aristotle says that unmusicality is the contrary and the man underlies,[20] or that shapelessness etc. are the contraries and bronze etc. are the *ὑποκείμενον*. So it seems as though Aristotle, when referring to that which underlies in a change, uses the expression *τὸ ὑποκείμενον* in an ambiguous way: sometimes he uses it for the SGR, and sometimes for the complex of SGR+SG~R. This is a serious challenge for any interpreter, and we will have to look into those uses in more detail later, and ask ourselves whether Aristotle really is using the expression *τὸ ὑποκείμενον* ambiguously, and if so, how the expression could come to be ambiguous.

Aristotle tells us here that the *ὑποκείμενον* is one in number but two in form. So now we have a substantive and intriguing pair of theses on the table:

(1) in a change, there is always something underlying (*ὑποκεῖσθαι*), namely the item which does the coming-to-be (*τὸ γιγνόμενον*);

(2) that thing is two in *form*.

In other words, in a change there is always (1) an item which undergoes the change, whose nature makes the change intelligible, and (2) it is a complex item. Aristotle's explanation for his claim that the *ὑποκείμενον* is two in form is straightforward: he points out that he means the same thing by 'form' and 'account', and that what it is to be for unmusical and what it is to be for man,

[20] *ὑποκεῖσθαι δὲ τὸν ἄνθρωπον.*

i.e. the accounts of these two things, differ.[21] On the interpretation I have been advocating, the two things whose accounts differ are (i) the man (the SGR), and (ii) unmusicality (the SG~R). They are the two aspects or components of the complex γιγνόμενον.

Now, even though we have now understood Aristotle's two claims, Aristotle still needs to offer us some reasons for *believing* these two claims. And this is what we get in the next lines: 190a17–31 offer us reasons for the claim that the ὑποκείμενον must be two in form—i.e. that we really must discern in the item which undergoes the change a *dual* aspect, and 190a31–b10 offer us reasons for the claim that there always is some such ὑποκείμενον, even in the *prima facie* difficult case of substantial change.[22]

> [190a17] And the one remains behind, whereas the other does not remain behind. The thing which is not an opposite remains behind (for the man remains behind), whereas the thing which is not musical, i.e. the [190a20] unmusical thing, does not remain behind, nor does the thing composed from both, such as the unmusical man. But 'coming to be something from something' (as opposed to 'this coming to be something') is rather said in reference to those things which don't remain, such as coming to be musical from not musical, but not [coming to be musical] from man. Not to overlook that sometimes this is said even with reference to the things which remain: [190a25] for we do [then] say that something comes to be a statue from bronze, not that the bronze becomes a statue. But the case from the opposite and non-remaining thing is said in both ways, both something comes to be that from this and this comes to be that. For it's the case that both the unmusical comes to be musical and from unmusical something comes to be musical. And this is why things are the same in the case of complex items: for both from being an unmusical [190a30] man something is said to become musical and an unmusical man is said to become musical.

Aristotle continues by pointing out that the two items just distinguished, i.e. the unmusical and man, play quite different roles in the change, despite being united together as the ὑποκείμενον. One of the two is the simple γιγνόμενον which remains and the other is the simple γιγνόμενον which does not remain. In more abstract terms, the thing which is *not* an opposite (e.g. man, the SGR) remains, whereas the thing which *is* an opposite does not remain (e.g. unmusical, the SG~R), and these are the two parts of the ὑποκείμενον, a composite thing,

[21] The structure of 190a16–17 looks a little strange at first with two occurrences of 'for': the ὑποκείμενον is one in number but two in form '*for* I mean by "in form" and "in account" the same thing. *For* being for a man and for an unmusical thing are not the same'. The second of these 'for' sentences hardly looks like the explanation of the first. How then should we understand them? Answer: the two γάρ sentences jointly explain the initial claim about the ὑποκείμενον. See Denniston 1954, 64, the section entitled 'Successive γάρ's have the same reference', with examples.

[22] This interpretation of the subsequent lines, as offering reasons for the second and then the first claims that Aristotle makes in 190a14–6, was urged on me by Adam Crager and derives from Kelsey (2008, 193). I'm indebted to Crager for discussion, and also to Kelsey's insightful article.

which itself does not remain (190a20–1). One thing which is new here is the introduction of talk of opposites, so far absent from the chapter: Aristotle distinguishes the SGR from the SG~R by pointing out that one is an opposite whereas the other isn't. Of course, opposites played an important role in the preceding discussion, so here Aristotle is showing the connection between the insights he has had in this chapter and the insights garnered through intense scrutiny of his predecessors and his own philosophical discussion in chapters 5 and 6. One part of the ὑποκείμενον is an opposite, and the thing which is opposite to that is obviously the simple ὃ γίγνεται.

Where is Aristotle going with all this? Well, he is saying this because he is now going to show that the distinction between SGR and SG~R, the two components of the ὑποκείμενον, can be put to explanatory work. Recall that earlier, Aristotle had said that of the two simple γιγνόμενα, one of them is such that you can always say '*it* comes to be such-and-such' and 'from *it* something comes to be such-and-such', whereas the other is such that you can always say '*it* comes to be such-and-such' but you *can't* always say 'from *it* something comes to be such-and-such' (190a5–7). He didn't say there how you can tell which γιγνόμενον is which. Here, he explains that it is the SGR which doesn't always support the locution 'from *it* something comes to be such-and-such'. So, to give an example, in cases of the schema 'from . . . , something comes to be musical', we cannot fill the dots with a reference to the SGR, namely 'man'. We can have a reference to the unmusical (the SG~R), but not a reference to man (the SGR).

Now, Aristotle immediately points out that there are nonetheless *some* cases where you can say truly 'from . . . , something comes to be such-and-such', filling in the gap with a reference to the SGR. The example he immediately gives is when you say that there comes to be a statue from bronze: 'Not to overlook that sometimes this is said even with reference to the things which remain: for we do [then] say that there comes to be a statue from bronze, not that the bronze becomes a statue' (190a24–6). The second half of this assertion has puzzled commentators. Aristotle is often thought to be saying that we don't *ever* say 'the bronze becomes a statue'. But manifestly, that is a perfectly fine thing to say: a true thing, and a kosher Aristotelian thing to say, as Code (1976) points out (360). Indeed, Aristotle said as much in 190a5–6, when he said that for both the SGR and the SG~R, we *can* always say that they come to be SHG. Code proposes that we understand 190a25–6 in this way: 'we say that a statue comes to be from bronze, not ONLY that bronze becomes a statue'. But this move, even if it has Aristotle expressing a true thought, is not necessary. The key to understanding Aristotle's phraseology here was suggested to me by Gina White. Aristotle's complete thought in 190a24–6 is this: *sometimes* (ἐνίοτε, 190a24) in the case of SGR we do say that the product comes to be from it, for *on those occasions* we say that a statue comes to be from bronze, not that the bronze becomes a statue. This is hardly to deny that *on other occasions* we might say the latter! Aristotle is just saying that we sometimes say the former and in so doing, *on that occasion*, we are not saying the latter. In other words, Aristotle is not issuing a blanket denial of the acceptability of the phrase 'the bronze becomes a statue'. Rather he is saying that on some occasions we employ another locution ('a statue comes to be from bronze'), and not it.[23]

After this, Aristotle refers us back to his claim from earlier (190a5–6), which is that in the case of the SG~R, we can use both locutions, both 'from this, that', and 'this comes to be that', because you can say the man becomes musical from unmusical (the SG~R), and you can say that the unmusical (the SG~R) comes to be musical (190a26–9).

Then there follows an important new case (190a29–31), not discussed earlier, and that is the question of which of the two locutions you can use with the composite or complex γιγνόμενον, i.e. the ὑποκείμενον. Aristotle tells us that we can use both locutions. We can say:

> 'From being an [unmusical man], someone comes to be musical'

and

> 'an [unmusical man] becomes musical'.

Why is it that this case follows the pattern of the SG~R and not the SGR? Answer, because—as Aristotle said in 190a20–1—the composite *also* does not survive the change. Thus, the composite is *like* the SG~R, and the same range of locutions apply to it as to SG~R. Once more, the distinction between the simple γιγνόμενον which remains and the simple γιγνόμενον which does not remain is key for explaining which locutions are applicable, thus showing the fecundity of making the distinction between the SGR and the SG~R, and giving us a reason for thinking that we must distinguish the role of the SGR and the SG~R: in other words, that the ὑποκείμενον really is two in form. Once more we see Aristotle focusing on the ἐκ-locution as a way of bringing to light the items playing the role of a *principle*. Notably, when we are referring to the complex items in a change, we can discern that the complex γιγνόμενον is

[23] There is one complication though. I have assumed that Aristotle would regard it as kosher to say that bronze becomes a statue, as he implied at 190a5–6. In this, I am joined by Code (1976) and Burnyeat (2003): 'Sometimes [Aristotle] will change the subject and speak of matter coming to be a human being' (*Apology*, 19). However, it is actually very hard to find places where Aristotle does actually *assert* something of the form '[Matter] becomes [composite]'. The closest case I know is *Physics* III 1, 201a29–30, which uses the verb 'to be' not the verb 'to become': 'bronze is potentially a statue' (ἔστι γὰρ ὁ χαλκὸς δυνάμει ἀνδρίας).

something *from which* the complex ὃ γίγνεται comes to be, and hence that the complex γιγνόμενον is a principle for the complex ὃ γίγνεται, as is made clear in the second half of the chapter when Aristotle at 190b29–30 says that the principles of natural substances are in a way two, counting as one of those two principles the complex ὑποκείμενον (analysed in 190b24–7 as consisting of matter and privation), with form as the other principle.

> [190a31] Since coming to be is spoken of in many ways, and in some of them it's not that this comes to be, but rather this comes to be something, whereas coming to be unqualifiedly applies only to substances, it is clear on the basis of the other cases that it is necessary for something to underlie, namely that which does the coming to be (for a thing comes to be quantified, or qualified, or [190a35] related to something, or at some point in time, or somewhere, with something underlying, since only substance is said of no other underlying thing, [190b1] whereas all other things are said of the substance in question).

Aristotle turns now to explaining why it is that in *all* cases of change there must be something underlying. The explanation comes in two parts: 190a31–b1, which explains why it must be so for non-substantial change (i.e. change where a substance is the persisting thing), and 190b1–10, which explains why it must be so for substantial change (i.e. a change where a substance comes into being).

In this passage, then, Aristotle points out that where there is a change whose outcome is a substance quantified in a certain way (where at the outset it lacked that quantity), then there must be something *underlying* of which that quantity and its lack are predicated. Likewise for quality, relation, place. I assume Aristotle is here referring us to the doctrine of subjects (ὑποκείμενα) of predication in the *Categories*, in particular the idea that accidents must belong to a primary substance. It is therefore worth noting that Aristotle is appealing to a fact about *subjects of predication* in order to argue his point about what underlies in a change. How then does the argument work? And what does ὑποκείμενον actually mean at its various occurrences?

Here is one suggestion, focusing on the case of quality for the moment. For something to come to be qualified in some way or other, it is obvious that there must be something underlying that quality (i.e. the quality at the *outcome* of the change), because of the doctrine of subjecthood from the *Categories*. But we know that there must be a privation of that quality at the start of the change (because change involves contraries, as *Physics* I 5 argued), and so now we have the conceptual resources to find a complex ὑποκείμενον in the change, the thing which underlies *in the change*, namely the combination of the *predicational* subject and the privation of the relevant quality.

Why put the argument in this way? Well, when Aristotle says 'since only substance is said of no other underlying thing, whereas all other things are said of the substance in question' (190a35–b1), he is obviously referring to the doctrine in the *Categories*, and hence using the word ὑποκείμενον in the sense of predicational subject. But his argument cannot be 'there's a ὑποκείμενον for the quality (in the sense of predicational subject) *hence, without further ado*, there's a ὑποκείμενον (in the sense of ὑποκείμενον in a change) in cases of change of quality', since this argument, without adding in more premises, would just be an equivocation on the word ὑποκείμενον: 'there's a ὑποκείμενον in one sense, so there's a ὑποκείμενον in the other sense'. The missing step, I contend, is one which employs the idea that change is from contraries to contraries, so that the complete argument is this: in cases of change of quality, there must be an underlying subject of predication for the outcome quality, and *hence for the privation of that quality at the start*, and so the combination of this subject and the privation is the ὑποκείμενον in a change. Understood this way, when Aristotle says 'it is necessary for something to underlie, namely that which does the coming to be' (190a33–4) he is referring to what underlies *in a change*, as you would expect, given the identification of it with the γιγνόμενον, as was asserted back at 190a14–15. But when he gives, as a reason for this, that 'a thing comes to be quantified, or qualified, or related to something, or at some point in time, or somewhere, with something underlying, since only substance is said of no other underlying thing, whereas all other things are said of the substance in question' (190a34–b1), his use of the word ὑποκείμενον is in the sense of the subject of predication, as you would expect, given the obvious recall of the doctrine of the *Categories*.

This interpretation relies on the assumption that the notion of the ὑποκείμενον in a change is different from the notion of a ὑποκείμενον for predication. This is a substantive assumption on my part. For it is a natural thought that Aristotle in *Physics* I 7, when trying to identify the ὑποκείμενον in a change, is simply trying to identify *the subject for predicates of change*. On that understanding, Aristotle argues as follows in this section: the subject for the predicates 'musical' and 'unmusical' is the man, *hence* the subject of the predicate 'changes from unmusical to musical' is the man (the persisting thing, i.e. the SGR). My motivation in this essay for resisting that move is Aristotle's clear identification of the ὑποκείμενον in a change as a complex item, and it seems to me unlikely that Aristotle thought that the man (in the case under discussion) is one in number, but two in being. The thing which is one in number yet two in being is *the unmusical man*, an item whose being is complex, and whose being neatly reveals its aptness as something, as we might put it, *subject to change*. So in my interpretation of the current passage, I am trying to show a different way in which Aristotle might leverage his notion of a subject (ὑποκείμενον) of predication, to help show what the ὑποκείμενον in a change is.

Now, the list of respects in which there is non-substantial change, namely the respect in which something is 'quantified, or qualified, or [190a35] related to something, or at some point in time, or somewhere' is strange in two ways. First, there is the inclusion of 'at some point in time' (ποτέ), and second, there is the inclusion of 'related to something' (πρὸς ἕτερον). To see why the inclusion of these two predicates is problematic, consider the following natural thought: *if category R is a genuine respect in which there is change, then whenever two (incompatible) predicates in R hold of one thing at different times, there must have been a change.* This holds for the unproblematic cases we have here: quantity, quality, place. If two different qualities hold of something at different times, then it must have changed between them, etc. However, this natural thought about respects of change does not hold for predicates of 'when' and 'in relation to something else'.

(1) 'When'. We are supposed to envisage something coming to be at some time or other, where 'being at some time or other' is supposed to indicate a new feature that the thing takes on in a change. So we need to find temporal predicates such that something could *come to have* that predicate true of it, where previously that predicate was false of it. So we need to find something which comes to be *on Thursday*, having previously had the feature (say) that it was *on Wednesday*. One option for a subject which could have those predicates at different times would be an event like a meeting: our meeting was to be on Wednesday but then was moved to Thursday.[24] But it seems highly unlikely that Aristotle would consider things such as *meetings* as subject to change; at any rate, I can think of no other text which suggests this. Rather, for ποτέ to indicate a respect of change, we must envisage that in such changes, a *substance* has the property of *being at one time* and then comes to have the property of *being at another*. Perhaps age-predicates could be one such group of predicates. Theaetetus *was* 18 years old on Monday and then later that week turns 19 years old. The problem is that such a description need not necessarily describe a change! Imagine that I have a small statue of Aristotle on my desk. It is now 5 years old (measuring from the date of its creation), and next year it will be 6 years old. Will it *thereby* undergo a change? Intuitively, it is possible for the statue to age without changing. In general, cases where a substance is said to be at time t1, and then to be at t2, are just not *necessarily* accompanied by change. All we have done is attribute *persistence* to it. No doubt this difficulty is why Aristotle nowhere else includes ποτέ as one of the respects in which something can change, and so it is understandable why Ross wants to excise it (*ad loc.*). If we do that, we need to be able to tell a story as to how it came to be included in the text, and such a story is not hard to tell: the list of predicational categories lulled the scribe into familiar patterns of thought and he unthinkingly wrote down the category of when

[24] I'm grateful to Hendrik Lorenz for suggesting such an example.

alongside the category of where (as Ross (1936) says: 'These words, if genuine, must be due to inadvertence', 492).

(2) 'In relation to something else' (πρὸς ἕτερον) has aroused less attention, but it still poses problems. For just as Aristotle's canonical list of the respects of non-substantial change nowhere else includes ποτέ, so it nowhere else includes πρὸς ἕτερον. See for instance *Physics* III 1, 200b33–4; VII 2, 243a35–7; VIII 7, 260a26–8. And this seems to be with good reason (as with ποτέ). For if Socrates is taller than Theaetetus and then at a later date smaller than Theaetetus, Socrates need not have undergone a change of any kind. (These are the notorious cases of mere Cambridge change.) Thus, the inclusion of πρὸς ἕτερον is problematic for the exact same two reasons as the inclusion of ποτέ: (i) neither ποτέ nor πρὸς ἕτερον feature in any other list of the respects of change, and (ii) in neither case is it true that *if* a predicate of that category is true of X at t2 where it was false of it at t1, *then* X changed.

Of course, *some* cases of being qualified relatively in one way at one time and then in another at another will be cases of genuine change, e.g. if Theaetetus is shorter than Socrates at one time and then taller at another (Theaetetus grew). And some cases of being qualified temporally in one way at one time and then in another at another will at least also be *accompanied by* cases of genuine change, e.g. if Socrates is (i.e. exists) on Monday and is (i.e. exists) on Wednesday: Socrates has engaged in all sorts of bodily changes, breathing, digesting, etc., and indeed we do talk of people *getting older*, which sounds like a change. This may be enough to let both categories, ποτέ and πρὸς ἕτερον, stand here. The thought would be that Aristotle is not committing himself to saying that difference in predicate in those categories *always* amounts to a change, but is rather being loose when specifying respects in which something can change in order to make the point that changes in respect of these categories, when they occur, demand recognition from the theorist of a subject for the relevant predicates, thus revealing the presence of a complex ὑποκείμενον. But then it seems to me that the two categories stand or fall together in this list; if you are in the business of excising, you should excise both.

> [190b1] But that substances too, i.e. all [other] things which have being without qualification, come to be from some underlying thing, will be clear to someone who examines things. For there is always something which underlies, from which that which comes to be comes to be, such as plants and animals from seed. [190b5] The things which come to be simply, come to be, (i) some of them by reshaping, such as a statue, (ii) some by addition, such as things which grow, (iii) some by subtraction, such as the Hermes from the stone, (iv) some by composition, such as a house, (v) some by alteration, such as things which turn in respect of their matter. All the things which come to be in this way [190b10] clearly come to be from underlying things.

As mentioned above, here Aristotle argues that the thesis that there is always something underlying in a change even applies to substantial change, even though it is not as easy to identify the ὑποκείμενον in those cases, since there is not an easy subject of predication we can identify and then link up with the relevant privation to form the complex ὑποκείμενον. (Indeed, there is not even an obvious candidate for what the privation is.) By looking at things in Aristotle's way, we are supposed to *come to see* that there must be such a thing: it will *become clear* to us.

Aristotle, in other words, is inviting us to extend our results from the previous discussion of non-substantial change (such as when a person becomes musical from being unmusical) to substantial change. The identification of a complex ὑποκείμενον in substantial change is going to be a very big step for him in overcoming the problem of change with which he is grappling in *Physics* I as a whole. For it will enable him to find an object in the world which is subject for such changes, namely things like seeds and the other items he will mention here.

In these lines, Aristotle's first explicit candidate for such a ὑποκείμενον is a seed, out of which animals and plants grow. During his list of five ways in which things might come to be, he also names a second example, the stone (out of which the Hermes is chiselled). His reasoning is that we can identify these as being the things *from which* the animals, plants, and herm come to be. Since they are there at the beginning of the change, and are the things which undergo the change in question, which results in the existence of a substance which was not there before, they are natural candidates for being the *subject* in the change, in the sense that the chapter has been attempting to analyse.

The first example of the ὑποκείμενον, the seed, is perfectly apt, given the thesis of this chapter that the ὑποκείμενον in a change is a complex (συγκείμενον, 190a3; a20) or compound (σύνθετον, 190b11) item, one in number but two in form (190a15–16; b24), in the sense of being an entity composed out of the SGR (matter, 190b25) and the SG~R (privation, 190b27). A seed, plausibly, should be analysed as matter characterized by a privation, the relevant privation being the absence of the form of whatever animal or plant the seed will grow into. Indeed, this is what a seed *is*, something which *will* grow into that animal or plant, all being well. The seed itself does not survive the change into the relevant animal or plant, but that is no surprise since we have been told already that the γιγνόμενον which is συγκείμενον—the one identified as the ὑποκείμενον—does not persist (190a20–1).

The privation of the relevant form is part of the being or definition or account of the seed, as you would expect given that it is being identified as a ὑποκείμενον in a change, which (as we have been told) has privation as one of its components. It is important to remember that when Aristotle talks about privations, he means more than just the lack of some form or other. Timothy Clarke has it just right when he describes a privation as 'the absence or non-existence of a form, *in the sort of thing which is capable of receiving that form*' (2015, 142, my emphasis).[25] The key thing to observe here is that the privation which is part of the being of the seed (lack of being an oak, or whatever) is more than just a mere absence of being an oak. I am not an oak tree, either, but I do not have as one of my predicates the *privation* of oak. The acorn is some matter which is *capable* of receiving the form of an oak, but is currently deprived of that form. The matter will take on that form when the acorn sprouts into a sapling.

In short, this is why the seed is such an important part of the ontology of natural science. It is a very special sort of item. It is a *ὑποκείμενον* for a change: the sort of thing which stands on its own as an item in the world, but which is ready or ripe to change into a substance. It is a thing whose being renders change intelligible, because it is an item with the requisite metaphysical complexity: it has matter in its definition (matter which will persist through the change, thereby giving the change its requisite unity) and it has privation in its definition (this privation will be lost in the course of the change, thereby ensuring that there is a difference between the beginning state and the end state of the change). It is a perfect candidate as Aristotle's first example of a *ὑποκείμενον* for a natural change, because it is a familiar item in our ontology, and one of which a student of nature must obviously take full account.

Moreover, it is a perfect candidate for being the sort of item we need to identify to overcome the Eleatic puzzle of change, discussed at the beginning of *Physics* I 8. As Clarke (2015) reconstructs it, that puzzle concerns the difficulty of identifying what is present at the start of a substantial change: 'Did the generated thing come to be from what it—that is to say, from its previously being in a state of existence? Or did it come to be from what is not—that is to say, from its previously being non-existent?' (136). The first option is impossible because if the generated thing was there to begin with, it can't come into being, and the second option is impossible 'because there must be something that underlies' (*ὑποκεῖσθαι γάρ τι δεῖν*, I 8, 191a31). This form of words deliberately echoes Aristotle's claim in I 7, 190a14–15, that if one examines change closely enough, one will see that *δεῖ τι ἀεὶ ὑποκεῖσθαι τὸ γιγνόμενον*. That point in I 7 is where Aristotle identifies the *ὑποκείμενον* as the complex of that which persists and that which does not persist, i.e. (for substantial coming to be) the complex of matter and privation. So Aristotle's point in I 8 seems to be that his solution to the Eleatic puzzle is the successful identification of an object in the world which is suited to be the initial object at the start of a substantial change.

[25] For a good text which makes this point about a different privation, see *Physics* VI 3, 234a32–3: 'we say that something is at rest when it is of such a nature as to be in motion but isn't in motion' (*ἠρεμεῖν γὰρ λέγομεν τὸ πεφυκὸς κινεῖσθαι μὴ κινούμενον*).

The aptness of the example of the seed is completely lost on interpreters who insist that the ὑποκείμενον in a change is the *matter* which persists during the change. Those who Interpret Aristotle in that vein have worked hard to make sense of Aristotle's example, an example which, remember, was supposed to be especially clear, since it is the very first example Aristotle gives of how there *must* be a ὑποκείμενον even in the case of substantial change. The problem comes about because interpreters are committed to the following view about what a ὑποκείμενον is: 'For any change, there is a subject (ὑποκείμενον) of that change, that subject persists, that subject remains the same kind of thing, and that subject along with some form makes up the product of that change' (O'Connor 2015, 7; cf. 9). The seed does not meet those conditions for subjecthood, since the seed does not persist and remain the same kind of thing during the change, nor does it make up the product of that change along with form (because it is not present in the final product of the change). The most common solution in the literature (see O'Connor 2015, 14 n. 20 for references) is to accept that the ὑποκείμενον of the change is the seed (at the beginning of the change), but that the description 'the seed' does not remain true of it during the change, even though the very thing which the description 'the seed' picked out is present at the end. This is just to invoke strategy (e) from above in this instance. In other words, proponents of this interpretation think that Aristotle treats 'the seed' in the same way as he treats—according to them—'the unmusical'. The thing which is unmusical, on that interpretation, is the man, and the *man* does persist (and is the subject of the change). 'Unmusical' may be a true description of the man at the start of the change, but it is not the privileged description under which the subjecthood of the man is made manifest. Similarly, the thing which is seed is some matter, and that *matter* does persist (and is the subject of the change). 'Seed' may be a true description of that stuff at the start of the change, but it is not the privileged description under which the subjecthood of the matter is made manifest.[26]

Thus, interpreters who adopt that strategy think that the very first example Aristotle gives of a ὑποκείμενον in a substantial change is somehow defective, because he has not actually pinpointed with an accurate description what the subject of the change really is. Although one must applaud this approach for its ingenuity, it does seem legitimate to wonder whether an interpretation which instead highlights the aptness of the example might be worth pursuing. This essay has been pursuing just such an interpretation, by insisting that the ὑποκείμενον in a change *does not* persist in the change, but instead has a dual nature (as Aristotle outright states so often in the chapter), one part of which persists, and the other

[26] Waterlow 1982, 47 suggests that 'Aristotle is used to ordinary vocabulary running dry when it comes to providing names for all the facts that his analyses enable him to distinguish'.

doesn't. Of course, we don't know what this persisting part of the σπέρμα is. That will depend on which kind of σπέρμα we are dealing with, of course, plant or animal. But Scott O'Connor, in his recent insightful paper, convincingly points out that 'some uterine blood, must, according to Aristotle, persist through the generation of the animal' (blood from the uterus of the mother, that is).[27] If this is right, then uterine blood looks to be a good candidate for the persisting matter of the σπέρμα, where the privation which is part of the being or definition of the σπέρμα is obviously the lack of whatever the relevant animal form is. The key point is that if the ὑποκείμενον is indeed a complex entity, matter plus privation, then σπέρμα is an ideal example of it.[28]

As for Aristotle's other, explicit, example here of a ὑποκείμενον, the stone, we should notice that the expression 'the stone' at 190b7 is ambiguous: it could be a reference to the particular lump of stone that the sculptor has before him at the outset of the process of bringing the herm into being, or it could be a bare reference to stone, the stuff or matter out of which the herm is made. To bring the example into line with the seed, however, it seems to me best to interpret it here as a reference to the lump that the sculptor has in front of him (below I discuss the reference at 190b16 to stone, where it seems Aristotle does mean to refer to the stuff, and not a particular lump). When the sculptor chooses such a lump of stone at the quarry, he is choosing a lump which is characterized by the fact that it lacks the form of statue, and *is suitable to be turned into* a statue (this obviously puts constraints on which lump he should choose: it should be firm, not cracked, soft enough to respond to chiselling, large enough, etc.). This is what statues come to be from: stones of *that* sort.

What is the point of the list of five ways in which substances can come to be? Why should having those processes enumerated help us in our discernment of the ὑποκείμενον in a substantial change? Aristotle presumably means to focus our attention on these cases because we will find it easier to pinpoint what the ὑποκείμενον in a change is if we are attuned to looking for whatever object is present at the start of one of these processes. Reshaping involves X being reshaped into Y; addition involves X having stuff added to it to become Y; subtraction involves X having stuff taken away from it to become Y; composition involves X being put together to become Y; alteration involves X changing its qualities to become Y.[29] In each case, X is the ὑποκείμενον.

[27] O'Connor 2015, 69.

[28] I am grateful to Scott O'Connor for help in these paragraphs. In his paper, he argues that σπέρμα *is* the uterine blood, which is a slightly different position from the one I take. In the framework of my chapter, σπέρμα would be uterine blood *plus* privation of animal form.

[29] The last example is a perennial puzzle for Aristotelian commentators, and I have nothing new to add to the discussion. I assume he is referring to wine turning to vinegar, or milk turning into butter or yoghurt. Why he should describe these cases using the term ὕλη is mysterious to me, as it is to Ross,

> [190b10] So it is clear from what has been said that that which does the coming to be is always composed, i.e. there is something which does the coming to be and something which this thing comes to be, and this thing is double, for it is either the underlying thing or the contrary. I mean that unmusical is opposed, and man underlies [in the case of non-substantial change], and [190b15] unshapedness and formlessness and disorder is the contrary whereas bronze or stone or gold underlies.

Aristotle wraps up the first section of the chapter by revisiting what he said about the γιγνόμενον. He had already told us (190a1–3) that there are two simple γιγνόμενα, the privation and the thing that remains. The complex combined out of them is the complex γιγνόμενον, and it is this which was identified with the ὑποκείμενον back at 190a13–16.[30] Aristotle here describes the two aspects of the complex γιγνόμενον as 'the underlying thing or the contrary', and in so doing, he appears to use the term 'underlying thing' to pick out the matter only. If the thesis of this essay is correct, then this comes as a shock. How can Aristotle now use the word ὑποκείμενον for the matter alone, given that up until now in the chapter he has used it to refer to the *complex* of matter and privation?[31] To answer this, we must ask what sort of thing the complex ὑποκείμενον is. In I 9,

given that this is the first occurrence of ὕλη in *Physics* I 7: 'Since Aristotle is carefully working up to his conception of ὕλη, the words κατὰ τὴν ὕλην are (as Mr. Hardie observes) an unfortunate anticipation' (Ross *ad loc.*).

[30] Aristotle's formulation of this point here has caused some trouble. He writes: ἔστι δέ τι γιγνόμενον, ἔστι δέ τι ὃ τοῦτο γίγνεται, καὶ τοῦτο διττόν. Earlier, Aristotle had used the phrase τὸ γιγνόμενον to refer to what does the coming to be, and the phrase ὃ τοῦτο γίγνεται to refer to the outcome of the change. Ross thinks that here, Aristotle reverses these terms. It's not hard to see why Ross thinks this. He thinks that the phrase καὶ τοῦτο διττόν must pick up the noun phrase τι ὃ τοῦτο γίγνεται. But then this noun phrase must refer to the start of the change, since it is the item at the beginning of the change, the complex ὑποκείμενον, which Aristotle thinks is double. So the phrase would have to mean 'something which becomes this', where 'this' picks up τι γιγνόμενον, which in turn would then have to mean 'that which comes to be' (as opposed to 'that which does the coming to be'). Ross says that these phrase are 'ambiguous in Greek' and he is right about that, but he is wrong that here we *have* to translate in the way he suggested. One could translate in the way I did: 'for there is something which does the coming to be and something which this thing comes to be, and this thing is double, for it is either the underlying thing or the contrary'. Here, 'this thing is double' refers back to the γιγνόμενον, which is also the reference of the first occurrence of the phrase 'this thing'. So the word τοῦτο in καὶ τοῦτο διττόν does not, *pace* Ross, pick up the whole noun phrase τι ὃ τοῦτο γίγνεται, but rather refers to the same thing as the τοῦτο does in τι ὃ τοῦτο γίγνεται. (I'm grateful to Adam Crager for pointing this out to me.)

[31] Code (1976) seems to think that Aristotle sometimes calls the privation alone the ὑποκείμενον: 'both the man and the unmusical are the substratum of the change (190a13–7; cf. 190b12–19)' (358). The second reference here seems to misfire completely, but on the next page, Code elucidates how he understands 190a13–17: 'Since the substratum of the change is to be identified with that which comes to be the product, both the matter and the lack are the substratum, and so the substratum is one in number, but two in form' (359). Code seems to interpret Aristotle as saying that *each* of the persisting thing and the lack is a ὑποκείμενον. But in the earlier part of this chapter, I gave reasons for thinking that at 190a15, Aristotle is referring to the *complex* of the persisting thing and the lack (the complex γιγνόμενον) as the ὑποκείμενον.

Aristotle helps us understand the nature of the two items composing this complex: 'we hold that matter and privation are different, and that one of them—namely, the matter—is non-being κατὰ συμβεβηκός, whereas the privation is non-being καθ' αὑτήν; also that one of them—namely, the matter—is nearly substance in a way, whereas the other is not so in any way' (following Broadie's translation in her contribution to this volume) (192a3–6). Presumably, privation is non-being in itself in virtue of containing negation in its very definition.[32] Given its lowly status, and matter's comparably lofty status (it's nearly substance!), Aristotle here chooses to refer to the complex of matter plus something which is a *per se* non-being as, simply, the matter. After all, when we call a man unmusical, we are saying what he isn't, rather than what he is. Or when we call bronze unshaped (cf. 190b15), we are saying what it isn't, rather than what it is. So in a way we add nothing to our description of the matter when we say which privation it has. In another way, of course, we *do* add something to our description when we attribute this privation to the matter: we describe it in such a way that we manifest clearly the role it is going to play in a change, namely the change where the matter takes on the form of which the privation is a lack. (And of course, our attribution of the privative term 'unmusical' to a man is supposed to capture the fact that it is the sort of thing that could be musical. A number is not 'unmusical' in the way a beginner violin player might be.) So when we are being precise about what metaphysical item the ὑποκείμενον is, we should specify the complex item. But this complex item, matter plus privation, is in a way the same thing as the matter itself, since attributing privation to it in a way does not add any being to the matter (since privation is non-being *per se*).

What Aristotle is doing in these last lines is to expand the analysis of the dual aspect ὑποκείμενον from cases of non-substantial change to the case of substantial change. So he reprises his example of the unmusical man, analysing him into a contrary and the (simple) ὑποκείμενον, and then gives a list of three contraries relevant to cases of substantial change (unshapedness, formlessness, and disorder), and three (simple) ὑποκείμενα (the bronze, the stone, and the gold which feature in a substantial change). In offering these two lists of three items, Aristotle makes good on his promise to clarify for us how it is that the identification of a ὑποκείμενον which is *complex* is possible even in the case of substantial change.

With this, the first part of our chapter ends. Aristotle has undertaken a study of the principles of change (albeit without actually employing the vocabulary of principles in this part of the text) in order to discern the presence of, and characterize, a complex ὑποκείμενον of change, present in all cases of change, both substantial and non-substantial. This complex ὑποκείμενον consists of the

[32] Cf. Clarke 2015, 142, who says privation is 'a particular kind of non-being: the absence or non-existence of a form, in the sort of thing which is capable of receiving that form'.

simple *γιγνόμενον* which remains and the simple *γιγνόμενον* which does not remain. Aristotle is justly proud of this achievement, as his remarks in I 9 indicate. This elaboration of the nature of the complex *ὑποκείμενον* of change sets the scene for his subsequent identification of the principles, not merely of substantial change, but of the being of natural things whose trademark feature is that they *come into being*. The focus on substantial coming-to-be in the second part of the chapter helps to explain why Aristotle is so intent, in the first part of the chapter, on getting the analysis of substantial change right, even though the discernment of the various elements present in substantial change is not immediately obvious, as he points out at 190b1–3.

8

The Principles of Natural Things—Two or Three?

Physics I 7, PART 2

HENDRIK LORENZ

Introduction

The section builds on the results of the first half of I 7 (189b30–190b17). It takes over from the preceding discussion of change the concepts of that which underlies, or the substratum, in any given change, and of the opposites between which any given change takes place, in such a way that something that undergoes the change passes from being characterized by the one opposite to being characterized by the other. The section is emphatically not a mere summary or recapitulation of the first half of the chapter, which introduced and explained these concepts. Rather, it presents a number of new and important thoughts.

The following are especially prominent among these. In the present section, Aristotle connects the concepts of what underlies and of the opposites with the question of what the principles of natural things are. He articulates the concept of the opposites (ἀντικείμενα) or contraries (ἐναντία) involved in any given

This is a revised version of a paper I read at the 20th Symposium Aristotelicum in Delphi. It greatly benefited from discussion at and around the Symposium, and from comments I received subsequently. I would like especially to thank Andreas Anagnostopoulos, Timothy Clarke, Alan Code, Sean Kelsey, André Laks, Benjamin Morison, and Oliver Primavesi. I also would like to thank Daniel Kranzelbinder and Samuel Meister for critical comments on an earlier draft that helped me in clarifying some of my views on *Physics* I 7.

change in terms of the distinction between form and privation. He privileges form over privation, treating only form and matter as truly indispensable principles of natural substances. Furthermore, Aristotle focuses in on the principles of the being and coming-to-be of natural substances, thus in effect focusing in on generation, the coming-into-being of substances, at the expense of the non-substantial forms of change such as alteration and locomotion. Lastly, the present section introduces the idea that form is not only a principle of natural substances once they have come to be, but also a principle from which they come to be.

Having introduced these new thoughts, Aristotle employs them in stating his own view of what the principles of natural things are; and he connects the resulting view with the earlier discussion in chapters 5 and 6 of *Physics* I, noting especially that he has now clarified the difference between the contraries of form and privation, the relationships that obtain among the principles, and the nature of what underlies (191a17–19).

Before beginning my detailed discussion of the section, I want to comment on my procedure in what follows of occasionally bringing in texts from the *Metaphysics* to explicate ideas from my section. My general view is that issues about the metaphysical status of the principles here introduced are very much in Aristotle's mind in composing this text. In fact they sometimes come to the surface, as when he notes that it is not yet clear whether form or the substratum are cases of substance (191a19–20), or claims that matter is more a case of some *this* (τόδε τι) than privation is (190b25–6), or that privation is in no way substance (chapter 9, 192a3–6). But it also seems clear that these issues are not fully entered into in *Physics* I or for that matter elsewhere in the *Physics*. However, Aristotle in *Physics* I does make some tentative, preliminary moves towards resolving at least some of these issues, and it does seem to me that the moves he makes point in the direction of the general kind of metaphysical picture presented and argued for in the *Metaphysics*, especially in *Metaphysics* Λ. And so it might be the case that calling attention to some relevant passages in the *Metaphysics* occasionally sheds light on the philosophical significance of ideas that are presented but not very fully developed in my section. It is in this spirit that I will occasionally refer to passages and ideas from Aristotle's *Metaphysics*.

1. The Principles and Causes of Natural Things

It is clear then, if there are causes and principles of things that are by nature, from which as first things they are and have come to be not incidentally but what each thing is said to be in substance, that everything comes to be from both that which

> underlies and the shape. For the educated person is in some way composed from person and being educated. For you will break down the accounts into the accounts of those things. It is clear then that the things that come to be would come to be from these. And that which underlies is one in number, but two in form (for the person, the gold and, in general, the matter is countable;[1] for it is more some this, and what comes to be comes to be from it not incidentally; by contrast, the privation, i.e. the contrariety, is an attribute); on the other hand, the form is one, e.g. the order, education, or any of the other things that are predicated in this way. (190b17–29)

It is supposed to be clear from what has preceded (οὖν, 190b17) that everything—meaning, I take it, every natural thing—comes to be from that which underlies and shape or form. The antecedent of the conditional at II 190b17–19 restricts the claim being made to things that are by nature. At the beginning of *Physics* II 1, Aristotle will give examples of such things: animals, their parts, plants, and the simple bodies, earth, fire, air, and water (192b8–12), adding later that the things that have natures are substances (192b32–4).

There are two other restrictions that we should note: the antecedent of the conditional limits the claim being made to being and coming to be what each thing is in its substance, and so the claim being made seems to apply only to substances, such as animals and plants.[2] Furthermore, the causes and principles that are under consideration are first or primary things from which natural things *are* and come to be. In the preceding discussion earlier on in chapter 7, Aristotle talks often about things coming to be from other things. The things from which X comes to be are the starting points, or aspects of the starting point, from which the change that terminates in X begins: the aspects so far introduced are that which remains and that which is replaced. One idea that is new, then, is that the causes and principles that are at issue are also things from which the natural thing in question *is*. What are we to make of this idea? In chapter 5, Aristotle speaks of one thing being from another, saying, for instance, that things that are intermediate between contraries are from the relevant contraries, in the way the colours other than white and black are from those contraries (188b23–6; cf. 188a27–30, where the idea is that it is not the case that contraries are from one another). So, to say that X is from A and B would seem to mean, in the context, that X is composed or constituted by A and B.[3] If so,

[1] This translation was suggested to me by André Laks. Alternatively: 'for there is the person, the gold and, in general, the matter that is countable'.

[2] In light of the distinction between substance and the non-substantial entities that Aristotle has just made at 190a31–b3, it seems difficult to construe οὐσίαν in 190b19 as meaning anything other than 'substance'.

[3] Sylvain Delcomminette, in his chapter on *Physics* I 5 (this volume, pp. 163–4), notes helpfully that there is good reason not to think that the things from which X *is* are, on Aristotle's view, limited to *material* causes of X. That he is right about this is also clear from the second half of I 7, given that Aristotle here treats both the form and the matter of X as things from which X is. However, Delcomminette's proposed

the causes and principles of natural things that Aristotle is now meaning to bring to light are first or primary things from which natural things come to be and which also are constituents of those same natural things.[4]

What Aristotle is claiming, then, is that the preceding discussion makes clear that natural substances like animals and plants come to be from and are constituted by two causes and principles, namely that which underlies and form. He immediately illustrates this by saying that an educated human is in a way composed of a human being and educatedness: 'for you will break down the accounts[5] (sc. of the composites) into the accounts of those things (sc. the constituents)' (190b22–3). It may seem awkward that Aristotle is apparently illustrating a claim about the generation and constitution of substances by reference to a human being who has become educated, which is the result of a non-substantial change, an alteration. However, we should recall how it is that in the first half of chapter 7, he illustrates the general claim that everything that undergoes coming-to-be is always a composite, being composed of that which underlies and that which is opposed. This is a general claim that is meant to apply to all cases of coming-to-be, prominently including the generation of substances. It is illustrated by reference to an uneducated human being, where uneducatedness is the thing that is opposed and a human being is the thing that underlies (190b10–17). This composite of course undergoes an alteration, not a substantial change, when the person in question comes to be educated. So there is a precedent in the first half of the chapter for using a case of alteration to illustrate cases of substantial change.

Aristotle is now applying the model of that which underlies and that which is opposed to the specific case that is primarily relevant to the *Physics*, natural things. He is also switching his attention from the starting point of change, which is the substratum characterized by absence of form or order, to the terminus of change, which is the substratum characterized by form (note that at 190b20–2 he is analysing the in-formed substratum into a substratum-element and a form-element, illustrating this by reference to being human and being educated). Moreover, he is introducing a notion of cause and principle which is at once a notion of what something comes to be from and a notion of what

translation of the phrase εἶναι ἐκ + genitive as 'to derive from' is too general, since it fails to capture the idea that the items denoted by the genitive expressions are conceived of as constituents of whatever is said to be from them. Things can, I think, be said to derive from their moving causes and perhaps from their final causes as well, but neither moving causes nor final causes are things from which the effects in question are. If my translation 'to be from' seems insufficiently smooth English, how about 'to be constituted from'?

[4] Compare the 'elements' of *Metaphysics* Λ 4, 1070b22–7.

[5] Reading τοὺς λόγους with mss. FIJ, against Diels' deletion, accepted by Ross. E has τοὺς ὅρους. For the address to the reader in the future tense, cf. *Categories* 2a24.

something is constituted by. This notion rules out absence of form, or privation, as a cause and principle of natural things, because privation is not a cause and principle that is a constituent of the natural substance that has come to be. Only substratum and form are.

The passage introduces the idea that form is something from which natural things not only are but also come to be (190b19–20). That Aristotle does mean to say this in our section will become even clearer as we move on (190b23). In the first half of the chapter, in saying that X is something from which Z comes to be, Aristotle has in mind that X is a starting point from which the change by which Z comes to be begins. Does he now mean to apply this idea to the form in question, in addition to applying it, of course, to the substratum? It seems that way, and I see no good reason to reject the impression. The idea is, then, that the form in question is not only a constituent of the relevant object or substance once it has come into being, but also something that serves as one of the starting points of its coming into being, as an element that has a role to play in making this change intelligible as the kind of change it is. This of course raises the question how the form of X can act as a starting point in the coming into being of X. Before a given statue or a given tree has come into being, the relevant substratum is of course characterized not by the form but by the appropriate kind of privation. So how can the relevant form be a starting point from which X comes to be?

Nothing that Aristotle says in our chapter, I think, answers this question or even points clearly in the direction of an answer. Perhaps Aristotle means this issue to remain somewhat indeterminate for the time being. He might mean to rely on an implicit principle to the effect that if A and B are constituents of X and explain what X is, they will also in some ways contribute to and explain X's coming into being.[6] However, it seems to me that against the background of chapters 5 and 6 a more determinate interpretation is rather plausible.

The first thing to note is that if we allow ourselves the notion of Aristotle's moving cause, we can answer the question how the form of X serves as something from which X comes to be.[7] In *Physics* II 7, Aristotle explains how it is that the form or essence of a given natural substance also in a way is its moving cause, in that the form and the moving cause are the same in kind, as when a human fathers another human (198a24–7), or an artisan or, rather, the art in question generates an artefact (*Physics* II 3, 195b21–5, together with the idea that the art in question is the relevant form present in the soul of the artisan, in

[6] I owe this suggestion to Caleb Cohoe.

[7] For the moving cause as something from (ἐκ) which something or other comes to be, cf. *Generation of Animals* I 18, 724a28ff.

Metaphysics Z 7, 1032a25–1032b14; cf. Λ 4, 1070b30–5). The relevant form, e.g. that of a statue or a tree, preexists the generation of the new substance, Aristotle thinks, in that it is present in the moving cause, the creator or parent of the substance that comes into being. In fact, it turns out that it is the form in question that is most of all the moving cause of the new substance.

Now, on the one hand the moving cause is almost completely absent from the entirety of chapter 7 (meagre exceptions being 190b33 and, I will argue, 191a6–7). But on the other hand, ideas about what will be presented as the moving cause are present in chapter 5 and, much more prominently, in chapter 6. In chapter 5, there is a hint that contraries are not only relevant to questions about what comes to be from what, but also to questions about what acts on or is affected by what (188a31–6). In chapter 6, the notion of something that underlies the contraries is motivated in important part on the grounds that contraries do not act on one another (189a22–6), whereas a trio of two contraries and something that underlies is sufficient for making intelligible the notion of one thing being affected by another (189b18–19). On the basis of these ideas presented in the immediately preceding chapters, it is not an enormously difficult task for thoughtful readers of I 7 to work out for themselves that when, for example, an uneducated person comes to be an educated person, this is a process in the course of which the uneducated person is affected, and improved, by an educated person. In this way, the relevant form, that of being educated, is present at the outset of the change and plays an important role in making the change intelligible. So the idea of the relevant form as something from which the object in question comes to be is not, in the context, completely mysterious. Nevertheless, it is introduced in the chapter without explanation and with remarkably little emphasis. Its introduction in the chapter, I suggest, is proleptic: it will be explicated later on in the *Physics*, when Aristotle discusses the moving cause and clarifies its relationship to the formal cause.

Having briefly reflected on the terminus of change, the substratum characterized by form, Aristotle then returns to the things from which things that come to be do so (190b23). They come to be from *these* things, he says, meaning no doubt the substratum on the one hand and form on the other. However, the shift from considering the terminus of change to considering the starting point of change brings with it a shift in how the substratum is conceived of. This is because at the outset of the change, the substratum is characterized by privation. Accordingly, Aristotle now analyses the substratum, conceived of specifically as something from which natural things come to be, into two components or aspects, namely into (what he now calls) a matter-component and a privation-component, saying that the substratum is one in number but two in form. He illustrates the matter-component by referring to the familiar examples of a human being and of some portion of gold (ὁ χρυσός, 190b25; cf. 190b16) and then offers the general formulation 'the matter' for this aspect of the substratum. He also claims that the matter—that is, the matter in question—is countable.

Why is he saying that the matter in question is countable? Presumably he means to justify or explain his doing so by noting that the matter is to some extent a case of some *this*, of a τόδε τι. It comes at least somewhat close to being a determinate, reidentifiable something or other.[8] This makes good sense, if Aristotle conceives of the matter in question as the proximate matter for the relevant object: the portion of gold needed for the production of a gold coin, say, or the amount of bricks and timber needed for building a house. A sculptor who wants to make three herms might obtain three lumps of rock of the right dimensions for turning them into herms. In fact, Aristotle presumably takes the view that it is the proximate matter of X from which X comes to be non-incidentally. By contrast, herms come to be from remoter kinds of matter—say, earth, air, fire, and water—only incidentally. And while the proximate matter of an object is countable, the stuffs that constitute the remoter kinds of matter need not be. So Aristotle's train of thought seems to be this: the matter in question is countable, because (i) the proximate matter of a given substance is in a way a determinate, reidentifiable something or other, and is countable by reference to the substance for which it is the proximate matter, and because (ii) we are in fact dealing with the proximate matter, since we are dealing with the matter from which the thing in question comes to be non-incidentally. In saying that the matter is 'more a case of some *this*' (τόδε ... τι μᾶλλον) Aristotle might mean to compare the proximate matter to some non-proximate matter. Perhaps more probably, he means to compare it to the privation-component of the substratum.

The countability of the matter-component of the substratum is important, I suggest, because the substratum is meant to be, in each case, numerically one thing. The unity of the substratum can hardly derive from its lack of form, so it should derive in some way from the matter-component. It makes sense, then, for Aristotle to point out that the matter-component of the substratum is countable and somewhat substance-like or object-like. We might add that it has unity and substance-likeness only derivatively, in relation to the substance or object that it is the proximate matter for. By contrast with the countability and substance-likeness of the substratum's matter-component, its privation-component, Aristotle points out, is an attribute.[9]

[8] I am presupposing an interpretation of what it is to be a τόδε τι along the lines of that offered by Beere 2009, 278–81.

[9] Note the similar contrast in chapter 9, where Aristotle says that matter (we say) is nearly substance, whereas privation is in no way substance (192a5–6).

This, I think, is the force of saying about the privation and contrariety[10] that it is a *συμβεβηκός*. The point is not, I think, that whereas what comes to be does so non-incidentally from the matter-component of the substratum, it is only incidentally (*κατὰ συμβεβηκός*) that it comes to be from the privation-component.[11] First of all, Aristotle is not here saying, on any tolerably literal construal of his language, that what comes to be does so only incidentally from privation and contrariety. Furthermore, Aristotle earlier on made it clear that it is perfectly appropriate to specify the relevant contrary as the starting point of a given change, as when you say, for instance, that someone becomes educated from being uneducated (190a21–3). So it would be perplexing if we were now told, without any explanation, that it is only incidentally that the thing in question comes to be from the privation-component of the substratum.[12] By contrast, it makes good sense for Aristotle to call attention to the fact that the privation-component is an attribute (that is to say, a mere attribute). The point is that it adds complexity in form, but that it depends metaphysically on the matter-component of the substratum, which, being itself countable and somewhat substance-like, accounts for the numerical unity of the substratum.

Aristotle adds that the form, the other thing from which things that come to be do so, is also some one thing from which things come to be. His examples of form are order (meaning, I take it, some specific order, such as that characteristic of a statue) and being educated. He then adds the general formulation 'or any of the other things that are predicated in this way' (190b28–9). The reference to a certain way of being predicated in this formulation is somewhat unclear. I suggest that the idea is that of a positive characteristic, such that its acquisition is a case of coming into being (though not necessarily a case of generation in the technical sense as a substantial change), rather than a case of destruction or loss: e.g. knowledge rather than ignorance, virtue rather than vice, the components of a whole being composed properly rather than being in disarray, etc.[13]

[10] I take it that by *ἡ ἐναντίωσις* Aristotle means to denote the member of the pair of contraries that is contrary to the form. For similar uses of the noun, cf. Horstschäfer 1998, 345–7.

[11] *Contra* (for instance) Horstschäfer 1998, 331–2.

[12] Nor is it part of what Aristotle says in *Physics* I 9 that what comes to be does so only incidentally from the privation or contrariety in question (*contra*, among others, Anagnostopoulos 2013, 264–7, 273). The correct interpretation of the key passage in I 9 (191b13–17) is presented and explained in Clarke 2015.

[13] Aristotle operates with just such a notion of coming into being (*γένεσις*) at *Gen. et Corr.* I 3, 319a14–17, relying on the notion of a column of opposites (*συστοιχία*) which arranges positive terms such as being knowledgeable in opposition to the relative privative terms, featuring in a co-ordinate column: *Οὐ μὴν ἀλλ' ὁμοίως ἐν πᾶσι γένεσις μὲν κατὰ τὰ ἐν τῇ ἑτέρᾳ συστοιχίᾳ λέγεται, οἷον ἐν μὲν οὐσίᾳ ἐὰν πῦρ ἀλλ'οὐκ ἐὰν γῆ, ἐν δὲ τῷ ποιῷ ἐὰν ἐπιστῆμον ἀλλ' οὐχ ὅταν ἀνεπιστῆμον* ('nevertheless, in all changing things alike, we speak of coming to be when the thing comes to be something in one of

2. Counting the Principles

> Which is why there is a way in which one should say that the principles are two, and there is a way in which one should say that they are three. And there is a way in which one should say the principles are the contraries, as when someone were to speak of the educated and the uneducated or the hot and the cold or the fitted together and the not fitted together. And there is a way in which one should not say that the principles are the contraries. For it is impossible for the contraries to undergo change by one another. This too is resolved because of the fact that the underlying thing is another thing. For it is not contrary. So that the principles are in a way no more than the contraries, but they are, so to speak, two in number. However, they are not completely two but three, since a difference in being belongs to them. For what it is to be for a human being and for being uneducated are different, and what it is to be for being unshaped and for being bronze. (190b29–191a3)

Aristotle now refers to the items just identified as ἀρχαί, and notes that because of what has just been said (διό) one should in one way say that the ἀρχαί are two, and in another way, that they are three. What does he have in mind in calling the items in question ἀρχαί: is he meaning to call them starting points of change, or principles of natural things? They are of course starting points of the changes by which the relevant objects or substances come to be: they are things from which what comes to be does so. But there is good reason also to think of them as the principles of natural substances. As we have seen, Aristotle begins the second half of the chapter (that is, the subsection I just discussed) by introducing the notion of causes and principles of things that are by nature. And in what immediately follows the present subsection, he will summarize it by saying that he has stated 'how many are the principles of the natural things that are to do with coming into being, and how it is that they are of that number' (191a3–4). So it is clear, I think, that Aristotle means in our present section to derive from the preceding subsection his answer to the question introduced in chapter 1 and discussed in detail in chapters 5 and 6 as to what the principles of natural things are, and in particular what their number is. They are two in one way, he says, and three in another.

This flows nicely from the preceding subsection: count the substratum as one principle, so as to reflect its numerical unity, and form as the other, and you get two principles; count the substratum as two principles, so as to reflect its duality in form, and you get three principles of natural substance—matter, privation, and form. In this way, then, Aristotle proposes to resolve the *aporia* stated at the end of chapter 6—namely, whether the elements of natural things are two or three. Chapter 5 articulates reasons for thinking that they are two; chapter 6 offers

the two columns—e.g. in substance, if it comes to be fire but not if it comes to be earth; and in quality, if it comes to be learned but not when it comes to be ignorant').

reasons for thinking that they must be three. He now resolves this apparent conflict by calling attention to the fact that the substratum can be counted as one principle but can also be counted as two, given its metaphysical complexity.

He adds that there is also a way in which the principles are the contraries, and a way in which they are not.[14] He illustrates the idea that in a way the principles are the contraries by saying that one might offer 'the educated' and 'the uneducated' as principles (presumably of being an educated person), or the hot and the cold, or again that which is fitted together and that which isn't. In support of denying that the principles are the contraries, he offers the consideration that it is impossible for contraries to be affected by one another. This latter consideration is presented in chapter 6 as a reason against thinking that the principles are only two, namely contraries such as the dense and the rare, love and strife, and the like (189a21–6). What does Aristotle have in mind in now saying that in a way the principles of natural things are the contraries? The idea is, I suggest, that there is a way of thinking of 'the fitted together' (*τὸ ἡρμοσμένον*) and 'the not fitted together' (*τὸ ἀνάρμοστον*) so that these turn out to be form on the one hand and the matter-privation composite on the other. Just as 'the hot' can denote, on the one hand, the quality of heat and, on the other, some stuff that is hot (metaphysically speaking, a bearer-quality composite),[15] so 'the not fitted together' can denote the privation of order, but it can also denote a matter-privation composite. The trick, I suggest, is to construe the positive members of the pair of contraries that Aristotle mentions as denoting just the forms in question, and the negative members as denoting matter-privation composites. In fact, in chapter 5 Aristotle does conceive of the contrary from which an artefact like a house comes to be as a matter-privation composite, saying that a house comes to be 'from *these things* not being put together but being separated in this way': the contrary from which a house comes to be is bricks and timber being characterized by the privation of house-form.[16]

Alternatively, one might construe both members of the pairs of contraries as composites: the positive member (e.g. the educated) as a matter-form composite, the negative one (e.g. the uneducated) as a matter-privation composite. This would allow the contraries to act on and be affected by one another, as Aristotle seems in chapter 5 to present them as being able to do (note again 188a31–6, where the contraries are presented as being relevant not only to the question of what comes to be from what but also to the question of what acts on and is

[14] I take it that in II 190b30–2, Aristotle is syntactically relying on the preceding clause: *ἔστι μὲν ὡς δύο λεκτέον εἶναι τὰς ἀρχὰς, ἔστι δ' ὡς τρεῖς· καὶ ἔστι μὲν ὡς τἀναντία* (sc. *λεκτέον εἶναι τὰς ἀρχάς*) ...*ἔστι δ' ὡς οὔ* (sc. *τἀναντία λεκτέον εἶναι τὰς ἀρχάς*).

[15] As pointed out by Socrates at *Phaedo* 103 b–c, in a strikingly similar context that clearly is relevant historical background to the development of ideas in *Physics* I 5–7.

[16] 188b17–9: *ἥ τε γὰρ οἰκία γίγνεται ἐκ τοῦ μὴ συγκεῖσθαι ἀλλὰ διῃρῆσθαι ταδὶ ὡδί.*

affected by what). An educated person can make an uneducated person educated, something hot can heat up something cold, and so on. However, this alternative construal does not fit the text as well. This is because it introduces the idea that one of the principles of, say, a house is the matter-form composite of the appropriate materials characterized by the form of a house. But this isn't a principle of a house; it just is a house. Aristotle has introduced the notion of form as a principle from which what comes to be does so, but he has not introduced the notion of a matter-form composite as a principle of a matter-form composite. Furthermore, in what follows very shortly Aristotle says that the principles of natural things, if their differences in being are taken into account, are three, namely form, matter, and privation (190b35–191a3): he is counting the matter-privation composite as two things, given their complexity in being, whereas he is counting form, conceived of as a principle of natural things, as just one, metaphysically simple, thing.

So it seems to me that when Aristotle says that in a way the principles are the contraries, it is preferable to construe the contraries he is envisaging in the lopsided way I have suggested, with the negative members of the pairs denoting matter-privation composites and the positive members denoting just the forms in question.

We can see, then, why Aristotle now says that the principles of natural things are in a way the contraries. It is easy to see, on the other hand, why he wants to say that in another way, it is not the case that the principles are the contraries. He evidently takes it that the principles of natural things account not only for the being but also for the coming-into-being of natural things. But no change can be explained just by reference to suitable contraries, such as the qualities of heat and coldness, or of being educated and being uneducated, because contraries, conceived of as metaphysical simples rather than composites, cannot be affected by one another. As Aristotle points out at 190b33–5, the difficulty of how to employ the notion of contraries so as to make change intelligible[17] is resolved on the basis of the fact that what underlies the contraries, or the relevant contrary, is not itself a contrary. Aristotle is suggesting, I think, that his notion of the substratum, conceived of as distinct from contrariety but also as in a way metaphysically tied to privation, resolves two difficulties that he has articulated in chapter 6:[18] first in the order in which they are attended to in our present section is the difficulty that there seems to be good reason for thinking that the principles of natural things are just two (189b27–9, with 189a9–10 in the background), but also that there seems to be good reason for thinking that they must be three (189b16–18); and, second, that while it is clear that the notion of

[17] The first *τοῦτο* in 190b34, I suggest, refers to this difficulty.

[18] I thus think that *καί* at the beginning of 190b34 means 'also' rather than, say, 'in fact'.

contraries does have an important role to play in making change intelligible, it is difficult to see how precisely this notion should be employed in explaining change, especially the kind of change in which natural things come into being.

Aristotle can now summarize the section by reiterating that the principles of natural things are in a way two and in a way three. A detail in this recapitulation that we might want to note is an asymmetry between form and privation. Aristotle presents the principles as 'so to speak' (ὡς εἰπεῖν) two in number, but not completely two, since a difference in being belongs to them, by belonging, I take it, to the principle that is the substratum. (The substratum is once again analysed into a matter-component and a privation-component, illustrated by being a person and being uneducated and by bronze and lack of shape.) The detail I mean to call attention to is that Aristotle seems to be presenting the principle that is form as one in number, so that the principles of natural things turn out to be, 'so to speak', two in number, whereas he is not treating privation as one in number. Privation, in his analysis, just shows up as an attribute of the pre-change matter in question; it has unity in form or account but not in number. Aristotle elsewhere runs unity in number together with being a particular rather than a universal (*Metaphysics* B 4, 999b33–1000a1).[19] He has just before introduced the matter-component of the substratum as countable, and has said that it is more a case of some *this* (τόδε τι), whereas privation is an attribute. Against this background, the suggestion that form, as a principle from which natural objects come to be, is numerically one in the way that the substratum is and privation is not suggests that Aristotle is here thinking of form as a countable, reidentifiable something or other, a case of being some *this*, and so as substance-like rather than as a mere attribute.[20] By contrast, privation is a mere attribute on the picture Aristotle is presenting, and we might want to fill in the picture by supposing that privation here is treated as just a repeatable universal.[21]

We may wonder why Aristotle qualifies the claim that the principles are two in number by saying that this is 'so to speak' the case, thus, I take it, indicating caution that this may not be precisely the right thing to say on a fully worked-out view of the matter. Why the caution? Here is a suggestion. Aristotle has focused almost completely on the substratum, rather than on form, as a principle

[19] 'For there is no difference of meaning between "numerically one" (ἀριθμῷ ἕν) and "individual" (καθ' ἕκαστον). For this is just what we mean by the individual, and by the universal we mean that which is predicable.' While this is, I think, the prevalent notion of numerical unity in Aristotle's writings, it is not the only one. In *Topics* I 7, Aristotle employs a notion of numerical unity in which universals like human being, two-footed terrestrial animal, and what can acquire knowledge are numerically one. Thanks are due to Timothy Clarke for calling my attention to this passage.

[20] Cf. *Metaphysics* Λ 3, 1070a11–12; H 1, 1042a28–9. In both texts, substantial form is treated as some *this*.

[21] I owe this suggestion to Benjamin Morison.

of natural things. Furthermore, he has mainly been concerned with the substratum and, to a lesser extent, form as principles from which natural substances come to be, rather than as principles from which they are. So far as form is concerned, he has stated that form, alongside the substratum, is a principle from which natural things come to be, but, as I have noted, he has done remarkably little to explain how form operates as such a principle. Moreover, he evidently wants to employ substratum and form not just as principles from which natural things come to be but also as principles that constitute them, once they have come to be. As for form, Aristotle will want to say that the form from which a natural substance comes to be, its moving cause, is numerically distinct from the form that is a constituent of it, once it has come to be. This, after all, is the view stated in *Metaphysics* Λ 5 (1071a27–9) and also, though not as directly and explicitly, in *Physics* II 7 (198a24–7).

So, although form, conceived of as a principle from which natural things come to be, is in each case numerically one, form conceived of as a principle from which natural things both come to be and are, turns out on proper analysis to be two in number, but one in kind. Our chapter is not the place in which Aristotle wants to enter into these complexities about the principle that is form. And so it seems reasonable for him to indicate in the way that he does the somewhat tentative, preliminary status of the claim that the principles of natural things, substratum and form, are two in number. Understood as a claim about the principles from which natural things come to be, this is correct. Understood as a claim about the principles from which natural things both come to be and are, it stands in need of revision.

3. Form, Substratum, and Substance

> It has been said then how many are the principles of the natural things that are involved in generation, and in what way they are this many. And it is clear that something needs to underlie the contraries and that the contraries need to be two. But in another way this is not necessary. For one of the contraries will be sufficient to produce the change by absence and presence. And the underlying nature is knowable by analogy. For as bronze is related to statue; or wood, to bed; or to anything else among the things that have shape, the matter and what is unshaped, prior to receiving the shape in question—so this [sc. the underlying nature] is related to substance and some this and the thing that has being. One principle, then, is this, not being one nor being a being in the way that some this is, another principle is that which is the account, and furthermore what is contrary to it, the privation. It has been explained in what precedes how these things are two and how they are more than two. First, then, it was said that only the contraries are principles, and later it was said both that something else must underlie and that the principles must be three. From the things that have now been said it is clear what differentiates the contraries, how the principles

> relate to one another, and what the underlying thing is. It is not yet clear whether the form or that which underlies is substance. But it is clear that the principles are three, and how it is that they are three, and in what way they are principles. As to how many principles there are and what they are, let our study be based on these points. (191a3–22)

As I anticipated already, Aristotle begins the present subsection by saying that he has now stated the number of the principles 'of the natural things that are involved in generation' (τῶν περὶ γένεσιν φυσικῶν).[22] As is clear from my translation, I take the reference to γένεσις to be meant as a reference to unqualified coming into being, i.e. the coming-into-being of a new substance. The focus on the coming-into-being of substances, and of natural substances, is introduced at the beginning of my section, when Aristotle presents the causes and principles of natural things as things from which each thing is and comes to be what it is said to be in substance. The natural things that Aristotle has in mind as the subject of his investigation are substances that are crucially characterized by the fact that they come into being, and so it makes good sense that their principles must be able to explain not only their being but also their coming-into-being.

It is clear, Aristotle adds, that something must underlie the contraries and that the contraries are two. The former formulation, that something must underlie the contraries, seems to me to be helpful in making sense of the fact that Aristotle in our chapter occasionally conceives of the substratum not as the matter-privation composite but as a metaphysical simple, namely just the matter in question (190b10–17, 19–22, and, I think, also 191a7–12). Recall Benjamin Morison's nice point about making changes intelligible as changes: specifying what underlies in a given change as a matter-privation composite is best suited to making what happens in a change intelligible as a case of *change*—it makes perspicuous both the aspect of the thing that is being replaced, and the aspect of it that remains, doing justice to two crucial facts about change, namely that when a change is complete, something *has changed*, and some *thing* has changed.[23] So, conceived of as what underlies in some given change, as what it is that undergoes the change in question (cf. 190a13–16, 33–4), the substratum is best specified as the relevant matter-privation composite. Conceived of as what at

[22] Ross proposes to delete περὶ γένεσιν, and Charlton (1992^2) accepts the suggestion. But it is present in all of our mss., and after 190b17–20 it shouldn't come as a surprise that Aristotle has in mind specifically natural things that come into being, to the exclusion of natural things that are everlasting. One might also, of course, think that Aristotle means here to use the term γένεσις loosely, to denote change in general, as he does at the beginning of the chapter. But in that case περὶ γένεσιν is redundant. *De Caelo* I 3 suggests that contrariety plays no role in grasping the nature of the heavenly bodies, and so it might well be the case that the discussion in our chapter of the principles of natural things, in which ideas about contrariety of course feature prominently, is meant to apply specifically and exclusively to natural things that come into being.

[23] Morison, this volume.

different times underlies the contraries of both form and privation, the substratum is just the matter in question, rather than the matter-privation composite. One might also note that as Aristotle's focus shifts from analysing change to the question of the principles of not only the becoming but also the being of natural things, the notion of the substratum as just the matter that underlies form (or its lack) becomes more prominent. After all, it is only the substratum conceived in this way that is a constitutive principle of natural substances, something from which those substances not only come to be but also are.

Aristotle qualifies the claim that the contraries needed for the explanation of natural things are two by adding that 'in another way this is not necessary', as one of the contraries is sufficient to produce any given change by absence and presence. The contrary that is sufficient to produce change plainly is form rather than privation, and one idea Aristotle is expressing in making this comment is that form and privation are not on a par with one another with respect to explanatory power. After all, privation is lack of form and so is conceived of in relation to form, whereas form is not conceived of in relation to privation. The comment may look like an aside, but it might offer a clarification that in fact is of structural importance. As we saw, at the beginning of my section Aristotle presents substratum and form as principles from which natural things both are and come to be, illustrating the relationship between substratum and form by reference to the way an educated person is composed of being a person and being educated (190b17–23). So this is in important part a point about causes or principles of natural substances that are also constituents of those substances, as I noted when discussing my first subsection (see section 1). The principles that feature in this analysis are form and substratum, with the substratum being conceived as not being characterized by privation. After all, once the natural thing in question has come into being and *is*, the substratum that partly constitutes it is not then characterized by privation.

However, Aristotle is conceiving of substratum and form not only as principles that constitute the given natural thing once it has come into being, but also as principles from which the natural thing in question comes to be. Now as we have seen, when you consider the substratum and form as constituents of the thing that has come into being, privation drops out of the picture. The substratum that constitutes the thing in question is the bare substratum: the matter, not the matter-privation composite. Can we conceive the substratum of X both as a principle from which (along with the form) X comes to be and as a principle that (along with the form) constitutes X, without switching from one way of conceiving the substratum (namely, as matter-privation composite) to another (namely, as matter)? Aristotle is now showing us how to do this, namely by thinking of the form as a principle that can, along with the substratum, account for change by absence and presence, without any need for privation to be appealed to as an additional explanatory factor.

We may wonder how precisely form can *produce* a change by absence and presence. Perhaps the thought is just that one can describe any given change in a way that makes it intelligible as the change it is by saying that some given substratum goes from absence of the form in question to its presence. But this suggestion does not do justice to the idea that, on this way of looking at things, it is form that produces or brings about the change in question (ποιεῖν τὴν μεταβολήν, 191a7).[24] It seems more plausible that Aristotle is thinking of form as productive of change by being somehow present as an agent and by being at the same time absent from some suitable substratum. Change is produced when form acts on a substratum which is suited to being acted on because the relevant form is absent from it. The change is produced not only because the form is present as an agent but also because the form is absent from the patient. If this is along the right lines, Aristotle is now picking up the lopsided conception of the contraries as form and matter-privation composite that he introduced at 190b30–2. In effect, he is adding to the earlier picture that the matter-privation composite is just the matter characterized by the absence of form, and that form and absence of form from matter are sufficient to produce change. If so, the present passage constitutes a second reference in the second half of I 7 to the moving cause, alongside 190b33. Furthermore, it presents form as an agent of change and so, in effect, as a moving cause.

After commenting on how change can in a way be explained by reference to only one contrary—that is, without reference to privation—Aristotle now turns to the other ineliminable principle of natural things, namely the substratum. He says that 'the underlying nature' (ἡ ὑποκειμένη φύσις) is knowable by analogy, and immediately offers examples of things that are related to other things as the substratum is related to 'substance, some *this* (τόδε τι), and what has being'. The examples are bronze in relation to a statue and wood (or timber) in relation to a bed, and Aristotle adds the generalization that this kind of relationship is exhibited by the matter in question in relation to what has the relevant shape, glossing the matter as what is shapeless before receiving the shape. These few lines (i.e. 191a7–12) raise a number of questions, including the following. First, are Diels and Ross right in bracketing the reference to matter [ἡ ὕλη καὶ] in 191a10? Second, what does it mean for the substratum to be knowable by analogy? And third, is the substratum now conceived of as a metaphysical simple (just as matter) or as the matter-privation composite?

[24] I am grateful to Alan Code for impressing this consideration on me.

The first question can, I think, be dealt with rather quickly. Ross is for deleting the reference to matter on stylistic grounds, claiming that it 'spoils the proportion ἄμορφον:τεχνητόν = ὕλη:οὔσια'. But first, all our manuscripts seem to include the reference to matter in this line.[25] Second, it makes good sense for Aristotle to generalize from bronze (in the case of a statue) and wood (in the case of a bed) to the matter in question; after all, he has done pretty much the same thing at 190b24–5, where he generalizes from man (in relation to educated man) and gold (in relation to something made of gold) to the matter, conceived of as countable. Third, Ross's stylistic point, apart from not being compelling, is not even correct. That is because the pronoun αὕτη in 191a11 picks up, not matter, but the underlying nature (in 191a8), and even though 'what underlies X' can refer to the matter of X, this is no good reason to think that 'the underlying nature' and 'matter' should be treated as synonyms. So, the proportion is ἄμορφον/ὕλη:ἔχον μορφήν = ὑποκειμένη φύσις:οὔσια.

What does it mean for the substratum to be knowable by analogy? This is not the place for a comprehensive study of analogy in Aristotle, but it might still be useful to offer a few comments on how the notion is put to use here.[26] Aristotle is assuming, I take it, that for something to be an object of scientific knowledge (ἐπιστήμη), it must be in some way unified. For the substratum as such to feature as a single entity in a body of scientific knowledge, there must be some systematic way in which all the things that count as substrata hang together so as to form a unified kind of thing. A random collection of unrelated things, such as the things denoted by a chance homonym like 'bank', is not an object of scientific knowledge.

Now, Aristotle recognizes a number of ways in which things that have a name in common may form a unified kind of thing. It might be that they belong to a single species, in that all of them are (for instance) tigers, or that they fall under a common genus, such as the genus of birds.[27] Things that belong to a single species or a single genus have a nature or essence in common, namely that of the species or genus in question. However, the things that count

[25] Ross's apparatus claims that Simplicius in his commentary on the passage omits the (relevant) reference to matter. It is true that Simplicius at 226.6–9 does not include a reference to matter in the subordinate clause (ὡς γὰρ πρὸς ἀνδριάντα χαλκὸς ἢ πρὸς κλίνην ξύλον ἢ πρὸς ἄλλο τι τῶν ἐχόντων μορφὴν τεχνητῶν τὸ ἄμορφον ἔχει πρὶν λαβεῖν τὴν μορφήν, οὕτως ἡ ἐν τοῖς φυσικοῖς πρώτη ὕλη πρὸς οὐσίαν ἔχει καὶ τὸ ὄν). But it seems that Simplicius means to paraphrase rather than to quote *verbatim*. If so, the lack of a reference to matter in Simplicius' subordinate clause provides no basis for thinking that Simplicius worked with a text of the *Physics* that did not include the reference to matter in l. 10.

[26] Some relevant discussion may be found in Beere 2009, 178–208; in Makin 2006, 128–35; and in Wilson 2000.

[27] Cf. *Parts of Animals* I 5, 645b28–9: 'some things have something in common by analogy; others, by genus; yet others, by species' (τὰ μὲν γὰρ ἔχουσι τὸ κοινὸν κατ' ἀναλογίαν, τὰ δὲ κατὰ γένος, τὰ δὲ κατ' εἶδος).

as the underlying nature of something or other do not share an essence that captures what it is for each of them to be the underlying nature of something. Very different kinds of things can be the underlying nature of something or other: there are the underlying natures of all the kinds of natural things that are involved with coming into being, e.g. earth, air, fire, and water, rocks, plants, and sublunary animals. There are also the underlying natures of artefacts. Given the enormous plurality and diversity of different forms that the underlying nature can take in all the relevant cases, it seems reasonable to think that there is no such thing as a single nature or essence of being the substratum of something or other. After all, it is not the case that earth, air, fire, water, or some combination of them are the underlying nature of all things that have an underlying nature. First of all, none of these candidates could, for Aristotle, plausibly be the underlying nature of earth, air, fire, or water themselves. And second, Aristotle in our section is clearly treating as the underlying nature of X what proximately underlies X, rather than what underlies X in some remote and indirect way: for example, the underlying nature of a bronze statue is bronze, not the earth, air, fire, and water that he takes to be the ultimate material constituents of bronze.

If there were a single essence of being the underlying nature of something that all relevant cases had in common, one could grasp the underlying nature by specifying the essence in question, just as one can grasp the kind *human being* by specifying its essence, e.g. rational biped (or whatever). Given the very different forms that the underlying nature takes, and the lack of a single essence of being the underlying nature that they all have in common, the best one can do, Aristotle seems to think, is to grasp it by attending to different kinds of things that constitute the underlying nature of this or that kind of thing, for example bronze in the case of a statue, timber in the case of a bed, and so forth.[28] These are all analogues of one another, and they can be organized into analogies of the form A:B = C:D. For example, as bronze is related to a statue, so is timber to a bed, whatever the matter of water is to water, and whatever the matter of a human being is to a human being. Such analogies, Aristotle seems to think, articulate how it is that bronze, timber, and so forth hang together so as to constitute a unified kind of thing, namely that of the underlying nature.[29]

Lastly, how should we construe the substratum in this passage: metaphysical simple or matter-privation composite? The question is difficult, and perhaps not in the end very important. Nevertheless, it seems on consideration preferable to think that Aristotle is here treating the substratum of X as just the

[28] I agree with Charlton, *ad locum*, that these in fact are meant to be examples of substrata of things.

[29] Cf. *Metaphysics* Λ 4–5, where the principles of matter, form, privation, and the moving cause are presented as each being the same by analogy in all cases.

matter of X, rather than as the matter-privation composite from which X comes to be. In support of the matter-privation composite reading, one should call attention to the fact that in Aristotle's general formulation of what it is that plays the role of substratum in relation to something else, he says that it is 'the matter and that which is shapeless before taking on the shape'. One can construe this temporally, so that it is a reference to the substratum before the change in question has taken place. This of course is the substratum characterized by privation. But even on this reading, the matter-privation composite reading is not compulsory. For one can use the expression 'the shapeless thing' to denote the bearer of shapelessness, i.e. the matter in question, rather than the matter-privation composite, just as one can use the expression 'the limited thing' (for example) to denote an entity that excludes the relevant limits, rather than one that includes them. Furthermore, we can perhaps construe the word *πρίν* in 191a10 not temporally but logically, so that Aristotle is referring to what is shapeless independently of taking on the shape. In this case, Aristotle's idea is just of what in and of itself is shapeless, but such as to have or lack the shape in question. In other words, he means to refer to the matter by itself, not to the matter-privation composite.

In favour of the metaphysical-simple reading, one should note that the immediately preceding reference to something that underlies is at 191a4–5, where the expression denotes something that underlies *both* contraries. What underlies both contraries is of course the matter in question, not the matter-privation composite. Moreover, the passage that immediately follows, 191a12–15, seems to me to militate rather strongly in favour of the metaphysical-simple reading, since it says that this thing (*αὕτη*), i.e. the underlying nature, is one principle, that form is another principle,[30] and that furthermore (*ἔτι*) that which is contrary to form, privation, is a principle (alternatively: 'furthermore, there is that which is contrary to form, privation'). So it seems to me that while 191a7–12 by itself is indeterminate between the two construals of the expression 'the underlying nature' that I have indicated, in the context of the passage the metaphysical-simple reading seems more likely. If so, Aristotle is best interpreted as here operating with a conception of the substratum as what underlies the natural substance in question, in such a way as to be a constituent of it.

Now, one might think that this is a very important point, on the grounds that on this interpretation Aristotle is here presenting the substratum in or of some change as just the matter in question, and that he is doing this in a passage that is rather prominent and that one might think is meant to provide Aristotle's

[30] I propose reading *μία δὲ ἧ ὁ λόγος* (with Torstrik) in l. 13, meaning: 'another principle is that which is the account'. This is not an emendation but just a different accentuation of a reading that has a great deal of manuscript support, namely *μία δὲ ἡ ὁ λόγος*.

authoritative characterization of the principle that is the substratum. So, one might think that the passage settles the question what Aristotle's preferred characterization of that principle is, in favour of thinking that the substratum in or of any given change just is the matter in question, rather than the matter-privation composite. But it would be rash to think this. This is because it is not clear that in this passage Aristotle means to provide a characterization of that which underlies in change—that is, of the substratum conceived of specifically as a starting point from which change proceeds or as that which undergoes change. In fact, it seems more plausible in light of the context that in this particular passage he means to characterize the substratum as one of the principles of natural substances (191a3–5; 12–13). And conceived of as a principle of natural substances, as something from which natural substances both are and come to be (190b18–19), the substratum just is the matter, because the privation-aspect of the pre-change substratum is not a constituent of the given natural substance: it is not something from which the natural substance *is*. However, to characterize the substratum, conceived of as a principle of natural substance, as just the matter in question is perfectly compatible with thinking that conceived of as what underlies in change, the substratum is most fully and most perspicuously characterized as a matter-privation composite.

Before we move on from the present passage, it is worth calling attention to terminology employed in the passage that might seem more at home in Aristotelian first philosophy than in a treatise on natural philosophy. He is specifying what the substratum underlies as substance, some *this*, and that which has being. He adds that the substratum is neither a single thing nor a being in the way that some given *this* is. These comments connect with the earlier claim that the matter of some given thing is some *this* more than (I take it) the relevant privation is (190b24–7), and with the observation, made near the end of the chapter, that it is not yet clear whether the form or the substratum are cases of substance (191a19–20). On the overall picture that emerges in the section, privation is not a substance but an attribute, and is not some *this*, or at least is a *this* even less than matter is.[31] The matter-form composites that are natural things are cases of substance, some *this*, and things that have being (191a8–12). The matter of any given natural thing is some *this* to some extent (190b25–6), but is not without qualification a case of being some *this* (191a8–12). It also is neither a unified thing nor a being in the same way and to the same extent that some given matter-form composite such as a tree is (191a12–13). As for form—the

[31] This is helpful for resolving a difficulty indicated in chapter 6, to the effect that contraries can't be substances because 'we claim' that one substance is not contrary to another (189a32–4). If privation does not have the status of substance, form can have it: there will in a way be a contrary to form, but that contrary will not be another substance.

forms of natural things such as plants and animals—its status with regard to substancehood, unity, being some *this*, and being is very much left indeterminate. But we have noted a hint that such forms are cases of some *this.* As we saw, Aristotle in our section seems to associate being countable and numerically one with being at least to some extent a case of being some *this*, and at the same time he ascribes numerical unity not only to the substratum but also to form—though not to privation. We also recall the thought, expressed in the difficulties of chapter 6, that it is problematic for substance to be from non-substances, or for non-substance to be prior to substance (189a33–4). Given that form is one of the principles from which natural substances *are*, this gives us reason to expect that form will in some way have the status of substance.

One last passage in my section that I think calls for comment is 191a15–19, in which Aristotle relates the work done in the present chapter to the discussions in chapters 5 and 6. 'First it was said', Aristotle reminds us, 'that only the contraries are principles, and later it was said both that something else must underlie and that the principles must be three. From what has now been said,' he adds, 'it is clear what differentiates the contraries, how the contraries relate to one another, and what it is that underlies.' It is, I think, clear that Aristotle now presents chapter 5 as introducing the idea of contraries as principles of natural things, and chapter 6 as arguing for a threefold conception of the principles that includes something that underlies the contraries. It is not immediately clear how the word μόνον in 191a16 should be construed. Depending on how you construe it, Aristotle is saying either that it was said that only the contraries are principles, or that it was only said that the contraries are principles.[32] On the face of things, the latter construal has the merit that it avoids making Aristotle present what was said first (namely in chapter 5) as on his own considered view false. On his considered view, after all, it is not the case that *only* the contraries are principles of natural things.

However, the word order of Aristotle's Greek favours the construal according to which what was said first is that only the contraries are principles. Moreover, Aristotle concludes chapter 5 by saying that 'it is clear that one should say that the principles are contraries' (189a9–10), rather than, say, that the principles include contraries. As I have noted already, a substratum for the contraries is implicitly in play already in chapter 5, when Aristotle says that a house, for example, 'comes to be from *these things* not being put together but being separated in this way' (188b17–19). We should also note the implicature in chapter 5 that contraries not only come to be from one another but also act on and are affected by one another (188a31–4). This, too, suggests that suitable substrata underlying contrary characteristics such as hot and cold, white and black,

[32] Cf. Ross's paraphrase: 'We showed first merely that contraries are principles'.

educated and uneducated, are already implicitly part of the picture presented in chapter 5. Perhaps, then, the *aporiai* of chapter 6 serve not to introduce the substratum as a completely new idea, but to recognize its distinctness from the contraries. As we have seen, in our present section Aristotle revisits the idea that in a way the principles of natural things are the contraries (190b30–2), and he makes sense of it, I suggested, in terms of the idea that form and the matter-privation composite, i.e. the principles from which natural things come to be, are in a lopsided way contraries of one another. So Aristotle seems to think the claim that the principles of natural things are only the contraries to be in a way not false but nevertheless importantly unclear, in that it fails to make conspicuous the important role of the substratum as a principle of natural things.

A second question that this short passage raises is what Aristotle has in mind in saying that from what has been said it is clear what is ἡ διαφορὰ τῶν ἐναντίων. Ross paraphrases this as concerning the difference between the contraries. Charlton complains that this construal is 'extremely weak' and claims that the literal meaning of the phrase is 'what the differentiating feature of the contraries is' (Charlton 1992[2], 47). On Charlton's construal the contraries are treated as a unit, and their διαφορὰ is what differentiates them from other things, for instance the substratum. I suggest that Aristotle's formulation is loose enough to express both the idea that the contraries of form and privation have been differentiated from one another, as in fact they have been, and the idea that the contraries have been differentiated from the substratum, as in fact they have been.

Concluding Remarks

In the second half of chapter 7, Aristotle connects the concepts of the substratum of change and of the contraries between which change occurs with the question of what the principles of natural things are. I want to close by briefly revisiting some of the interpretive difficulties that my discussion has brought to light.

One such difficulty is the question what to make of the fact that τὸ ὑποκείμενον sometimes in the chapter denotes the matter-privation composite, while on other occasions it denotes matter by itself. I have suggested that we should ascribe to Aristotle an implicit distinction between the substratum of change and the substratum of the relevant contrary/opposite. I have also suggested that as Aristotle turns in the course of the chapter from analysing change to the question of what the principles of natural things are, the conception of the substratum as just the matter of the object in question becomes more prominent. Furthermore, the seamlessness with which he shifts between these two ways of conceiving of the substratum suggests that he is thinking of these two ways as aspects of a unified conception: the matter-privation composite from which X comes to be and the matter of X once X has come to be are in a way the same entity, namely the principle of X that is its substratum, but the same entity in different conditions. This seems intelligible and even reasonable to me. What makes it intelligible and even reasonable is the fact that Aristotle is conceiving of matter as substance-like and of privation as a mere attribute of it, something in whose absence it remains what it is.

Another difficulty raised by my section is the question what to make of the presence of both the idea of a principle from which X comes to be and the idea of a principle from which X *is*. Aristotle runs both ideas together at 190b17–20, but they clearly are distinct ideas, as one can see by noting that privation is a principle from which things come to be but not a principle from which things *are*. I have suggested that Aristotle takes both ideas to be importantly connected in that the kinds of things he is interested in, natural things that are involved in generation, are things that are crucially characterized by the fact that they come into being, and so the principles that explain the nature of these things should also be explanatory of their coming-into-being. The principles from which the being of these natural things derive should then also somehow be the principles of their coming-into-being. Combining these two closely related conceptions of principles is also facilitated, I suggested, by the fact that the two indispensable principles that Aristotle is introducing, form and substratum, really are principles that things both come to be from and are from. They both are pre-existing starting points from which the coming into being of X proceeds and constituents of X once it has come to be.

A third difficulty that the section raises is that it introduces or hints at a number of philosophically important ideas that are not explained or developed either in chapter 7 itself or elsewhere in *Physics* I. An idea that is presented but not explained in the chapter is that form is a principle from which X comes to be, so that in some way it preexists the coming into being of X. One could develop a philosophical picture on which this is true on a rather slim textual basis by noting that chapter 5 suggests that contraries not only come to be from one another but also act on and are affected by one another, as when something hot heats up something cold or an educated person imparts education to someone else who at the beginning of the process is uneducated. On this picture, the form that is a principle of the thing that has come to be (e.g. heat) in a way preexists the process of coming to be. A picture along these lines will of course be presented in *Physics* II.

An idea that is only hinted at in the present section is that form is in some way substance-like and perhaps in some way and to some extent has the status of being some *this*. Aristotle begins to substantiate this hint in *Physics* II 1, when he runs together talk of the substance of X and talk of the nature of X (193a9–11), and then argues that both the form of a given natural substance and its matter in different ways count as its nature (193a28–31). So if the nature of X is also its substance, then both the form and the matter of a given natural object count, in different ways, as its substance.

The section, then, has a somewhat proleptic character. To some extent, issues that it raises are taken up shortly in *Physics* II. But it seems that some of the issues receive detailed and direct discussion only in the *Metaphysics*, for example the question whether the forms of natural objects are cases of substance. On the other hand, *De Anima* II 1 begins with a short, dogmatic statement that substance is spoken of as matter, as form, and as the composite (412a6–9). Aristotle seems to think that it is appropriate for natural philosophy to presuppose, or accept on a rather incomplete basis, that both the matter and the form of natural substances themselves have the status of substance, whereas it falls to first philosophy properly to clarify and justify these claims.

9

The Route to Solving the Eleatic Puzzle

Physics I 8

MARISKA LEUNISSEN

1. Introduction (191a23–24): Context and General Purpose of *Physics* I 8

That in this way alone (μοναχῶς) also the puzzle of the ancients is solved, we should discuss next. (191a23–24)

Up until this chapter, Aristotle has accomplished at least two important things. First, he has presented a general methodology for the study of natural things and marked off the tasks of the natural scientist from those engaged with first philosophy or the study of being qua being (in *Physics* I 1–2). Second, after critically reviewing the opinions of his predecessors (in *Physics* I 3–4), he has also identified—at least at a general level—the principles of natural things, and, most importantly, has introduced his notion of an 'underlying nature' (*Physics* I 7, 191a8–9: ἡ δὲ ὑποκειμένη φύσις), a role performed by matter in the case of things that are, and by the matter-privation composite in the case of things that come to be (as defended in *Physics* I 5–7).[1] According to this analysis, then, natural change always occurs between two contraries (i.e. contrary qualities in the case of alteration, and form and privation in the case of substantial change) acting on an underlying nature, which constitutes—in a sense—a third principle. Technically speaking, this identification of the principles of nature concludes

[1] See the contributions by Morison and Lorenz on *Physics* I 7 in this volume.

the inquiry announced at the very beginning of book I of the *Physics* (*Physics* I 1, 184a10–16).

The next issue on the agenda, as Aristotle announces at the beginning of *Physics* I 8, is to offer an account showing that 'in this way alone'—that is, only by using Aristotle's own principles (such as, most importantly, the notion of an underlying nature in addition to the opposites form and privation as principles of natural things)[2]—'also the puzzle of the old [philosophers] is solved' (*Physics* I 8, 191a23–4). The purpose of *Physics* I 8 is thus to broadcast the correctness and benefits of the use of Aristotle's own principles in the analysis of natural change by showing how this can solve one of the oldest and most pernicious problems in the philosophical traditions preceding him.

The puzzle Aristotle has in mind here concerns the denial of coming to be as it is known best from the Eleatic monists.[3] As is clear from Aristotle's reformulation of the puzzle in 191a27–31, which I will return to below, it is based on the paradox that there can be no coming to be of what is, since what is already exists, and that there can also be no coming to be of what is not, since nothing can come from what does not exist—for there must already be some existing thing that underlies it.[4] Because the Eleatics believe that these two possibilities exhaust the realm of reality, that is, they believe that coming to be can only be from what is or from what is not, and from what is so without qualification, they are forced to deny the existence of generation. And, by drawing inferences even beyond this immediate conclusion that there is no coming to be (or, as Aristotle puts it, 'by exaggerating the immediate consequence of this'; see 191a32: τὸ ἐφεξῆς συμβαῖνον αὔξοντες), they also end up denying the plurality of things altogether and instead claim that being is one. In this opening section, Aristotle's use of the term μοναχῶς in 191a23 is a direct reference to Parmenides' characterization of the Goddess's account of the unity of the being and the denial of the existence of not being as the 'only roads worth thinking' (fr. 2, 7–9; 8: αἵπερ ὁδοὶ μοῦναι διζήσιός εἰσι νοῆσαι): Parmenides was right that there is only one correct way of solving the dilemma, but since he was wrong about the principles of natural things, his proposed solution is also wrong. Aristotle is about to provide the only 'true' route out of the *aporia*.

[2] The οὕτω in 191a23 picks up on the conclusion of I 7, and the καὶ suggests that just as the introduction of the notion of the underlying nature was necessary to solve the puzzle concerning the number of principles of natural things—namely whether there are two or three—it is also necessary to solve the infamous Parmenidean dilemma.

[3] He already explicitly discussed the positions of Parmenides and Melissus and their fallacies in *Physics* I 3. Aristotle does not mention the Eleatics explicitly when formulating the puzzle, but the references to the Parmenidean road metaphor and other references to the poem make sufficiently clear that it is this group of philosophers in particular that he has in mind.

[4] I am thus reading the Parmenidean dilemma as a puzzle about the existence of change, much in the line as has been suggested by Clarke 2015; I will return to the various lines of interpretation that have been proposed below.

The reason why Aristotle returns to this puzzle (which, although it has not been formulated explicitly in the earlier chapters, already surfaced in the discussion of the Eleatic monists in *Physics* I 2–3) in this chapter is not because establishing the truth of his identification of the principles of substantial change as form, privation, and underlying nature requires the refutation of the Eleatic position that being is one. For, as Aristotle had already argued in *Physics* I 2, 184b25–185a14, it is not possible for the natural scientist *qua natural scientist* to defend the first principles of his own science against someone who challenges those principles, since the latter denies the very foundation of every claim to knowledge the former can make.[5] Within natural science, the thesis that natural things undergo change simply has to be assumed as a first principle, and it is vindicated by observation (*Physics* I 2, 185a13–14: δῆλον δ᾽ ἐκ τῆς ἐπαγωγῆς; cf. *Physics* II 1, 193a1–9 and *Physics* VIII 3, 253b2–6). However, Aristotle also believes that it *is* a task for *natural scientists* to solve as many problems or falsehoods that someone might have inferred on the basis of any relevant first principles (*Physics* I 2, 185a14–17; a14: λύειν), such as in this case, the Eleatic axioms that what is, is, and that what is not, is not, and the false inferences they drew from these axioms pertaining to the possibility of coming to be, namely that there is no change and, overblowing this conclusion even more, that there is no plurality. Although the Eleatic monists are not explicitly talking about nature here, their puzzles do pertain to the natural scientific realm, and so Aristotle thinks they need to be addressed (*Physics* I 2, 185a17–20): for discussing them has philosophical relevance (*Physics* I 2, 185a20: ἔχει γὰρ φιλοσοφίαν ἡ σκέψις). Puzzling through the problems raised by the old natural philosophers in this way is thus one manner in which dialectic can be useful for natural scientists. For by explaining the cause(s) of errors in the views of others and by explaining what caused the puzzles, the true view—as perhaps already presented and demonstrated by oneself—becomes clearer,[6] more convincing,[7] and will be shown in the best way possible.[8]

[5] This kind of defence of first principles requires the second use of dialectic in science, as defined in *Top* I 2, 101a36–b4: 'In addition, [this is useful] for the first [principles] regarding each of the sciences. For it is impossible to say something about them on the basis of the ones [i.e. principles] that are proper to the science at hand, since the principles are prior to everything, and it is necessary to go through them via the reputable opinions about them. And this is what is proper or most fitting to dialectic. Being capable of examining it is the road towards the principles of all investigations'.

[6] For Aristotle's definition of the first use of dialectic in science, see *Top* I 2, 101a34–6: '[It is useful] for the philosophical sciences, because by being able to puzzle on both sides of something we will perceive with more ease what is true and what is false in every matter'.

[7] See e.g. *NE* VII 15, 1154a22–5: 'Since we must not only state the truth but also the cause of the falsehood: for this contributes to conviction (τὴν πίστιν)—for whenever the explanation for why something that appears true while not actually being true seems to be reasonable, that produces more conviction for the truth—therefore . . .'. Cf. also *DC* I 10, 279b4–12 and *Physics* IV 1, 211a7–11.

[8] See e.g. *Physics* IV 4, 211a6–11.

In what follows, then, Aristotle offers us precisely this kind of elucidation. For Aristotle first provides two explanations for why 'the old philosophers' made the mistake they made and for why the Eleatic thesis, understood in the way these old philosophers defended it, is indeed false (in 191a23–34 and 191a34–b13). However, understood *correctly*, that is, when analysed in terms of the principles and tools of Aristotle's own natural philosophy, the Eleatic thesis that if there is coming to be, it must be either from what is or from what is not, can be preserved in a manner that is no longer false and that solves the problem of coming to be (see 191b13 and 191b26–7: it is not Aristotle's intention to do away with the thesis altogether, since ultimately—and again, understood correctly—it is one he endorses himself). Thus, in the next sections, he provides his own solution to the problem that builds on the three principles of natural things—and in particular on the notion of the underlying nature—as defined in *Physics* I 7 (in 191b13–27), and hints at a *another way* in which Aristotle's own distinctions can be put to work to solve the Parmenidean dilemma—a solution that is developed more rigorously in a different treatise (in 191b27–9). And, finally, Aristotle concludes the chapter by summarizing its accomplishments (in 191b30–4). In the sections below, I will discuss each of these five sections in turn.

2. Aristotle's Account of the Old Philosopher's First Mistake (191a24–34): Inexperience

191a24: For the first [people] who inquired in a philosophical manner into the truth and into the nature of existing things were diverted—as if being pushed down onto some other road—by their inexperience,

191a27: and they say that of the things that are none come to be or pass away, because it is necessary for what comes to be either to come to be from what is or from what is not, and yet it is impossible to [come to be] from either of those. For there is no coming to be of what is (for it already is) and from what is not there could not be any coming to be either: for something must underlie. And in this way, then, by exaggerating the immediate consequence of this, they claim that there is no plurality but only what is itself.

191a33: These [people] acquired that opinion because of what was [just] said.

In formulating the problem raised by the old philosophers,[9] Aristotle describes their first mistake as a form of getting sidetracked from the proper road as a result of a certain kind of inexperience (191a26–7: *ἐξετράπησαν οἷον ὁδόν τινα ἄλλην ἀπωσθέντες ὑπὸ ἀπειρίας*).

[9] This formulation is itself possibly a paraphrase of the way in which Melissus formulates the dilemma in fr. 1: *ἀεὶ ἦν ὅ τι ἦν καὶ ἀεὶ ἔσται. εἰ γὰρ ἐγένετο, ἀναγκαῖόν ἐστι πρὶν γενέσθαι εἶναι μηδέν· εἰ τοίνυν μηδὲν ἦν, οὐδαμὰ ἂν γένοιτο οὐδὲν ἐκ μηδενός.*

Before turning to the nature of the inexperience that pushed the old philosophers down the wrong road, I would like to note that manuscripts EJS offer an alternative reading in a26–7, namely that the old philosophers were pushed off the road by 'an' or 'the *aporia*' (ὑπὸ ἀπορίας). However, I prefer to follow Ross's (1936) reading here for two reasons. First, it seems odd to me for Aristotle to identify an (or the?) ἀπορία as one of the causes for the Parmenidean dilemma which is itself presented as a kind of ἀπορία (an/the ἀπορία would both have to be the source and the result of the Eleatic mistake). I take it that instead, the Parmenidean dilemma or ἀπορία resulted from following the wrong road, which itself was due to a certain kind of inexperience (cf. διὰ ταύτην τὴν ἄγνοιαν in 191b11, which picks up on the existence of a certain cognitive lack in the predecessors). Second, the passage seems to be a play on and an implicit critique of Parmenides' own advice, formulated through the words of the Goddess, not to let 'habit, rich in experience' force one along the route that involves sense-perception (μηδέ σ' ἔθος πολύπειρον ὁδὸν κατὰ τήνδε βιάσθω) and to 'judge by reason' (κρῖναι δὲ λόγωι) alone her words (DK 28B7.3–6). If our text is indeed a reference to this passage in Parmenides, it is inexperience with a certain type of reasoning (and thus a failure to heed the Goddess's own words to use reason correctly), rather than too much—and allegedly misleading—empirical experience that pushed the Eleatics off the road to truth as Aristotle sees it.[10] For one, it was the wrong decision to leave the road of not-being 'unthought' (DK 28B8.15–18).

For Aristotle, properly carrying out an investigation requires one to stay on track (ὁδῷ) or to follow the proper manner of proceeding (μέθοδος):[11] one has to attend to certain norms of inquiry if one is to stay on track towards truth and towards the acquisition of scientific knowledge of the nature of things. For instance, in *Physics* I 1, 184a16–26 Aristotle had described the natural manner of proceeding (184a16: πέφυκε . . . ἡ ὁδὸς) in natural science as comprising (1) an advancement from what is clearer to us to what is clearer by nature; (2) an analysis of things that are clear to us but 'jumbled up together' into elements and principles; and (3) a procession from the general to the particulars, where the general is understood as the whole that is more knowable according to perception and that comprises many things as its parts. In other words, Aristotle generally believes that the only manner in which one can come to grasp the principles, causes, or elements of natural things is to start from a broad familiarity with what is available to *observation*—with 'what is clear to us'. That this is the road Aristotle follows himself, becomes clear from various explicit appeals to it,

[10] Cf. also 191a33–4: paradoxically, the road the Eleatics ended up taking leads to opinion—τὴν δόξαν—and not to truth.

[11] See e.g., *PA* I 4, 644b17–22 and *Meteor.* I 1, 338a20–339a9. For a discussion of Aristotle's discourse on *methodos*, see Lennox 2011, 28–32.

both in the *Physics* and in the other natural treatises,[12] and it is not uncommon for him to criticize predecessors for failing to follow it (see, e.g., *Gen. et Corr.* I 2, 315a29–b15 and *Gen. et Corr.* I 7, 323b15–17) or to praise the ones who in making their distinctions manage to stay 'on track most' and therefore postulate principles that are more in accordance with nature (see, e.g., *Gen. et Corr.* I 8, 324b35–325a2; b35: ὁδῷ μάλιστα).[13]

However, in our text, the inexperience that caused Aristotle's predecessors to have fallen off the right track is most likely—as both Philoponus and Themistius already pointed out in their commentaries—not inexperience with the relevant empirical data (although there is a sense in which 'observing the underlying nature' and thus analysing natural change correctly is empirical: see Broadie in this volume). Rather, it is inexperience with dialectic and Aristotle's method of making the appropriate distinctions.[14] Aristotle believes that his predecessors simply posited the first principles of natural things, based on a wrong and simplistic analysis of generation (see 191b10: οὐ διελόντες; cf. perhaps *Physics* I 5, 188b26–30, where Aristotle argues that his predecessors used the right name for their principles, but without any account or understanding, as if pushed in that direction by truth itself), and then mistakenly argued from those principles. The proper method, of course, is to provide an argument *to* the first principles first (*NE* I 4, 1095a30–b1), using a dialectical method in the broad sense (as opposed to a scientific or demonstrative method which cannot be used with regard to establishing the first principles of a science) that often involves working through the existing opinions (as opposed to directly appealing to empirical evidence) and that requires one to provide a proper conceptual analysis.[15] The inexperience diverting the predecessors from the right road is thus a methodological one; again, only experience with and knowledge of the proper method allows one to stay on track (ὁδῷ).[16]

[12] See e.g. *Physics* I 7, 189b31–2: 'for it is in accordance with nature—after having first talked about what is in common—to then investigate the particulars regarding each'; *Gen. et Corr.* II 9, 335a27–8; *DA* II 2, 413a11–13; *HA* I 6, 491a7–26; and even *Pol.* I 1, 1252a17, *Pol.* I 2, 1252a26, and possibly *NE* II 7, 1108a2–4 with 1107a28–9.

[13] Lack of empirical familiarity with the phenomena of nature can result in 'speaking in non-natural scientific ways' (*DC* III 1, 298b14–18), in being less able to posit more comprehensive principles (*Gen. et Corr.* I 2, 316a5–10), or even in positing limited or false principles (*GA* II 8, 747b27–748a16). For mistakes made by his predecessors which according to Aristotle resulted from this kind of inexperience, cf. also *DC* III 7, 306a5–17; *GA* II 8, 747a34–b10; *Gen. et Corr.* I 8, 325a13–15; and *Gen. et Corr.* II 9, 335b17–24. For Aristotle, natural scientific principles—or even whole theories—that do not harmonize with what can be observed (and, as stated above, that there *is* natural change and plurality Aristotle believes is evident to perception and ἐκ τῆς ἐπαγωγῆς) need to be rejected (*GA* III 10, 760b28–33 and *NE* I 8, 1098b11–12).

[14] Philoponus, *In Physica* 169.15–16: ὑπὸ ἀπειρίας διαλεκτικῆς and 170.22: διὰ τὴν ἀπειρίαν τῆς διαιρετικῆς μεθόδου; Themistius, *In Physica* 29.25: ὑπὸ ἀπειρίας τοῦ διαλέγεσθαι.

[15] Cf. *Met.* Γ 4, 1005b35–1006a11.

[16] Cf. *Met.* Γ 3, 1005b1–4: 'As for the attempts of some of those who speak about the truth, as to the way it should be accepted, they do this because of a lack of educatedness in analytics. For people should

How, then, are we to understand the puzzle itself? Given the ambiguity of the Greek verbs τὸ εἶναι and τὸ γίγνεσθαι, which can be taken either existentially (without a complement) or predicatively (with a complement),[17] the puzzle could be read as either pertaining to how things can come into existence (as is suggested by its pairing with the verb for passing away in 191a27 and Aristotle's conclusions that it led these philosophers to deny plurality and change altogether),[18] or about how things can become something. In the latter case, Aristotle's formulation of the puzzle is elliptical and requires a complement, such as a general quality F[19] or a substance.[20] Under this interpretation, the puzzle is about how a something can come to be F from what is or is not F, or about how something can come to be from what is a substance or from what is not a substance. The reading I follow here is the existential reading as proposed by Clarke: what comes into existence must do so from what exists or what does not exist, where what comes into existence refers not to the precursor or the source of coming to be, but rather to the generated thing itself. The problem, then, is as follows: the generated thing cannot come into existence from its being something that exists, for it already exists, nor can it come into existence from it being something that does not exist, for how could a non-existing thing become something existing, without it having an already underlying—existing—subject? It would appear, then, that nothing can come into existence, and that there is no change and no plurality—which are the opinions embraced by the Eleatic monists.

3. Aristotle's Specification of the Old Philosopher's Mistake (191a34–b13): Conceptual Confusion

191a34: [a] We, on the other hand, say that to come to be from what is or from what is not—or [saying] that *what is not* or *what is* 'does something' or 'undergoes something', or that it 'becomes some particular this'—in one way does not differ from a physician doing or undergoing something, or from being or becoming something from a physician. Because of this, since this [i.e. the statement about the physician] is said *in two ways*, it is clear that also 'from what is' and 'what is does or undergoes' [are said in two ways].

come with prior scientific knowledge of analytics, and not look for it while listening [to lectures in the science they are learning]'. Transl. Reeve 2016.

[17] I am setting aside veridical readings ('that something is the case'), since those only apply to the verb τὸ εἶναι and are clearly not at stake in this context.

[18] For a defence of the existential reading, see Loux 1992 and, most recently, Clarke 2015.

[19] For a defence of the predicative reading of the puzzle as concerning the coming to be of F from what is F or not F, see Ross 1936 and, most recently, Anagnostopoulos 2013.

[20] For a defence of the predicative reading of the puzzle as concerning the coming to be of substances, see Kelsey 2006.

> 191b4: [b] Thus a physician builds a house, not qua physician, but qua house-builder, and he comes to be pale, not qua physician, but qua dark. But he acts as a physician or comes to be a non-physician qua physician.
>
> 191b6: [c] And since we say that a physician does or undergoes something or comes to be something from being a physician most appropriately when he undergoes that or does that or comes to be that qua physician, it is clear that also 'to come to be from what is not' means [exactly] this, namely '[to come to be from what is not] qua what is not'.
>
> 191b10: [d] These [old philosophers] precisely failed to make this distinction, and gave up, and on account of this ignorance they gained such a large additional ignorance (*διὰ ταύτην τὴν ἄγνοιαν τοσοῦτον προσηγνόησαν*), that they believed that there is no coming to be or existence of other things [other than being itself], and instead did away with all coming to be.

In what follows, Aristotle clarifies how *he* understands the puzzle and explains that, if one makes the appropriate kind of distinctions, part of the puzzle (namely the impossibility of there being coming to be from what is not)—and thereby the need to deny all generation—is solved. The specific mistake, then, that Aristotle attributes to the old philosophers is that of an imprecise conceptual analysis in their unpacking of the puzzle due to their inexperience with the proper method (191b10–13): they failed to draw the correct distinctions with regard to what it means for something to 'come to be from what is not' and also to what it means for something to 'come to be from what is' (191b17–18). By not speaking correctly and by failing to analyse existing ways of speaking about coming to be and thereby neglecting to draw the correct distinctions, the Eleatics failed to get a correct grasp of the phenomena under investigation.

For Aristotle, natural or ordinary language often carries ontological presuppositions, which, when analysed correctly, can reveal something about reality itself. On the assumption that the structure of language often mirrors the ontological structure of nature, Aristotle believes that one should be able to discover these ontological structures by carefully and methodologically examining the way things are said. Thus, for instance, Aristotle claims that people were right in keeping water-dwelling and flying animals separated and in not grouping them together by giving them one higher, more general name, because these animals are in fact separate greatest kinds (*PA* I 4, 644a12–16). And elsewhere in the *Physics*, he suggests that by surveying the way in which people ordinarily speak about causal explanation yields—again, when analysed correctly—a distinction between four types of causes in reality (*Physics* II 3, especially at 194b24; 194b34–5; 195a3–4; and 195a15). On the other hand, 'by not making distinctions concerning in how many ways necessity is said' (*οὐ διελόμενοι ποσαχῶς λέγεται τὸ ἀναγκαῖον*), Aristotle's predecessors—for instance—failed to recognize that necessity belongs to natural things in different ways (i.e. as material and as conditional necessity, with the former being a type of unconditional necessity that belongs most properly to the heavenly realm),[21] and because of this failed to offer the correct causal explanations of natural phenomena (*PA* I 1, 639b21–30). Similarly, by failing to recognize that 'perception' and 'objects of perception' are said in two ways, namely in actuality and potentiality, they ended up offering wrong theories of sight and taste (*DA* III 2, 426a20–5). Their mistake in these cases, as Aristotle puts it, is to speak simply about things that cannot be said simply (*DA* III 2, 426a26: ἀλλ' ἐκεῖνοι ἁπλῶς ἔλεγον περὶ τῶν λεγομένων οὐχ ἁπλῶς); they failed to analyse things further.[22] For Aristotle it is thus always important to investigate or analyse in how many ways something is said,[23] for that is part of the path 'to truth and scientific knowledge'.

The distinction his predecessors failed to make and that Aristotle has in mind here is, of course, between two ways in which things can come to be from what is and from what is not. In his rendition of the puzzle, Aristotle slowly builds up to his own distinction by first [a] illustrating how things are often said in two ways by using an example that is presumably meant to be 'clearer to us', then [b] by spelling out these two ways while using the qua-locution (ᾗ), by which it is possible to pick out something under its causally relevant or most appropriate description, and finally [c] by applying the more appropriate of these two ways of speaking to the case of change in general, which is presumably the case that is 'clearer by nature'.

In section [a], which runs from 191a34 to 191b4, Aristotle first unpacks 'coming to be from what is or what is not' in terms of his own categories of relevant types of changes, namely those of qualitative change (I assume that action and passion refer to alteration here) and substantial generation. He then explains that in one way these more general descriptions of change are no different from the particular example of a physician being the subject of those qualitative changes or generations: that is, although there is a difference between the level at which those descriptions are provided (i.e. general versus particular), the changes themselves are the *same in kind* and therefore in this one way identical. And finally, he explains that since in the case of the particular example it is clear that something 'which is' is the subject of change in two ways, we are justified in concluding (191b3: ἐπειδή) that the same differentiation in speaking must also belong to the general description of which the particular case is an instance.

[21] The exact interpretation as to what types of necessity are distinguished here is controversial: see Lennox 2001, 127–8 and Leunissen 2010, 100–5.

[22] Cf. also *Gen. et Corr.* I 2, 315a29–b15 where Aristotle praises Democritus for having investigated change less superficially and in a manner that is more appropriate to natural science than Plato, but criticizes both for not having made any distinctions concerning different types of change.

[23] For more examples in the *Physics*, see *Physics* I 2, 184b20–1; I 5, 188a30–1; II 2, 193b22; III 4, 204a2–3; IV 3, 210a14; and IV 13, 222b28.

The two ways in which the physician is a subject of change in the particular example is explained in section [b], which runs from 191b4 to 191b6, with the help of the qua-locution: in some cases we say that a physician is the source or subject of change not qua physician, that is, *not qua* or *in virtue of what he [essentially] is* and hence *in that sense* is a source or subject of change '*qua what is not*', while in other cases we say that a physician is the source of change qua physician, that is, he is such a source *qua* or *in virtue of what he [essentially] is* and hence in that sense '*qua what is*'. In other words, when someone says that 'a physician builds a house' he is—at least in a sense—saying that 'there is coming to be from what is not', as the house does not come to be from the physician qua physician, whereas when he says that 'a physician acts as a physician', he is—again, in a sense—saying that 'there is coming to be from what is' since it is qua physician that he is the source of change when acting as a physician.

The application of the analysis from the particular case back to the general case is provided in section [c], which runs from 191b6 to 191b10. There Aristotle explains that since in the case of the physician we say that the physician is the subject and source of change *most appropriately* (μάλιστα . . . κυρίως) when he is such *qua* physician—that is, when we say that there is coming to be from something in virtue of what that thing is—it is clear that 'coming to be from what is not' means 'coming to be from something qua or in virtue of what that thing is not'. In other words, 'coming to be from what is not' is said most appropriately when we say that something comes to be—not from something that does not exist in an absolute sense (such as a house coming to be out of nothing, or any kind of *creatio ex nihilo*)—but from something that is not something in a certain (causally) relevant sense (such as a house coming to be out of the building activities of a physician, who in that case is acting, not qua physician, but qua house-builder; for it is the art of house-building that has as its proper product houses, not the art of medicine). The parallel between speaking about changes pertaining to the physician and speaking about coming to be from not being, then, is that in both cases there are two ways of speaking, but only one is most appropriate (and is the one the Eleatics should have employed), namely the one that picks out the source of change in the way that is causally relevant to the kind of change discussed.

So again, Aristotle's criticism of the Eleatic monists pertains to their lack of conceptual accuracy in the formulation of that puzzle, with the subsequent failure of getting the phenomena right. That is, the puzzle results from an initial linguistic confusion about how people ordinarily speak about coming to be, followed by a failure to analyse these ways of speaking correctly: for such an analysis, if done well, can make evident the true structure of ontology. According to Aristotle, making the correct distinctions in how one speaks about coming to be reveals that there are more and less appropriate ways of saying that something comes to be from what is or from what is not and this, subsequently, provides an important step towards dissolving the puzzle. For, as it turns out, it is only because existing things can be picked out *qua* their *non* (causally)-relevant descriptions, that there is a sense in which they can be the subject and source of generation qua what is not. Having thus outlined this distinction between the two ways in which 'coming to be from what is and what is not' is said and having made clear that speaking in one of these ways is more appropriate because it tracks nature, Aristotle can now turn to his own solution of the puzzle, which requires the introduction of his own technical distinctions and principles.

4. Discussion of Aristotle's Solution (191b13–27): Distinguishing Between 'Absolutely' versus 'Incidentally'

191b13: We ourselves also say that there is no coming to be from what is not in an unqualified sense; however, there is coming to be from what is not in a way, such as [coming to be from what is not] *accidentally* (for [there is coming to be] from a privation, which in itself is something that is not [i.e. it is the absence of form in something capable of possessing that form],[24] since it [i.e. the privation] does not persist when something comes to be.[25] This causes surprise, and it seems that it is impossible for something to come to be in that way, from what is not).

191b17: In the same way there is also no coming to be neither *from* what is nor *of* what is, except accidentally. In this way [i.e. accidentally] there is also coming to be of this [i.e. of what is], in the same way as for instance if an animal were to come to be from an animal and a certain animal from a certain animal: for instance, if a dog were to come to be from a horse.[26] For [in such a case] the dog would not only come to be from a certain animal [i.e. a horse], but also from [the genus] animal, but [he would] not [come to be from animal] qua animal, for that [namely, being an animal] already belongs [to it].

[24] Cf. Aristotle's discussion of privation in the previous chapter, *Physics* I 7, 190b17–191a3.

[25] There was some debate at the Symposium about how best to read the elliptical phrase ἐκ γὰρ τῆς στερήσεως, ὅ ἐστι καθ' αὑτὸ μὴ ὄν in 191b15–16, especially because one would expect an ἥ rather than an ὅ to refer back to the privation. I here follow the interpretations already given by Simplicius (*In Physica* 236.12–237.2) and Philoponus (*In Physica* 178.1–11): on this reading, Aristotle claims that there is coming to be from not being, in a sense, since there is coming to be from a privation, which is itself a per se non-being, but this coming to be is not itself per se from what is not, since the privation does not persist in the thing that comes to be; what comes to be thus comes to be from what is not in an accidental way (the only per se coming to be is from matter).

[26] I deleted Ross's addition of [ἐκ κυνὸς ἢ ἵππος] in 191b20–1 (yielding 'for instance if a dog would come to be from [a dog or a horse from] a horse'), based on the suggestion by Laas 1863, and preserved the manuscript reading of EJP, which is also followed by the commentators and by most of the recent scholarship on Aristotle: see e.g. Anagnostopoulos 2013, 269 n.42; Clarke 2015, 143; and Kelsey 2006, 351 with n.35.

191b23: And if something is going to become an animal in a non-accidental way, then it will not [come to] be from animal, and if [something is going to become] something that is [in a non-accidental way], [it will not be] from something that is, nor [will it come to be] from what is not. For it has been said what 'from what is not' means to us, namely 'qua what is not'.

191b26: In addition, we also do not do away with [the claim that] everything is or is not.

Aristotle's discussion of his own solution of the ἀπορία—i.e. the solution that is supposed to provide the only way out of the puzzle—falls apart into two separate, but allegedly parallel parts (see 191b17: ὡσαύτως),[27] where Aristotle first deals with the question of 'coming to be from what is not' (in 191b13–17) and next with that of 'coming to be from what is' (191b17–26). Aristotle concludes with a statement highlighting a second point of agreement with his predecessors (191b26–27: ἔτι δὲ καὶ . . .).

In the first part of Aristotle's solution of the puzzle, he first confirms that he agrees with the statement of the old philosophers that there is no coming to be from what is not if what is not is *said simply* or *in an unqualified, absolute sense* (191b14: ἁπλῶς). That is, Aristotle agrees that there is of course no coming into existence from what really does not exist. However, he denies that there is no coming into existence from 'what is not' if 'what is not' is said *in a certain sense* or *in a certain qualified way* (191b14: πώς), such as for instance when 'what is not' is said in an accidental way (191b15: κατὰ συμβεβηκός). Aristotle provides an example of such coming to be from what is not that builds on distinctions he made in the previous chapter concerning the coming to be of natural substances (*Physics* I 7, 190b17–27): for there is coming to be *in an accidental way* from what is not from a privation (cf. *Physics* I 7, 190b27: ἡ δὲ στέρησις καὶ ἡ ἐναντίωσις συμβεβηκός), as a privation is something 'which in itself is not' (191b15: καθ' αὑτὸ), in the sense that it is the lack or absence of the relevant form of something, and in the sense that it is something that does not persist through the change, as it is replaced in the change by its contrary (see *Physics* I 7, 190a9–13 and 190a18–21). In this way, Aristotle further specifies the distinctions that he had made earlier with the help of the qua-locution in the ordinary uses of the language of 'coming to be from what is or what is not' in terms of his own technical distinctions between what is the case absolutely speaking or in an unqualified way versus what is the case in a qualified way, accidentally, or in itself.[28] Identifying 'what is not qua what is not' as the privation thus solves at least one part of the puzzle (which itself is explained in 191b16–17 in terms of the surprise that results from 'thinking that things cannot come to be in that way, from what

[27] On the importance of reading the two arguments as parallels, see Kelsey 2006, 344.

[28] Cf. *Met.* H 2, 1042b7–8.

is not') by reference to one of his own principles of change, namely that of privation: for something can come to be accidentally from what is not, namely from a privation, but since this privation always belongs to some underlying nature, this coming to be will not be 'coming to be from what is not' in an absolute, unqualified sense. Aristotle's solution thus meets the challenge raised in 191a31 that 'there must be something underlying' (ὑποκεῖσθαι γάρ τι δεῖν) the coming to be of something.

In the second part of his solution of the puzzle, Aristotle states that the same distinctions also apply to the 'coming to be from what is or of what is',[29] namely that there is only such coming into existence accidentally. In other words, while Aristotle agrees with the Eleatic monists that there is no coming into existence from what is in an absolute, unqualified sense—for what is already exists or is the thing it is and therefore cannot come into being (this is the objection raised in 190a30: εἶναι γὰρ ἤδη), he denies that things cannot come to be *in a certain way* from what is. And, presumably, given Aristotle's presentation of the second part of the solution as a parallel for the first part, and given that his general aim in this chapter is to show that only by using his own principles of change the puzzle of the old philosophers can be solved, the kind of 'coming to be from what is' Aristotle will want to hold on to must be a coming to be from the underlying nature and form (cf. *Physics* I 7, 190b19–20: ὅτι γίγνεται πᾶν ἔκ τε τοῦ ὑποκειμένου καὶ τῆς μορφῆς). If this is correct, then Aristotle's argument must show that coming to be from the underlying nature and form are—not cases of accidental coming to be (for Aristotle ruled this out in *Physics* I 7), but—cases of accidental coming to be *from what is*.

Aristotle uses a somewhat odd thought experiment to illustrate this point. For one might think that the coming to be of animals from animals at the genus level, or the coming to be of a certain animal from a certain animal at the species level, are not at all accidental coming-into-beings from what is. Surely it is not an accident that 'humans beget humans' as Aristotle states himself repeatedly.[30]

[29] The shift in the formulation of the puzzle from 'coming to be from what is' to 'coming to be from what is or of what is' need not be taken as a break with the parallel argument concerning coming to be only 'from what is not' (where 'from what is not' is most naturally read as the source for the coming to be) and not also concerning 'of what is not' (which would most naturally be read as the outcome of the coming to be). For, as Kelsey 2006, 352 explains, Aristotle's use of the expression 'coming to be of what is' also refers to the source and not to the outcome of the process of coming to be: see *Physics* I 7, 190a6: λέγεται τόδε γίγνεσθαι ἀλλὰ καὶ ἐκ τοῦδε, and *Physics* I 8, 191a30–1: οὔτε γὰρ τὸ ὂν γίγνεσθαι (εἶναι γὰρ ἤδη) ἔκ τε μὴ ὄντος οὐδὲν ἂν γενέσθαι.

[30] See, e.g., *PA* I 1, 640a22–7: '[Empedocles] failed to see, first, that the seed previously constituted must already possess this sort of potentiality, and, next, that the producer was prior not only in definition but also in time; for it is a human being who generates a human being (γεννᾷ γὰρ ὁ ἄνθρωπος ἄνθρωπον), such that it is because the one *is* such, that the other's coming to be happens in that way', and *PA* II 1, 646a33–5: 'for a human begets a human and a plant a plant from the underlying matter of each'. See also *Physics* II 2, 194b13; *Physics* II 7, 198a26–7; *Met.* Z 7, 1032a25; *Met.* Z 8, 1033b32; *Met.* Λ 3, 1070a8 and

What Aristotle is trying to make clear, however, is that these are *accidental coming-into-beings from what is*. Take the hypothetical (and impossible)[31] case of a dog coming to be from a horse: in that case, it is immediately clear that, at the level of the species at least, the coming to be of the dog is an accidental coming to be from a horse, since the dog is not the horse (the offspring is not identical, in any way, to its parent).[32] Similarly, but more difficult to see as it requires the correct application of the qua-locution, the coming to be of animal from animal in that case is also an accidental coming into being from what is, for the way in which animal comes to be is not identical to the way in which animal already belonged to what is coming to be. In neither case, then, there is coming to be from what is *simply*.[33]

Aristotle concludes this section by pointing out that his solution to the puzzle, namely that there is only coming to be from what is not or from what is accidentally 'does not do away with the claim [of the old philosophers] that everything is or is not': for by correctly distinguishing between how what is or

1070a27–8 ('for a human begets a human, and an individual [begets] each individual'); *Met.* Λ 4, 1070b34; and *Met.* N 5, 1092a16. Cf. also Plato, *Crat.* 393b7–c6.

[31] Technically speaking, this kind of 'kind-crossing' is contra to nature according to Aristotle. See e.g. *Met.* Z 8, 1033b29–1034a2: 'Indeed in some cases it is even obvious that that which generates is of the same kind as that which is generated—not however identical with it, nor numerically one with it, but the same in species (τοιοῦτον... τῷ εἴδει), such as in the natural [generations]—for human begets human—unless something happens contrary to nature, such as when a horse [begets] a mule. And even these cases are similar (καὶ ταῦτα δὲ ὁμοίως): for that which would be common to both horse and ass does not have a name, namely the kind that is closest to them, but it would likely be both, just as that of the mule [is both]'.

[32] Cf. perhaps *PA* I 1, 641b33–6, where Aristotle makes a distinction between two ways of speaking about seed, one as the seed 'from which', such as from a horse, and one as the 'of which' or as what will be from it, such as a mule: in this example, the seed from a horse is what underlies the substantial change, but the substantial being that results is a mule, such that there was coming to be from what is, but not qua what the resulting substantial being already was.

[33] I would like to thank Marwan Rashed for help with this difficult passage. Pace Kelsey 2006, 355–66, who explains this example as being a parallel case of air coming to be from water, where air comes to be from something that is air-material but also happens to be water, so that a dog comes to be from something that is dog-material but also happens to be a horse. The point of this example is, according to him, that while substances (such as air) *can* come to be from substance-materials that are a substance (such as water) incidentally, substances *cannot* come to be from substance-materials that are the same substance of the specific kind they are to become.

This reading may be foreshadowed by Philoponus, who takes the example to be entirely about the material causes of natural generation (2009, *In Physica* 179.3–10): 'For, he says, we say that animal comes from animal, but here we use the expression "from animal" to mean from the material cause (ὡς ἐξ ὑλικοῦ αἰτίου), since that is the topic of discussion in this case. For we do not mean it in the manner of "human being from human being" (meaning as efficient cause), but rather in the manner of wasps [coming] from a horse, and bees from a bull. For horses are the source of wasps and bulls of bees. So whenever we say that animal comes from animal, e.g. wasps from horses, obviously we are not saying the following: that qua animal the horse changes into animals, namely wasps (since the horse does not serve as the underlying matter for the wasps). Rather, we are saying the following: that from something of which being is predicated, another thing, which also exists, develops *per accidens*.'

what is not is said and by recognizing that not everything is or is not in an absolute sense, it becomes clear both that and how there is room for genuine coming into existence and passing away in the sublunary, natural world.

5. A Variation of the Same Solution (191b27–9): Distinguishing Between Potentiality and Actuality

> 191b27: This, then, is one way [of solving the puzzle], and another is that it is possible to speak about the same things in accordance with potentiality and actuality. This has been defined with more rigour in other [treatises].

The other way in which the Eleatic dilemma can be solved involves another way of introducing technical Aristotelian notions that allows Aristotle to show that there is coming to be from what is and from what is not in a qualified way, namely by differentiating existing and coming to be in potentiality and in actuality. Aristotle is thus not introducing a completely different solution, thereby going against his earlier claim at the opening of the chapter that there was only one way of solving the problem: both solutions rely on a more careful conceptual analysis of the process of coming to be than his predecessors had engaged in.

The more rigorous discussion of the distinction between potentiality and actuality can presumably be found in Aristotle's *Metaphysics* Z–Θ, but for the way in which this distinction helps to solve the puzzle of the old philosophers, the most relevant discussion is perhaps the one Aristotle offers in the methodologically later treatise *On Generation and Corruption*. There Aristotle returns to this very same puzzle (*Gen. et Corr.* I 3, 317b13–15): although it was solved and discussed 'at greater length' in another work, he claims that the current account requires a concise (συντόμως) discussion of it as well. By drawing a distinction between 'being in potentiality' and 'being in actuality', Aristotle is able to clarify or reformulate the claim that 'coming to be is always from what is or from what is not' as 'coming to be is always from something which is in potentiality, but in actuality is not' (*Gen. et Corr.* I 3, 317b16–17: δυνάμει ὂν ἐντελεχείᾳ δὲ μὴ ὄν), such that the underlying matter from which there is coming to be can be said to be both 'what is' and 'what is not'. Even so, Aristotle says, the puzzle of whether there can be coming into existence without qualification at all—that is, whether substance comes to be at all—is extraordinarily difficult (*Gen. et Corr.* I 3, 317b18–22; 317b18: θαυμαστήν). The solution Aristotle proposes, then, is—in a simplified version—that substances come to be out of what is that kind of substance potentially (and perishes into what is that kind of substance potentially) but not actually, and that a substance thus comes to be out of something that exists *in a sense* and that already is what it is going to become *in a sense* (such that there is something that underlies the coming to be and there is no coming to be out of nothing), but that is not yet a 'this' or an actual substance of that kind and *in that sense* is not (such that it is not *in actuality* already what it will become).

Both variations of the solution of the puzzle thus require distinctly Aristotelian ways of speaking and principles that are proper to his natural science.[34]

6. Conclusion (191b30–4): 'Observing This Nature'

> 191b30: So, as we said, the puzzles on account of which they were forced to do away with some of the things we talked about are solved: for it was because of this that also some of the earlier [philosophers] were diverted to such a great extent from the road (τοσοῦτον ... ἐξετράπησαν τῆς ὁδοῦ) towards generation and perishing and change in general. For if they had observed this nature (αὕτη γὰρ ἂν ὀφθεῖσα ἡ φύσις), it would have dissolved them from all their ignorance.

In the final section of the chapter, Aristotle reiterates that at this point he has solved the puzzles and that the reason for why the old philosophers lost track of the road of the truth was their failure to see 'this nature', which most likely refers to the 'underlying nature' (cf. *Physics* I 9, 192a10: ὅτι δεῖ τινὰ ὑποκεῖσθαι φύσιν and 192a30–2).[35] In chapter 7, this underlying nature had been identified with matter as the principle of being, and with matter plus privation as the principles of becoming, and in chapter 9, Aristotle criticizes other predecessors for having touched on this nature, but not in a sufficient manner, and then turns to their understanding of coming to be out of *what is not*, and finally clarifies the distinction *he* makes between the matter and the privation in coming to be. Had his predecessors been sufficiently experienced with the appropriate methods of investigation, they would not have let these puzzles drive them off the road of truth; and, as Aristotle hopes to have shown, his own technical distinctions are required to get us back on track and to solve the dilemma.

[34] See, e.g., *Gen. et Corr.* I 9, 326b29–31: 'In what way coming to be and action and passion belong to the things that are, let us speak about this taking the principle we have often stated (λέγωμεν λαβόντες ἀρχὴν τὴν πολλάκις εἰρημένην). For if to be such is on the one hand in potentiality and on the other in actuality'.

[35] *Physics* I 7, 191a7–8: ἡ ... ὑποκειμένη φύσις. See Themistius, *In Physica* 31.25–7; cf. also Charlton 1970, 81. Philoponus understand 'this nature' metaphorically as referring to the distinctions between 'absolutely' and 'incidentally', and between 'actuality' and 'potentiality' (*In Physica* 182.3–6).

10

Responding to the Platonists

Physics I 9

SARAH BROADIE

Introduction

In *Physics* I 9, Aristotle turns to a Platonist theory of coming to be. It recognizes just two principles as opposed to the three—subject, form, and privation—which he himself has argued for so far. Aristotle is hostile and dismissive towards the dyadic schema. He presents it as a confused deviation from his own view: the Platonists have come up with two principles because on the one hand like him they recognize both form and the 'underlying nature' but on the other hand they fail to distinguish the latter from privation. This mistake, in their case, has not led them to despair of making sense of coming to be, as happened with some of the earlier thinkers according to Aristotle (see especially *Physics* I 8). On the contrary, the Platonists are hospitable to the notion of coming to be, but have failed to analyse it correctly.

In assessing this Aristotelian criticism, we must keep in mind the extensiveness of the Platonists' dyadism. Its scope went far beyond the domain of physics and physical coming to be. But, as we shall see, Aristotle has reason to reject dyadism only in the domain of sublunary change (and in particular, as I shall argue, in the biological part of that domain). At a different level, that of the eternal motion of Aristotle's celestial spheres, his own approach will in fact turn out to look remarkably similar to the Platonist dyadic one.

My thanks to Andreas Anagnostopoulos, Sean Kelsey, Diana Quarantotto, and Barbara Sattler for valuable comments.

We shall come to these questions after presenting I 9 itself. The passage divides into two main discussions, 191b35–192a27 (which I call Argument 1) and 192a27–34 (Argument 2). These are followed by a coda, 192a34–b4. Argument 1 brings out differences between a Platonist account of coming to be and the triadic scheme as developed earlier in *Physics* I, and points to absurdities of the former. Argument 2—which may or may not be connected with the Platonists[1]—brings out a sense in which, in Aristotle's own scheme, the matter or 'underlying nature' is imperishable and ingenerable.[2] Section 1 provides a translation with comments on the immediate text, and the following sections discuss philosophical background and questions arising.

1. Translation and Comments

1.1. Translation

Argument 1, 191b35–192a27:

> Others too have touched on it, but not adequately. They first agree that a thing simply comes to be from not-being [192a1] (in this respect Parmenides in their view is right), and then it appears to them that if [something] is one in number, it is also only one in function (δυνάμει). But this is a very different point.
>
> Our view is that matter and privation are different, and that one of them—namely, the matter—is that-which-is-not in virtue of concurrence, [a5] whereas the privation is [not-being] in virtue of itself; also that one of them—namely, the matter—is near [to substance] and is substance (οὐσία) in a way, whereas the other is not so in any way. For them, by contrast, that-which-is-not is the Great and the Small, without differentiation (ὁμοίως)—whether what one takes is the pair together or each of the two by itself. So this latter way of having a triad is completely different from the previous one [sc. Aristotle's]. They did get far enough [a10] to see that a nature (φύσιν) of some kind must underlie, but they make it one (μίαν). For even if someone makes it a dyad, calling it Great and Small, he is still doing the same thing: he overlooked the other [nature]. For the one [nature] remains, and with the form (μορφῇ) is joint cause, like a mother, of the things that come to be; but the other part of the contrariety, if one focuses one's thought on its maleficence (τὸ κακοποιὸν αὐτῆς), might often (πολλάκις) [a15] give the impression of not even being at all.
>
> For granted that there is (1) something divine and good and desirable, we say that there is on the one hand (2) its contrary and on the other hand (3) that which by

[1] See Section 2, paragraphs 18–19.

[2] This division puts 'ὡς μὲν γὰρ τὸ ἐν ᾧ . . . στέρησις' into Argument 1 and the answering sentence 'ὡς δὲ κατὰ δύναμιν . . .' into Argument 2 (192a26–9). If this seems too awkward one can begin Argument 2 at φθείρεται (25), following Ross.

nature desires and reaches towards it in accordance with its own nature. But the implication for those [thinkers] is that the contrary reaches towards [a20] its own destruction. Yet it is impossible both that the form (εἶδος) desires itself—because it is not in need of anything—and that the contrary desires it—since contraries are destructive (φθαρτικὰ) of each other. Rather, that [which desires the form] is the matter, like female desiring male and ugly desiring beautiful—except that [the matter] is not something ugly nor yet female in virtue of itself but in virtue of [a25] concurrence. And in one way it perishes and comes to be, but in another way it does not. As that-in-which (τὸ ἐν ᾧ), it perishes in virtue of itself—for what perishes, the privation, is in this.

Argument 2, 192a27–34:

But considered in terms of its function (ὡς . . . κατὰ δύναμιν) it [sc. the matter or the underlying nature] does not [perish or come to be] in virtue of itself; it must, rather, be imperishable and ingenerable. For if it was coming to be there would have to be some primary underlying thing from which [a30] as constituent it comes to be: but this is the nature itself [i.e. the material principle], so that it would be before it came to be. For by matter I mean the primary thing that underlies in each case, from which as constituent something comes to be not in virtue of concurrence. And if it perishes, that is where it will finally arrive, so that it will have perished before it perished.

Coda, 192a34–b4:

As for the form-principle (τῆς κατὰ τὸ εἶδος ἀρχῆς): whether it is [a35] one or many, and which it is or which they are, are matters whose detailed determination is the task of first philosophy; so we may set them aside for that occasion. In the expositions which follow [192b1] we shall be speaking of natural (φυσικῶν) and perishable forms εἰδῶν) [or: kinds]. That there are principles, and which they are, and how many in number, can be taken as fully discussed. Let us then mark a new start in what we now go on to say.

1.2. *Comments*

191b35: Others too have touched on it, but not adequately.

'Others too': the Platonists or some of them. The referent of 'it' is 'this nature', 33–4, which must refer back to 'the underlying nature' (or: 'the underlying matter' on Simplicius's reading) at 191a8. Aristotle often speaks of 'touching on' a topic or theory, e.g. *Physics* III, 203a2; *Metaphysics* A, 988a23; 29; 32. The perfective conveys that these thinkers are current presences for Aristotle, although this does not imply chronological simultaneity; at *Metaphysics* A, 988a29 'have touched on this sort of cause' is said of Empedocles and Anaxagoras along with Plato.

> 191b36–192a1: They first agree that a thing (τι) simply (ἁπλῶς) comes to be from not-being (in this respect Parmenides in their view is right)

That is: they agree with Parmenides in making 'from not-being' a condition of coming to be, a condition which he, of course, held to be unintelligible. Does ἁπλῶς qualify γίγνεσθαι alone (this seems to be the understanding of Hardie and Gaye 1984), or γίγνεσθαι τι ἐκ μὴ ὄντος (cf. Charlton's 1970 translation), or ὁμολογοῦσιν (cf. Charlton's note, p. 47)? Logically, the last two options make the same point: 'they understand coming to be out of not-being as a simple issue' is tantamount to 'they understand coming to be from not-being in a simple—i.e. too simple—way'. This is the criticism developed in Argument 1; the ground for it has been laid in I 8. Hardie and Gaye's construal suggests the contrast between coming to be *simpliciter* by contrast with coming to be so and so, e.g. cultured; but, as Charlton points out, this distinction plays no part in the ensuing argument.

According to Ross (1936, 497; cf. Charlton 1970, 81) the view attributed here to Parmenides is the dilemma which according to Aristotle baffled the ancients: 'Necessarily, what comes to be comes to be either from being or from not-being, but both are impossible' (191a23–31). But why mention just one side of the dilemma? The allusion is more likely to be to Parmenides' argument that (a) coming to be is unintelligible because (b) it would be from not-being, and (c) not-being is unintelligible (fr. 8, 7–9). Aristotle says that the Platonists agree with Parmenides on (b); he implies that they reject (c) and therefore reject (a). Parmenides equated not-being with not-being anything at all, but in the light of Plato's *Sophist*, the Platonists would hardly have agreed to this. Rather, they agree with Parmenides in emphasizing the not-being aspect of the *terminus a quo* of becoming. This is by contrast with those natural philosophers who saw it as impossible that things should come to be from what is not and concluded that they must come to be from what-is, namely their imperceptible constituents (187a32–b1). The Platonists are right, Aristotle implies, to recognize the 'from what is-not' aspect of becoming, but he faults them for overlooking the matter-principle, which is not not-being *per se* but is only coincident with not-being (192a3–5; cf. 191b13–16); this simultaneously secures the 'from what-is' aspect in a non-paradoxical way.[3]

> 192a5–6: the matter is near [to substance] and *is* substance (οὐσία) in a way

On matter as nearly substance cf. *Metaphysics* H, 1042b9; 1044a15; Θ, 1049a36. In the Platonism targeted here, only the formal principle would count as οὐσία;

[3] I am especially indebted to Sean Kelsey here.

hence the material factor, although not absolutely nothing—it is the Great and the Small—is very far removed from οὐσία. Cf. *Timaeus* 52d3 where the formal principle is ὄν, whereas the Receptacle or χώρα is said to be πάντων ἐκτὸς εἴδων and intelligible only in the most puzzling way (51a7-b2).

> 192a6–12: For them, by contrast, that-which-is-not is the Great and the Small, without differentiation (ὁμοίως)—whether what one takes is the pair together or each of the two by itself. So this latter way of having a triad is completely different from the previous one [sc. Aristotle's]. They did get far enough to see that a nature of some kind must underlie, but they make it one (μίαν). For even if someone (τις) makes it a dyad, calling it Great and Small, he is still doing the same thing: he overlooked the other [nature].

In the view reported in the first sentence I take 'that-which-is-not' as grammatical subject, with the Great and the Small as complement (thus Ross 1936, 348). This way round (Charlton 1970 adopts the other) gives a better contrast with the first part of the previous sentence, 192a3–5, where *our* view lays emphasis on not-being, i.e. on the difference between the matter's incidental not-being and the not-being imported by the privation *per se*. ὁμοίως, then, means that there is no such differentiation of not-being in the Platonist theory (thus Charlton 1970, 21). This construal is clearer with Ross's comma after ὁμοίως. Without the comma the meaning could be that conflating that which is-not with the Great and the Small comes to the same thing whether one takes the Great and the Small as a single pair or separates them into two (thus Ross 1936, 497 and Hardie and Gaye 1984). In any case, 'this latter way of having a triad' refers to taking 'each of the two by itself', since only so can the Platonists have the semblance of three principles. The Great and the Small was often called a dyad, but its duality allows for no difference between the coincident and *per se* ways of non-being that distinguish matter from privation and, with form, give the three Aristotelian principles. Here Aristotle indicates that the number of things named as principles is unimportant compared to the number of distinct principial roles (cf. 190b29–30; 35–191a7). Whether you look at the Platonists' Great and the Small as *itself* single or double, it or the pair of them only ever has one and the same *role*. Thus even when the Platonists glimpse the underlying nature 'they make it one' (192a10–11), i.e. they make this the only principle additional to the form-principle. At these moments they either 'overlook' (12) the privation altogether or completely merge it with the underlying nature. In sum, this passage features two cases of the same thing appearing as both two and one. (a) The Great and the Small can be taken as double or single (two directions or one range)[4]; as double it counts in a way as two principles,

[4] Thus sometimes Aristotle speaks of the Great and the Small, e.g. 192a7 and 187a17, and sometimes of the Great-and-Small, e.g. 192a11–12; *Metaphysics* B, 998b10; M, 1083b23 and 32; N, 1087b8; cf. I, 992a11–13. Ross 1924 vol. I, lxi, traces the second locution to Plato, the first to Xenocrates.

which gives the Platonists three. However, *this* doubleness of the Great and the Small is unimportant for Aristotle here; it does not make the Platonist account any less inconsistent with his own. (b) Matter, or what underlies, is one factor and not-being or privation is a second; but the self-same Platonist Great and Small (whether seen as double or not) figures sometimes as underlier (9–10), sometimes as that which is-not (6–7), hence in effect it is both. This is the duplicity intolerable to Aristotle.

192a11: someone

i.e. Plato. At 187a17–18 Aristotle makes a similar observation naming him.

192a12: the other [nature]

This is the privation (called 'the other part of the contrariety' at 14). It may seem strange to say that a privation is a nature, especially given the near synonymy at times of φύσις with οὐσία (see, e.g., *Physics* II, 193a9; 20; cf. 192b32–34). But at 193b19–20 Aristotle says that privation in a way is form, on which see Simplicius's explanation reported by Ross (1936 *ad loc.*, 505–6).

> 192a13–16: For the one [nature] remains, and with the form (μορφῇ) is joint cause, like a mother, of the things that come to be; but the other part of the contrariety, if one focuses one's thought on its maleficence (τὸ κακοποιὸν αὐτῆς), might often (πολλάκις) give the impression of not even being at all.

Aristotle speaks *in propria persona* here, but 'like a mother' must be a deliberate echo of *Timaeus* 50d2–4 and 51a4–5 (cf. 49a6 and 52d5 on the 'nurse' of becoming). He is alluding to the Platonist doctrine that their second principle (the Great and Small or the Receptacle) is responsible for evil (*Metaphysics* A, 988a14–15; N, 1091b30–1092a5).

'the other part of the contrariety' refers to the contrary of the form. It is natural to take αὐτῆς at 15 as referring to this contrary.

'τὸ κακοποιόν' is in antithesis to 'something divine and good and desirable' at 17. Ross (1936, 497), explains τὸ κακοποιόν as destructiveness ('association with φθορά'). Perhaps it was a Platonist notion that merely excluding good causes evil.[5] One can only guess at Aristotle's exact turn of thought here. It may be this: if the Platonists look for the contrary of 'something divine and good and desirable' on the assumption that what they are looking for is something *maleficent*, they would in many cases (πολλάκις) miss the contrary and hence

[5] Thus the contrary is the formal cause of evil. On the use of -ποιός for the formal cause, cf. *Nicomachean Ethics* 1144a3–6.

conclude that it simply doesn't exist. (For in many cases a contrary of the good is not a positive evil (cf. *Categories* 12a14–25; *Metaphysics* Δ, 1023a5–7), but is neutral—as indeed the underlying nature is, as such, neutral.) Aristotle is trying to explain how the Platonists might have reached their dyadic position, or anyway the version that equates the Great and the Small with just the nature that remains.

> 192a16–25: For granted that there is (1) something divine and good and desirable, we say that there is on the one hand (2) its contrary and on the other hand (3) that which by nature desires and reaches out for it in accordance with its own nature. But the implication for those [thinkers] is that the contrary reaches towards its own destruction. Yet it is not possible either (οὔτε) that the form (εἶδος) desires itself—because it is not in need of anything—or (οὔτε) that the contrary desires it—since contraries are destructive (φθαρτικὰ) of each other.

The genitive absolute at 16–17 marks common ground between Aristotle and the other side. I follow Ross (1936, 498) in reading ὃ at 18. The pleonasm πέφυκεν . . . κατὰ τὴν αὑτοῦ φύσιν (18–19) stresses the fundamental difference of the third principle from the principle that is the contrary. Aristotle argues that since the Platonists in effect identify these two, they view the contrary of the good as by its very nature reaching towards the good, hence towards its own destruction: which Aristotle seems to think is obviously absurd (cf. *Metaphysics* N, 1092a1–3).

I understand οὔτε . . . οἷόν τε . . . οὔτε at 20–1 as claiming that just as the first limb is impossible, so is the second. No doubt it is undeniable that the form, which is good, cannot hanker for itself (cf. Plato, *Symposium* 200a–b), but it is harder to dismiss the idea that the contrary can hanker for the form, and so in a way for its own destruction. Treating this as absurd arguably begs the question when the passage of arms is with Platonists. If Aristotle simply assumes that nothing of itself can tend towards its own destruction, the other side could reply that this is false because their non-formal principle does precisely that. They could then point out that it makes all the difference whether this contrary-and-underlying principle is understood as able to attain the good at which it aims. If it does attain it, then as contrary it must be destroyed while as underlying it must remain, which is a contradiction given that this is a single principle. But there is no obvious contradiction in thinking of the contrary as necessarily always only *tending towards* the good. And if that is the case there is no reason to distinguish underlier from contrary since they never part company. It would again beg the question for Aristotle to insist that whatever tends towards some end *E* must in principle be able to arrive at *E* so as actually to become or possess it. (On sensible objects as only ever reaching towards the Platonic Form, see *Phaedo* 74d–75a; for a similar structure see *Philebus* 53c–54c; also *Symposium*

201a–c on ἔρως as a non-beautiful power reaching towards the beautiful.) Moreover, the premise 'contraries are destructive of each other' (21–2) is self-evident only in a sense too weak to support Aristotle's conclusion. Self-evidently, contraries are mutually destructive in the sense that *if* a subject were to acquire one it would necessarily lose the other, but it is not self-evident that contraries are in all cases actually realizable potentials to obliterate one another, since this depends on the same question-begging assumption, namely that the subject of one contrary is metaphysically able to become subject of the other. Furthermore, Aristotle himself elsewhere recognizes things that naturally tend towards their own destruction. (a) In *De Caelo* I, 281a28–33, he postulates that finite (in both temporal directions) states of being *P* and not-being *P* (or being not-*P*) have inherent spans of duration, such that each by its own nature brings itself to an end so as to give way to the opposite.[6] (b) In *Physics* III 1 he defines movement or change as something that both *is towards* a terminus and ceases to be once the terminus is reached. Moreover, being in *T*-ward movement is in a sense contrary to being at *T* (or being *T*) in that these are mutually exclusive properties of the subject. Aristotle, of course, emphasizes the differences between the being and actuality of three entities: the movement; the subject of movement which existed before the movement began and will do so when it is finished;[7] and the result, e.g. the built house, which exists when the movement is finished (201a29–b13). But it is perfectly in keeping with Platonism to view the very being of the subject as consisting precisely in an interminable becoming, i.e. an interminable reaching out towards the good or the form, and hence as essentially negative because always falling short of the end. Aristotle places himself in a somewhat analogous situation when conceptualizing the infinite as something that only ever potentially exists since its actuality can only be in an uncompletable succession (*Physics* III, 206a14–28). And what about Aristotle's endless celestial motion in *Metaphysics* Λ? Its ultimate cause is a presumably never-attainable object of desire and love (1072a25–b4).[8] It would seem, then, that Aristotle is over-ambitious in *Physics* I 9 if his aim is to show that an account of coming to be that integrates the underlying or material nature with the contrary, thus recognizing just one non-formal principle, is simply unintelligible.

[6] Cf. *On Generation and Corruption* II, 336b10–18, where the annual approach and recession of the sun provide the conditions for the inherent lifespans.

[7] Aristotle argues for the difference between a given movement *M* and its subject by pointing to cases where the identical thing has the potential to be subject of a movement incompatible with *M*. The Platonists would have focused on cases where the movement is the only one naturally available to the subject, e.g. the development of a growing thing. See Section 2, paragraph 8.

[8] See Section 4 for further discussion of this example.

> 192a22–25: Rather, that [which desires the form] is the matter, like female desiring male and ugly desiring beautiful—except that [the matter] is not something ugly nor yet female in virtue of itself but in virtue of concurrence.

Here as at 14 Aristotle speaks for himself while using imagery congenial to the other side. The female–male image is a powerful tool for them. Clearly, female is *both* contrary to male *and* is the male's συναιτία in reproduction: their contrariety is essential to their complementary cooperation. The Platonists can appeal to this familiar model to explain and defend their dyadic position about coming to be. For once form is identified with male the contrary of form is inevitably cast as female, yet the female is still the fellow-parent, i.e. the material principle. Thus the contrary fails to be distinguished from 'the nature that remains'.

Aristotle himself in different places treats female (a) as contrary to male, and (b) as recipient or matter in relation to the male. For (a) male and female as contraries see *Generation of Animals* IV, 766a21–2 and b16; cf. the Pythagorean list of opposite principles, *Metaphysics* A, 986a22–6. For (b) female as recipient and providing the material principle, see *History of Animals* V, 541a2–3 and *Generation of Animals* I, 729a9–11; b4; 738b20–36. These different characterizations present no logical problem for literal animal generation because, since female and male are different individuals, the contrariety of their natures does not put their cooperative com-presence into question. In the metaphysical analogue it is different: one and the same principle is supposed, as matter, to be cooperatively present alongside form, and, as privation, to exclude it. However, Aristotle does not directly attack this incoherence; instead he seems to display it by rolling from one side to the other. In our present passage he first (presumably speaking for the Platonists) identifies matter as that which desires form, illustrating this with female desiring male and ugly desiring beautiful; and promptly adds (perhaps now speaking in his own voice) that what desires form is not the ugly or the female as such, since these are only concurrent with the matter.[9] The upshot is that female is now being identified with the contrary or privation of beautiful form, even though back at 192a13–14 it was the remaining nature that was like the female parent. This swivelling about the female role—assigned first to matter as such, then to the contrary or privation—may be a deliberate parody of the Platonists' ambivalence about their non-formal principle.

> 192a26: As that-in-which (τὸ ἐν ᾧ), it [sc. the matter or underlying nature] perishes in virtue of itself—for what perishes, the privation, is in this.

[9] It is obscure whether ugliness is mere lack or a positive evil, κακοποιόν, 192a15. Another conceptual muddiness is that female desires (ὀρέγεσθαι, ἐφίεσθαι, 18–20) male in a different sense from that in which ugly desires beautiful: the ugly desires to become, itself, beautiful (cf. *Symposium* 200a–201c), whereas the female partner does not desire that she become male.

The word 'this' must refer back to 'that-in-which': but what does the latter refer to? It cannot refer simply to the subject underlying the whole process from privation towards form, because this subject does not, as such, perish. Some interpreters therefore take the sentence up to 'for' (γὰρ) to mean: 'the matter [sc. underlying the whole process], considered [just] as that-in-which-the-privation-is, perishes as such'. (Thus Ross 1936, 498, with Philoponus and Simplicius.) This gives the right logical sense, but now 'for' introduces a tautology: 'what perishes, namely *the privation, is in that-in-which-the privation-is*'. It is better to take 'that-in-which' as referring not to the matter underlying the whole process, but to the *contrary* itself positively described. The privation of the form is 'in' this contrary in the sense of being based on it or consisting in it: e.g. privation of sweetness in an apple is based on or consists in the positive sourness of the unripe fruit. (For this sense of ἐν see *Physics* IV, 210a20, where the example is 'health in hot and cold factors'.[10] The hot and cold factors are not themselves a subject that changes between being healthy and being unhealthy: they are what health, when present in its subject, consists in or supervenes on.) So at 192a26 Aristotle says in effect: 'the matter considered as positive contrary perishes as such'. On this reading 'what perishes, the privation, is in this' is not tautologous, since 'this' and 'that-in-which' refer to the matter considered as locus (not of privation but) of the *terminus a quo* positively conceived.

The consideration introduced by 'for' could be evidence for what precedes, or an explanation of it, or both. Does Aristotle think it more obvious that the privation perishes than that the positive contrary does, hence he uses the former claim to support the latter? Or does he think that the perishing of the positive contrary is due to or governed by the perishing of the privation? He surely at least thinks this to the extent that he agrees with the Platonists in putting explanatory weight on the divine quality and goodness and desirability of the form (192a16–17), since from this point of view the emphasis is on overcoming privation rather than displacing whatever positive condition may happen to underlie the privation.

192a27–34: this passage (Argument 2) explains a sense in which matter as persisting ὑποκείμενον of coming to be is 'imperishable and ingenerable'.[11] For ingenerability the argument seems to be this: insofar as *X* is coming to be *Y* (e.g. bronze coming to be a statue), *X* cannot be what is thereby coming into being, i.e. *X* cannot be identical with *Y*, the objective or *terminus ad quem*. For otherwise *X* would have come to be only when *Y* has come to be. Since *X* as matter for the coming to be of *Y* is the immediate constituent of *Y from* or *out of* which *Y* will come into being (cf. *Physics* II, 193a10–11), it follows that *X* exists before

[10] At the meeting Marwan Rashed also suggested this interpretation.

[11] On the relation of Argument 2 to Argument 1 see Section 2, paragraph 18.

Y comes into being. Hence if *X* is both matter and *terminus ad quem* (of a given episode of coming to be) *X* will turn out to exist (ἔσται, 30) before *X* comes into being (and so too for *Y*). Therefore the matter, as such, of a given coming to be is necessarily not what is thereby generated; hence in this logically limited context it is necessarily ungenerated.

On imperishability: if *Y* is primarily constituted from matter *X* (*Y* is the statue, *X* is the bronze), and if '*Y* perishes' means that *Y* perishes into its primary material constituent (the statue is melted back into a lump of bronze), the matter in this episode of perishing cannot be what perishes. For the primary constituent immediately survives, and retains *its* δύναμις (192a27) to be made again into a statue or into any object that might have been made from it instead of the original statue. The primary constituent's own perishing (as when this piece of bronze decomposes into water and earth) will be the last thing that happens to it. So *X qua* what *Y* immediately perishes into is necessarily not that which perishes in this episode of perishing. As such and within this context *X* is immune to perishing, hence imperishable; for otherwise it will have perished before it perishes.

It may seem to us that what Aristotle is arguing for here is too obvious for argument. Perhaps, however, by showing how his own account allows for a sense in which matter is ingenerable and imperishable, he brings out a sort of continuity between himself and certain predecessors, although their position is naïve compared to his. The theories of Empedocles, Democritus, and in a way Anaxagoras, discussed earlier in *Physics* I, all posit literally everlasting material principles (see also II, 193a16–28). It is not easy to resist the view that the principles of coming to be and perishing are themselves straightforwardly immune to coming to be and perishing. For if all principles come to be and pass away, (a) it is not clear how they deserve to be considered principles as distinct from things that principles are supposed to be principles *of*. And (b) since all comings to be and passings away demand to be explained by principles, if all principles come to be and pass away there will be an infinite regress of principles; hence arguably no principle is in the end a real starting point and principle (cf. *Metaphysics* α, 994a1–b31; Β, 999b5–9; 1000b23–9).

Against this background we can see Argument 2 as not so trivial after all. This is because it shows a new way of satisfying the virtually universal assumption that principles must be ingenerable and imperishable. According to this new way the ὑποκείμενον-principle is ingenerable and imperishable in the sense that, for any episode of change, the particular empirical item that plays the role of ὑποκείμενον cannot, as a matter of logical impossibility, play the role of what comes to be (what the change is a change towards) or the role of what passes away (what the change is away from). So: *qua* ὑποκείμενον of a given change, the item is indeed beyond reach of coming to be and passing away. In emphasizing this,

Argument 2 underpins the real possibility of coming to be and perishing. For if the bronze, in a given episode, could function both as ὑποκείμενον and as what the ὑποκείμενον is coming to be, it could be undergoing a change into what it already is, i.e. it could change without changing at all. Furthermore: the sense in which the ὑποκείμενον of a given episode is, as such, ingenerable and imperishable in no way precludes the item that plays the ὑποκείμενον role from playing, in other contexts, the role of what something else comes to be or ceases to be. The same thing-like item can play different principial roles in different comings to be. In other words, the principles in terms of which change is to be understood are not special pre-eminent everlasting things—whether corporeal, empirical things or incorporeal, non-empirical things—but essential roles or functions that structure all change and are everywhere the same even as their occupants differ at different times.[12] This is surely one great lesson of *Physics* I. (On this see also Section 2, paragraphs 21–2.)

At *Metaphysics* Z, 1033a24–b19 (cf. Λ, 1069b35–1070a4) Aristotle has an argument rather similar to Argument 2, with these differences: it makes the case about form as well as matter (in fact focusing more on form); it does not use the terms 'imperishable' and 'ingenerable', and perishing is not mentioned; it revolves round the notion of an agent's making *X* into *Y*, whereas Argument 2 speaks only of *X*'s becoming *Y*; and it explicitly invokes the absurdity of infinite regress.[13]

> 192a34–36: As for the form-principle (τῆς κατὰ τὸ εἶδος ἀρχῆς): whether it is one or many, and which it is or which they are, are matters whose detailed determination is the task of first philosophy.

To launch the science of nature ('second philosophy'), we need not have settled questions of first philosophy such as whether the form-principle involved in coming to be is a separate incorporeal substance as the Platonists assume.

> 192b1: the natural and perishable forms [or: kinds]

[12] This is the solution to the problem at *Metaphysics* B, 1000b29–31, of how perishable things can have imperishable principles. On the analogical nature of the solution see *Metaphysics* Λ, 1070a30–b21; 1071a24–35, with Crubellier 2000 on the identity of principles in *Metaphysics* Λ 4; also Frede 2000; also Stephen Menn's chapter in this volume on sublunary principles as types or variables (in the logical sense) made salient through drawing analogies between categorially and empirically different cases. The discussion by Wieland 1960–1 is seminal. Wieland sees Aristotle as proceeding by way of conceptual analysis and 'concepts of reflection' (cf. Bostock 1982). Moravcsik 1991 balances this by emphasizing that Aristotle is not seeking linguistic clarification as such, but explanation in terms of the real configurations of things. See also Freeland 1991 on Aristotle's causal and explanatory realism. My own stress is on a different point: that Aristotle's move to the meta-level, and his focus on omnipresent analogy-patterns, are his way of preserving for sublunary phenomena something of the intuition that a genuine principle is eternal-like in not being restricted to particular places or times.

[13] See note 43 for the argument.

Does this mean natural and perishable individuals according to their kinds, or does it refer precisely to the *forms* of such individuals, i.e. to their structures, patterns of organization, which in particular instances will cease to exist through natural causes? Philosophically it makes little difference. Nor at this stage does it matter whether one takes 'and perishable' as epexegetic of 'natural' or as an additional determination. Aristotle may have expected Platonists in his audience to take it epexegetically (everything natural is in principle perishable), while keeping in reserve his own theory of the heavenly spheres that are both natural and *not* perishable because everlastingly in movement. The proof that there even are such beings lies far ahead in the *Physics*, in book VIII.

192b2: the expositions which follow

According to Ross (1936, 498–9) this is a general reference to Aristotle's physical works, not just to the *Physics*.

2. Philosophical Background

Whether or not Argument 1 succeeds as a refutation of the Platonist position (for more on this see Section 3), it certainly brings out deep differences between the Platonists and Aristotle on the principles of coming to be. Most of this section will be concerned with Platonic sources and materials relevant to the Great and the Small in Argument 1. It will mainly focus on the Phileban Unlimited and the Timaean Receptacle (*Philebus* 23c–27c; *Timaeus* 48e2–53a7).[14]

(1) The Great and the Small in Argument 1 resembles the Receptacle insofar as the latter is one of just two principles of becoming, namely paternal form and that which underlies.[15] In the *Philebus,* by contrast, we have three principles: the Timaean paternal formal one has turned into two, namely Limit (τὸ πέρας) and Intelligence (νοῦς), the distinct efficient cause. These two along with the Unlimited give a triad of principles, but one very different from Aristotle's own form-matter-privation, which makes no reference to efficient causality. If we set aside νοῦς from the Phileban triad, what remains is a pair of principles closely approximating the Platonist pair targeted in Argument 1. I shall now consider the Phileban Limit and Unlimited in more detail.

[14] One of many uncertainties concerns the term ὕλη (for I 9 see 192a3, 6, 22, 31). We do not know whether Aristotle takes the Platonists to be using it too, and in something like a technical Aristotelian sense. At *Physics* I, 4, 187a17–18, where he says that Plato treats the Great and the Small as ὕλη, he could be describing the view entirely in his own terms, or reproducing something said by Plato outside the dialogues. See Kelsey 2010, on the term ὕλη in *Physics* I.

[15] On the relevance of the *Timaeus* to the Platonic theory targeted by Aristotle, see Johansen 2010.

(2) In the *Philebus* the Unlimited is not called 'the Great and the Small' in so many words, but it fits that description well. It is illustrated by the hotter and the colder, the stronger and the milder, the dryer and the wetter, the taller and the shorter, the high and the low in acoustics, the faster and the slower; and it is explained generically as the more and less (24b5; c2–3; e7–8; 25c10–11). A specimen of the Unlimited is as such without determinate poles: any notional point on it is more/less than something that is more/less than something else, and so on. Thus such a range, if we can even call it that, is not between contraries (understanding contraries not merely as mutually exclusive but as the extreme opposites determining a range (*Categories* 6a12–18; *Metaphysics* Δ, 1018a25–31; I, 1055a3–21)). Without the imposition of Limit (e.g. the equal or the double or some definite number), the range has neither poles nor intermediate positions defined by distances from poles or ratios thereof. At *Philebus* 25e1 Plato speaks of the imposition of a member of the class of Limit as ending the 'conflict' (διαφόρως ἔχοντα) between the more and the less.[16] Perhaps he means that in the raw Unlimited any given 'point' or 'stage' (if identifiable at all) is simultaneously both more (or more so and so) and less, which seems to be a contradiction because there are as yet no distinct definites *A* and *B* such that the stage can be seen to be more *than A* and less *than B*. It would be equally true to say that in the Unlimited or the Great and the Small as such there is no contrariety and no mutual exclusion of properties, nor even the appearance of contradictoriness, because there are no determinate properties to entail each other's negations. But when the accession of Limit has marked out the Unlimited into a series of positive determinate natures, these are mutually exclusive in the sense that no particular or individual subject can possess more than one at a given time. However, the Unlimited itself possesses or contains them all rather as a genus contains its species. Hence in the Unlimited *itself*, once it has been marked out into a series of positive determinate natures, there is no privation or absence of any of these natures: they are all in it. The Unlimited can be thought of as a ὑποκείμενον—a sort of substrate—of these specific determinations, but with respect to them it is not a subject of empirical change, nor of anything suitably analogous to empirical change; i.e. it is not a sort of place or seat such that the presence in it of one of the determinate properties excludes the presence of a contrary that could be present but is not.

(3) It seems that this theory of the 'generation' of determinate properties by Limit from the Unlimited could never have been intended to depict *change*, i.e. the successive appearance of mutually exclusive situations, or the replacement of one state of affairs by another.[17] It is not as if the range of determinates comes

[16] See Delcomminette 2006, chapter 4 for a recent full discussion.

[17] This formulation marks off change by the notion of 'replacement', which we probably think of as implying temporal differentiation. Hence the formulation may seem misguided given that Aristotle

into being at some temporal juncture not having been before. The range of determinates is a formal, purely intelligible, timeless entity, no less than Limit and Unlimited, and its being generated is a timeless relationship. This is a theory of the purely metaphysical origin of ranges of determinate properties—that is, properties in terms of which, at a later metaphysical level, empirical change and replacement are possible at all. We may imagine (or think we can) the Unlimited, on the 'accession' of Limit, as 'breaking out into' the plurality of determinate properties. One can invoke or seem to invoke the notion of a non-temporal *transition* between a purely metaphysical or logical before and after—a transition to the Unlimited's having (or even being) the set of mutually incompatible determinate properties from its lacking them all. But these 'moments' or 'phases' of the Unlimited cannot coincide, since they would cancel each other out. So must they in some unimaginable non-temporal sense 'take turns'? If we say 'Yes' it will be hard to show how this unimaginable scenario avoids being meaningless. It seems better to think of the 'moments' as corresponding to two perspectives that cannot be combined. In one there is the Unlimited considered apart from the range of determinates which it makes possible, and in the other there is the Unlimited considered as material of the, so to speak, actualized range. But without a way of making non-temporal sense of 'first one, then the other' we cannot integrate the data of both perspectives. In particular, we cannot coherently see identity or persistence running from the Unlimited in its purely negative aspect to the Unlimited considered as configured by Limit. We cannot really see the determinates as somehow formed out of the negative Unlimited through the imposition of Limit.[18]

(4) In face of the difficulty or impossibility of making sense of the non-temporal transition, an obvious move is to say that this talk is metaphorical: literal coming to be involves temporally successive change or replacement, whereas the determinates 'come to be' from Limit and Unlimited only in the sense of

defines *χρόνος* by reference to change (*κίνησις*). However, (a) *χρόνος* for him is much more than what we think of as bare temporal (as distinct from logical or mathematical) succession; and (b) he himself uses the language of temporal differentiation when explaining his definition of *κίνησις* (*Physics* III, 201b6–7: *τότε . . . ὅταν . . . καὶ οὔτε πρότερον οὔτε ὕστερον*; 11, *οὐκέτι*).

[18] Logically, the Platonists could have claimed a *Physics* I-style triadic analysis for their explanations of eternal things by eternal things, by assigning the role of privation to the counterfactual state of affairs. On this understanding 'indeterminate' describes how the Great and the Small *would have been* in the absence of the formal principle that configures it as a range of determinate properties. Thus the Great and the Small appears as a subject common to both *termini*, the real *ad quem* and the counterfactual *a quo*, hence as a third principle distinct from both. But there is no evidence that such a counterfactual approach played any part in Platonist thinking, although it would have been useful for those who refused to accept as literal the Timaean transition from pre-cosmos to cosmos. They could have claimed that the text really presents a contrast between how things are and how they would have been without the Demiurge. Had the Platonists made this move Aristotle would surely have registered it when he criticizes their anti-literalist interpretation (*De Caelo* I, 279b32–280a10).

being ontologically constituted from determinable and determinant factors. (There would then be a discussion to be had on whether these factors are in fact ontologically and explanatorily prior to the determinates, e.g. the genus to the species.) Aristotle would surely go along with the obvious move. But now one must ask: if the Phileban type of example is typical of the dyadic theory attacked by Aristotle in *Physics* I 9, how exactly is his attack relevant to its target? For him all coming to be is change (*μεταβολή*; cf. *Physics* III, 200b33–201a3), and whatever comes to be can pass away—whereas the set of determinates 'generated' in the Phileban sort of case is presumably eternal or timeless like Limit and Unlimited themselves. *Physics* I, by contrast, is concerned with the natural, empirical, changeful, coming to be of natural empirical items. If the Platonists focus only on timeless intelligible relationships then even though they talk about these in the language of 'generation', 'making', and 'coming to be',[19] and fail in good faith to realize that this language is deeply problematic—still, there would be nothing in their position to worry the writer of *Physics* I. Their dyadism would not be a competitor with, or a botched version of, Aristotle's triadism, for Aristotle's triadism is about literal coming to be.

(5) There is further reason to wonder just how Plato and the Platonists deserve Aristotle's attack in *Physics* I 9. Empirical, temporally successive, coming to be is something that Plato himself not only takes seriously but seems to understand in terms of the triadic framework. At *Phaedo* 102d–103c, in the course of disambiguating the idea of change among determinate contraries, he shows that: (a) one contrary as such, e.g. the hot as such, or the Hot, cannot become the cold as such; (b) the hot 'in us', e.g. in Socrates, cannot become cold;[20] and (c) a concrete (as we might put it) entity such as Socrates that has one of these qualities can come to have the other albeit not at the same time: when one approaches the other 'flees and retreats, or is destroyed' (102d5–103a2; d7–12). So here Plato recognizes a distinct subject which at a given moment receives (*δεξάμενος*) hotness (from having been cold, hence not hot) yet remains itself (102e2–5). Thus he clearly recognizes the differences between: (i) the receptive *ὑποκείμενον* that remains; (ii) the privation of heat, implied by cold, that disappears; and (iii) the heat that takes the latter's place. But these are precisely Aristotle's three principles.[21] Moreover, Plato's argument depends on understanding contraries as destructive of each other or driving each other out. On this basis one may wonder how Aristotle can in good faith accuse Plato of 'overlooking the other nature' (192a12).

[19] See, e.g., *Philebus* 25e2, *ἀπεργάζεται*; 4, *γενέσεις τινὰς*; 26a6–8, *ἐγγενομένη . . . ἀπηργάσατο*, and *Physics* I 9, 191b36.

[20] The example in the text is 'the tallness in us' (102d7–8), but I have changed this to avoid the complications of a relational property.

[21] Cf. Cherniss 1944, 90–1.

(6) Clearly for Plato different levels are in play. The *Phaedo* passage is about physical change of an individual subject from one to another of mutually exclusive mathematical or empirical contraries, whereas the Phileban Limit and Unlimited is about the timeless 'generation' of sets of such contraries whose determinateness is precisely what enables them to *be* mutually exclusive. The first set-up points towards a three-principle schema such as Aristotle wants for sublunary physics; the second is a two-principle one to be studied as a topic in the logic or metaphysics of intelligibles. The contraries 'generated' in the latter are, in that context, metaphysical products, not principles; they are thereby available to act as principles of empirical change in accordance with the triadic schema. So there is no contradiction in Plato's acceptance of both accounts.[22] Even if Plato or the Platonists insist that the talk of 'generation' in connection with the two-principle schema is to be understood somehow literally, albeit timelessly, the accounts operate at different levels, hence are not in competition.

(7) There is no discrepancy, then, between the Plato of the *Phaedo* passage and the Plato of the *Philebus.* But we still have the puzzle of explaining why, given Plato's *de facto* recognition of a triadic as well as a dyadic level, Aristotle in *Physics* I 9 ignores this differentiation, writing as if he identifies the Platonists with dyadism *tout court.* One might conclude that Aristotle is unfair or confused. But this is not unavoidable. Instead, the situation may be that some Platonists, perhaps including Plato at some stage, came to the view that everything they denominated 'coming to be', at every level whether timeless or changeable, should be understood in terms of a single set of principles, perhaps for the sake of theoretical simplicity and unity—and for some reason they chose the two-principle model as their candidate.[23] On that basis Aristotle has a reasonable complaint against them if the dyadic model fails for empirical, change-involving, coming to be, or for central cases of such coming to be.

(8) That it does so fail is not obvious. Let us consider the arguably most important type of physical coming to be, the genesis of natural organic substances. In the triadic account in the *Phaedo* the examples are of change from one quality or size to another. In typical cases both *termini* are equally positive determinations, or neither is clearly more positive than the other. This is not the case when a substance comes-to-be. Here, the natural (direct) way to characterize the *from*

[22] Somewhat similarly, the *Timaeus* has a dyadic schema for the metaphysical generation of the corporeal elements in the Receptacle, alongside what can be construed as a triadic one for transformation of one element into another. See paragraphs 14–15 below.

[23] This supposition is supported by *Metaphysics* A, 988a7–14, where Aristotle says without qualification that Plato uses only the principles of form and matter. *On Generation and Corruption* II 9, 335b5ff. is mainly about the Platonists' use of form, but may glance at matter as one of their principles for explaining sublunary coming to be. In many places Aristotle focuses on the Platonist obsession just with form as cause of sublunary coming to be (ibid. 9–16; *Metaphysics* A, 991b3–5; M, 1080a2–4).

which is as absence of the *to which*. Changes in quality to some given quality, say green, can and do have different positive starting points: white, red, yellow, etc. This is why a particular change to green is properly specified only when we also specify the *terminus a quo*. The same holds for changes of quantity and place. The journey from Thebes to Athens is quite different from the journey from Syracuse to Athens. But the 'journey' from not being, to being, a fully formed dog or oak tree is always of the same nature because in each case the journey always has the same sort of starting point. And what is salient about the thing at the beginning of the process, whether foetus or seed, is that (a) it is *not yet* a dog or an oak tree, and (b) it *comes from* an adult dog or oak.[24] It would seem that to characterize the process one need mention only the form of a complete member of the species and the absence of this complete form. So these cases do seem to invite dyadic analysis—so much so that Aristotle's efforts in I 7 to extend his triadic account to the coming to be of natural organic substances may begin to look rather artificial.[25] The Platonists, it seems, could reasonably say that even on the level of natural coming to be, dyadic analysis trumps the triadic one, given the centrality of organisms as paradigm natural *substances* (as distinct from qualities, quantities, etc.) to all parties in the debate.[26]

(9) These considerations would help explain why, if the Platonists (or some Platonists) sought a single model of coming to be, they would have embraced the dyadic account. For we need a reason why a quest for unitary explanation would have led them to opt for dyadism rather than triadism across the board. One might, after all, manage to persuade oneself that a triadic scheme is discernible in the 'genesis' of a set of determinate contraries from the Unlimited considered as lacking them to their existence in the Unlimited now considered as configured by them (see paragraph 3)—at least if sense could only be made of the timeless transition: and perhaps the Platonists had faith that it could. Thus I think it is worth asking why, if they wanted a single model, they didn't prefer the triadic. Well, one difficulty is that a fully triadic reading in the supposedly timeless case just mentioned would give us the set of contraries twice over, so to say. This can be seen by trying to construct an analogy with Aristotle's account of physical genesis. Here we have (a) the matterless form of the statue

[24] Something like this might also be said about the qualitative example by which Aristotle clarifies his triadic proposal, the uneducated man becoming educated (I 7, 189b34–190a31). The *terminus a quo* is saliently privative, and the *ad quem* reproduces, in an obvious way, the knowledge-form possessed by the teacher.

[25] Cf. Charlton 1970, 75–7.

[26] The line of thought in this paragraph was prompted by comments from Diana Quarantotto. It may help explain the fact that the Platonists were willing to concede that sense-perceptible human beings etc. *are* substances (see *Met.* Z, 1028b8–13; 1029a33–4; 1034a3–4; Λ, 1069a30–2). The concession would have given them a weapon for attacking Aristotle's general attempt to prise apart substrate and privation: for the attempt is obviously doomed if it fails for the central case of natural substances.

as *terminus ad quem*, and then (b) the materiate sensible statue (not principle, but product) that reflects that form in its shape etc. So analogously on the metaphysical level there should be (a★) something like the sheer form or Idea of the set of contraries playing the part of *what* the Unlimited 'seeks' timelessly to become, and (b★) something analogous to the materiate sensible statue. Now to unpack the difficulty. (i) It is natural to explain the *per se* indeterminate Unlimited, e.g. of acoustic higher and lower, as the genus of which the contraries are species (cf. *Philebus* 16c–19a; on genus as matter see *Metaphysics* Δ, 1016a25–8; 1023b1–2; 1024b8–9; Z, 1038a5–8; H, 1045a33–5; *I*, 1058a23–4). So (ii) it is natural to accept in general the converse point, i.e. that the contraries are species of the indeterminate Unlimited. And (iii) it is natural to see this as a definitional fact about them: the contraries are essentially species of the generic Unlimited. But then (iv) this definitional fact must hold of the set of contraries even in its role of terminal form or Idea. A terminal form is definable if anything is. But (v) in that case what is the analogue of the materiate sensible statue, the product of becoming? It was supposed to be the Unlimited considered as 'converted' (by the power of the Limit) into a set of determinate contraries. But this determinately differentiated Unlimited is nothing other than the set of contraries itself viewed as species of the generic Unlimited. Hence (vi) the set of contraries viewed as species of the Unlimited functions both as the form, i.e. *terminus ad quem*, that 'guides' the metaphysical 'process', and as its result or product. This absurdity results from trying to impose the triadic model at this level. Aristotle can use the triadic model for physical change because here the form *can* appear twice over, both as *terminus ad quem* of coming to be and then as informing the matter in the actualized product.[27] This is because this matter is *not* the genus of which the form is a species—it is not an element of the form's definition—and also because temporal differentiation allows for earlier and later contributions by the form. This argument shows that if the Platonists wanted a single scheme of principles for all kinds of genesis it could not be the triadic scheme. On the intelligible level they could not go beyond their two principles: the matter-like undifferentiated one, and the determining form.

(10) I turn now to the Receptacle. At *Physics* IV 2, 209b11–13 (cf. b33–210a2) Aristotle says that Plato in the *Timaeus* identifies χώρα (= place or space, another designation for the Receptacle at *Timaeus* 52a8 and d3) with matter. Certainly no such explicit statement appears in the *Timaeus*, and some scholars have thought the identification so plainly incoherent that Aristotle could only have ascribed it to Plato from confusion or perverse intention.[28] But others have held that some sort of *rapprochement* between the notions of χώρα and matter might not

[27] For more on this point see Section 3, paragraph 27.

[28] On the different interpretations see Algra 1995, chapter 3.

be such a gross confusion after all.[29] So whether the *Timaeus* implicitly assumes some sort of space–matter merger, or Plato retrospectively sanctioned it for the *Timaeus*, or Aristotle was the first to read it into the text, it is not clear that any of these intentions injected disastrous confusion. However, I do not think that anything relevant for our present passage, *Physics* 191b35–192a25, hangs on deciding such questions. The situation is this: Aristotle here presents a contemporary approach, related in some way to the Receptacle-theme of the *Timaeus*, that fails sufficiently to distinguish the role of matter on the one hand and that of non-being or privation on the other hand; and he is anxious to show that his own approach is genuinely different on this point. We can understand his concern and the way he works it out without lengthy speculation on who said what and when.[30]

(11) Let us note some differences between the Receptacle-motif in the *Timaeus* and the account which Aristotle is opposing here. One difference lies in the fact that in the *Timaeus* (as the dialogue's evidence very strongly suggests) the Receptacle is brought in not to help explain every process of coming to be, but for the narrower task of explaining the genesis of the four corporeal elements, earth, fire, water, and air.[31] The Receptacle only appears in the cosmology (see the 'second beginning', 48a7–b3; e1–2) at the point where Timaeus has just broken off from discoursing on the 'craft-works of Intelligence' (47e7ff.) and turns to the four basic corporeal materials. Up to now we have been shown the Demiurge and his ancillaries crafting various organic structures out of the materials, which were in disorder before construction of the cosmos began (30a4–5). So far their pre-cosmic presence and availability, and indeed their natures, have been taken for granted. The focus now shifts to the question of what these materials distinctly contribute to the realization of the organic structures.[32] In pursuing this inquiry, Timaeus does not merely investigate the natures of the four considered as in the raw (or, as he puts it, the natures and properties they each possessed even before the ordered cosmos came into being, 48b3–5; cf. 52d3–4; 53a7). He also, in fact first, turns his attention to the question of how these corporeally primary elements *themselves* came, or come, to be: for they do have an origin, but no one so far has shown what it is (48b5–c2). And it

[29] E.g. Zeyl 2000, lxi–lxiv; Johansen 2004, chapter 6; Zeyl 2010, 117–30; Sattler 2012, 161–2. If 'matter' meant only the ὑποκείμενον of all the materials collectively (a sort of medium in which they appear) there would be no difficulty. It is more problematic if, as Aristotle says, 'matter' means something constituting each parcel of fire and water etc., since then we have parts of χώρα moving about in χώρα. This may be implied by *Timaeus* 51b. Aristotle's interpretation may also rest on a misunderstanding of τὸ μεταλαμβάνον . . . τοῦ νοητοῦ at 51a7–b1; cf. *Physics* 209b12–13 with Ross's note *ad loc.* (Ross 1936, 565–6).

[30] For detailed discussion see Cherniss 1944, chapter 2.

[31] See Broadie 2012, chapter 6.

[32] These include the cosmos at large, since it is a living being (30b; 34bff.).

is at this point that the Receptacle is introduced as matrix and supporter-in-being of the four elements specifically (49a5–51b6) and also as somehow responsible for the separative movements whereby they tend to occupy different regions (52d4–53a7). It is never said with any precision that the Receptacle is an immediate principle for literally every process of coming to be, e.g. of plant from seed, or foetus from sperm and formed offspring from foetus (discussed at 91a1–d5), or blood from food ingested from outside the body (discussed at 80d3–81a2).[33] By contrast, the position which Aristotle confronts in 191b35–192a25 seems to abstract from the coming to be of any particular type of thing.

(12) A related difference between the Receptacle-motif and Aristotle's focus in Argument 1 is this: the principle of the Great and the Small belongs at a higher or more general level than that of the genesis of sensible entities, including even the four elements.[34] The Great and the Small was postulated as material principle (or something analogous) for the timeless and eternal generation of the numbers—whether mathematical or ideal numbers—by the One. Then various species of it were distinguished: the long and short for lines, the broad and narrow for surfaces, the deep and shallow for solids (*Metaphysics* A, 992a10–13; M, 1085a9–12; N, 1089b11–14). These types of magnitude too may have been variously understood as mathematical or ideal. The three-dimensional ones could be seen as being, in some way,[35] predecessors of sensible or physical objects; cf. *Metaphysics* Z, 1028b18–27.

(13) In trying to make out Aristotle's target in Argument 1 in the light of these differences, we can speculate whether the Timaean motif of Receptacle as matrix of the four materials was the origin of a more abstract Academic notion of the Great and the Small as matter for the numbers, or whether instead Plato came to the Timaean picture from a pre-existing theory of the principles of numbers and other incorporeals.[36] But, again, our ignorance of the historical evolution of Aristotle's target may make no philosophical difference for

[33] It is true that the Receptacle is said to be receptacle and as it were nurse of '*all* becoming' at 49a5–6; e7; 50b6–c2; e5; 51a4–7; 52d4–5; 88d6; cf. 52b1. See also 51a1–3 with Cornford's emendation of line 1 (1935). But the universal reference is either confined to the four elements or, if unrestricted, is due to the fact that the elements are constituents and necessary conditions for all organic phenomena. In the latter case the Receptacle is not a metaphysically immediate principle of every sensible phenomenon in the way it is of the four elements.

[34] See especially *Metaphysics* A, 987b14–988a1, with Ross 1924, vol. I, 166–77 and lvii–lxxvi. See also Ross's commentary on Theophrastus, *Metaphysics* 6a15–b17 (Ross and Fobes 1929, 50–9).

[35] 'In some way' is meant to allow for the difference between Xenocrates who derived everything from a single set of first principles, and Speusippus for whom there were independent principles for each level of object (numbers, magnitudes, sensibles), although he might have recognized a scientific or conceptual order of priority (Aristotle, *Metaphysics* N, 1090b13–20; Λ, 1075b37–1076a3; Theophrastus, *Metaphysics* 6a15–b17).

[36] *Timaeus* 48b5–d1 points to the notion of a principle or principles 'of everything', but refuses to discuss either them or the number of removes between them and the corporeal elements. The passage

understanding his present concerns and how he responds. Did the Platonists want just one set of principles for all coming to be? Or analogous ones for coming to be at different levels (cf. *Physics* I, 188b30–189a9)? Or did they take the principles operative at the most fundamental Platonist level and use these as somehow emblematic of coming to be in general? Or were they simply undecided between these options, or did some endorse one and some another (not to speak of different ones at different times)? Or is Aristotle's target a sort of abstraction skimmed or constructed by him from variants—an umbrella '-ism' (as in 'idealism', 'expressivism', 'consequentialism', etc.)? No doubt the most positive stance to take on this issue is that the target as Aristotle represents it is all that he needs for making his main point in Argument 1. At any rate I shall proceed on this assumption. His main point and objection against the view is simply that it assumes two principles where three are necessary. In this he adheres to the theme that has already organized so much of his discussion in *Physics* I, i.e. the question of how many principles there are (184b15ff.; 187a12ff.; 189a11ff.; b11–29; 190b10–191a22; 192b3; cf. a34–5).

(14) One aspect, however, of Plato's Receptacle-account may tend to obscure its affinity with the target of Argument 1. The Receptacle-account runs from 49a to 53a. Its main concern is with just two principles of coming to be and their offspring. The two principles are, of course, the Receptacle itself and certain separate forms of which the offspring are said to be 'copies' (50c5; 51a2; 52c2). But part of the passage seems to indicate a triad of principles similar to Aristotle's three. This part, 49b–50a, comes near the beginning. Plato introduces the Receptacle by emphasizing how the four materials seem to be caught up in a cycle of mutual transformations. They all, it seems, continually turn into each other through processes of solidification, melting, dispersal, conflagration, gathering, quenching, and packing together (49b7–c7). This is supposed to be obvious from experience, although afterwards we learn that appearances were misleading: earth, because of the triangle-geometry of its particles, cannot be part of the cycle (54b5–d2; cf. *δοκοῦμεν*, 49b8; *ὡς φαίνεται*, 49c7; cf. *οὐκ ὀρθῶς φανταζόμενα*, 54b8, harking back to *ἑκάστων φανταζομένων* at 49d1). Even so, the other three do inter-transform: water becomes air, air becomes fire, fire becomes air again, and air in the form of cloud and mist becomes water. So one might get the impression that the Receptacle passage presents (a) that which in any given case is coming to be (i.e. the *terminus ad quem*), say fire; (b) that stuff, say air—a sort of contrary, hence a privation, of fire—which turns into fire; and (c) the underlying and persistent Receptacle. This triad looks remarkably like

also says nothing on whether one of these ultimates is a matter-like principle analogous to the Receptacle. On the difficulties see Cherniss 1944, chapter 2, and Steel 2012, esp. 191–200.

the Aristotelian one (if for the moment we put on one side Timaeus's paternal formal principle).

(15) Even so, the Receptacle passage is not centrally *about* the transformation of the materials. This is not its topic-for-study and *explanandum*.[37] The focus on transformation is a preamble. The main body of the passage (50b–52d4) is not concerned to explain how a portion of fire, air, etc. serves as a source, i.e. principle, for the next element in the cycle, or how these materials come into being *from* one another in the Receptacle. Instead the Receptacle passage mainly presents just two principles: the paternal forms, changeless and intelligible, and the maternal Receptacle. The four empirical materials are shown simply as 'offspring' of these two (50d2–4). The materials are images of the forms, and these images successively come into and go out of *the Receptacle* (50c1–5; e1; 49e7–50a1). After 49e the passage never returns to the idea that the materials pass away into and develop out of *each other*.[38] Instead, the stress is on their collective and wholesale dependence on the two metaphysical principles. Moreover, although portions of diverse materials show up in a time-like sequence relative to each other, there is no suggestion that the father- and mother-principles first generate one portion of stuff and then another; the metaphysical generating seems to be of the entire spread and series of the materials considered as a collective offspring.[39] This genesis of the materials is thus on a par with the eternal genesis of numbers or ideal numbers through cooperation of the One and the Great and Small, or the genesis of determinate contraries discussed earlier.

(16) So the main topic of the Receptacle passage is not the elements considered as each a source or principle for the next member of the transformative cycle. The main point is their collective status as metaphysical products of metaphysical principles—specifically, just *two* metaphysical principles, the trans-natural forms and the trans-natural Receptacle. So we are not wrong to think that Plato's Receptacle stands for a theory of just the dyadic type that Aristotle targets at 191b35–192a25. The main Receptacle theory in the *Timaeus*

[37] By contrast, elemental inter-transformation is the *explanandum* of the distinct geometry-passage, 53cff. Here too the picture is triadic with the triangles serving as the distinct ὑποκείμενον.

[38] The image of gold moulded and re-moulded (50a4–b6) is not an appropriate illustration of one thing coming to be from another in the natural course of events. A triangular shape in the gold does not by its own nature turn into the next shape; the craftsman (τις, 50a5) fashions one shape to replace another.

[39] Consequently, in allowing earth to feature in the Receptacle-passage despite its later geometrical exclusion from the cycle of transformations, Plato is not confused or cheating or caving in to the influence of the Empedoclean four elements: For although earth will turn out incapable of being the *physical* offspring or progenitor of any of the other three, all four are on a par *qua* metaphysically generated from the same pair of parents. As Plato must have been aware, placing earth on its own outside the cycle of transformations is exactly the kind of move that, in an earlier tradition, would have proclaimed that earth is somehow the ultimate first principle of everything (cf. Aristotle, *De Caelo* III 306a17–20). Plato shows that this whole way of thinking is obsolete by making empirically indestructible earth no less metaphysically dependent than its empirically destructible siblings.

has exactly the structure of the doctrine which Aristotle ascribes to Plato by name at *Metaphysics* A 6:

> It is evident from what has been said that he has used only two causes, that of the essence (τοῦ τί ἐστι) and the material cause (for the forms are the cause of the essence of all other things, and the One is the cause of the essence of the forms); and it is evident what the underlying matter is, of which the forms are predicated in the case of sensible things, and the One in the case of forms, viz. that this is a dyad, the Great and the Small. (988a7–14, Ross's translation)

(17) Even so, the Receptacle passage as a whole should make us wonder, as we did with *Phaedo* 102d–103c, why Aristotle in *Physics* I 9 completely ignores the Timaean evidence that Plato recognized two levels, one of metaphysical so-called generation involving just two principles, and one of empirical coming to be where the Aristotelian triad is clearly discernible. Although both involve the corporeal elements, these levels are clearly distinguishable. As before (see paragraph 7), the explanation may be that Platonists moved towards extending their dyadic view to cover not only the timeless metaphysical 'becomings' that were no doubt their original inspiration but also the real (physical) changes that are the topic of *Physics* I. Or the explanation may be that Platonists hoped to extend their dyadic view to a certain central type of physical change, even if not to all types. I now propose that what Aristotle is warding off in *Physics* I 9 is a Platonist extension of dyadism to the biological domain and especially to biological reproduction.[40] This after all was the model for those timeless metaphysical figures of 'father', 'mother', and 'offspring', so the Platonists might reasonably have felt some impulse to apply the two-principle schema to the literal biological phenomenon. More on this in Sections 3 and 4.

(18) Now for some remarks on Argument 2 in *Physics* I 9. I take Argument 2 to begin at 192a27, at ὡς δὲ κατὰ δύναμιν (although see note 3). Its logical relationship to Argument 1 is unclear. Perhaps it is in no particular logical sequence to Argument 1, but instead they are coordinate reflections on the theme of coming to be and passing away. In that case, Argument 2 may have nothing to do with the Platonists in particular. On the other hand, perhaps it was Aristotle's view that they were misled by the logical 'fixity' of the matter in a given episode of coming to be into concluding that for *every* coming to be there is a fixed, in

[40] The reason for this more restricted hypothesis is that Aristotle's discussion of the *Timaeus* theory of elemental transformation by means of sub-elemental triangles shows that this theory is still very much a live option (*De Caelo* III, 298b33ff.; 299b24ff.; 306a1–6; 306b3–307b19; *On Generation and Corruption* II, 332a28–30; cf. *Timaeus* 53c4ff.) It is a triadic theory, with the triangles functioning as the distinct ὑποκείμενον. So the context in which *Physics* I was composed included Platonists who accepted triadism for some physical changes.

the sense of eternal, matter or ὑποκείμενον.[41] If this was the view of some Platonists, then the difference between them and Aristotle is even starker than has fully appeared so far. Not only would these thinkers be imposing the dyadic schema on every area of coming to be, temporal and eternal alike, but in every area they would be populating the schema with none but eternal principles both formal and material. What seems to be a perishable *substance*, a plant or an animal, would really be a transient configuration of underlying eternal matter, and there could be no true scientific understanding of the plant or the animal that did not refer back to this eternal principle (cf. 193a12–28).[42]

(19) One reason for thinking that Argument 2 has a Platonist target is that elsewhere (*Metaphysics* Z, 1033a24–b19) Aristotle gives a somewhat similar regress argument which seems to segue into an attack on Platonic forms (b20–9).[43] This argument too is about the 'fixity' of form and matter in a given coming to be, although the focus is more on form. Aristotle's point at 1033a24–b19 may be that misinterpretation of fixity helps bolster the doctrine of forms as eternal substantial entities (cf. Ross 1924, 188 *ad* 1033b19). Thus Argument 2 in *Physics* I 9 may be making a parallel point against some distinctively Platonist notion of the material principle.

[41] See the comments on 192a27–34 in Section 1.

[42] By another slide they might also have assumed a single eternal matter for every coming to be.

[43] 'Since anything which is produced (γίγνεται) is produced by something (and this I call the starting point of the production [γένεσις]), and from something (and let this be taken to be not the privation but the matter, for the meanings we attach to these have already been distinguished), and since something is produced (and this is either a sphere or a circle or whatever else it may chance to be), just as we do not make the substratum—the bronze, so we do not make the sphere, except incidentally, because the bronze sphere is a sphere and we make the former. For to make a "this" is to make a "this" out of the general substratum. I mean that to make the bronze round is not to make the round or the sphere, but something else, i.e. to produce this form in something else. For if we make the form we must make it out of something else; for this was assumed. E.g. we make a bronze sphere, and that in the sense that out of this, which is bronze, we make this other, which is a sphere. If, then, we make the sphere itself, clearly we must make it in the same way, and the processes of making will regress to infinity. Obviously then the form also, or whatever we ought to call the shape of the sensible thing, is not produced, nor does production relate to it—i.e. the essence (τὸ τί ἦν εἶναι) is not produced; for this is that which is made to be in something else by art or by nature or by some capacity. But that there is a bronze sphere, this we make. For we make it out of bronze and the sphere; we bring the form into this particular matter, and the result is a bronze sphere.... It is obvious then from what has been said that the thing, in the sense of form or substance, is not produced, but the concrete thing which gets its name from this is produced, and that in everything that comes to be the matter is present, and one part of the thing is matter and the other form. Is there then a sphere apart from the individual spheres or a house apart from the bricks? Rather we may say that no "this" would ever have been coming to be, if this had been so. The "form" however means the "such", and is not a "this"—a definite thing; but the artist makes, or the father generates, a "such" out of a "this"; and when it has been generated, it is a "this such".... Obviously then the cause which consists of the Forms (taken in the sense in which some maintain the existence of the Forms, i.e. if they are something apart from the individuals) is useless with regard both to comings-to-be and to substances; and the Forms need not, for this reason at least, be self-subsistent substances (οὐσίαι καθ'αὑτάς)' (*Metaphysics* Z, 1033a24–b29, tr. Ross; see also Λ, 1069b35–1070a4).

(20) At any rate, Argument 2 is Aristotle's explanation that positing the matter as 'fixed' in the only sense in which we are clearly entitled to do so is equivalent to rejecting the absurdity that something comes to be before it comes to be etc. This rejection, as I understand it, in turn amounts to acceptance of what, perhaps partly thanks to Aristotle's work, seems an utter truism: namely, that for any given coming to be there must be distinct occupants for the roles (a) of whatever it is that thereby comes to be and (b) of matter. So one point to be picked up from Argument 2 is that something has certainly gone wrong for anyone who draws heavy-duty metaphysical conclusions from the equivalent of that truism. If Platonists or others want a valid route to the conclusion that there is an eternal substrate, they will have to start from premises which they have no reason to assume would be readily granted. Thus we could read Argument 2 as intended to break down the misunderstanding of any thinkers, materialist or Platonist, who regard it as self-evident that coming to be as such involves eternal principles.[44] If Aristotle thought that the Platonists saw the logical fixity of matter in each single coming to be as implying an eternal substrate, he might equally have thought this about the materialists discussed earlier in book I, who postulated fire or water or some set of more or less empirical elements as imperishable matter for everything else and perhaps even as substrate of *every* change; again cf. *Physics* II, 193a16–27. On this reading Argument 2 rounds off the whole discussion of earlier thinkers in *Physics* I.

(21) One might however think that no one could be so stupid as to postulate an everlasting substrate on grounds equivalent to a failure to recognize that in every coming to be the matter must be distinct from what thereby comes to be. But the postulate is in any case tempting because it tends to be associated with a natural way of understanding a basic assumption of science. This is the assumption that the whole of nature is governed by *the same principles*—which for materialists would be naturalistic principles. On a simple understanding this becomes the view that change throughout nature depends on the involvement of a set of ultimate, self-standing, naturalistic *entities* or *things*.[45] Since a principle cannot be a principle of itself (185a3–5), these thing-principles of the natural world cannot themselves come into being since nothing is prior to them. So if nature and the things that depend on nature are all that there is, certain naturalistic entities—or perhaps one such entity—lies at the foundation of all that is.

[44] For Aristotle it is not coming to be as such that depends on eternal principles, but the endless succession of coming to be.

[45] This is meant to include such entities as forces, e.g. Love and Strife. In the present context what qualifies something to be *thing*-like is its being thought of as numerically identical through all its manifestations: it is not merely a pattern abstracted by us from repeated cases of, e.g., combination and dispersal. (The point does not depend on whether the Love and Strife of Empedocles were actually thing-like in this sense.)

As parts of nature such beings do natural, empirical, cosmological, work; but they also fulfil a metaphysical—on some constructions, a divine—function insofar as they constitute the basis of absolutely everything beside themselves.

(22) This pattern of thought is part of the cultural background to Aristotle's *Physics*. There were two very different reactions. According to one, which we meet full-blown in Plato, the natural world is not self-standing and nothing belonging to it is finally ultimate. The other, typically Aristotelian, response rejects that identification of principles of change with *things*. The same thing sometimes acts as a given type of principle and sometimes not—e.g. sometimes it is an agent of change, sometimes a substratum or patient, sometimes it is both in relation to different changes. The principles, i.e. the principial roles, are always the same, but occupants come and go in relation to them. Hence role-occupants may come into being and pass away without undermining the immutability of the principles. At any rate this is how Aristotle views sublunary change. In itself this approach leaves it open whether the natural world depends on an ulterior metaphysical reality. Whether it does is another discussion, indicated as such in the coda to *Physics* I 9, with its distinction between postponed questions of 'first philosophy' and the immediate topic of 'natural and perishable kinds'.

3. The Dialectical Situation in *Physics* I 9

(23) There are good reasons to include the Platonists in the survey of views in *Physics* I: they offer a definite set of principles; as with previous theorists and Aristotle himself, contrariety is important for them; they recognize a matter-principle, whether or not they use that term; their theory stands in interesting contrast with earlier ones in that earlier ones particularly struggled with the 'not-being' aspect of coming to be, whereas the Platonists are confident about including privation in the picture; and finally their theory emphasizes the notions of form and the good, themes which have played no part in the earlier chapters but will be so important in *Physics* II. On the other hand, when we look at the *arguments* of I 9, we cannot but think that they would hardly scratch the skin of a convinced Platonist. In Argument 1 Aristotle mainly just shows how their set and number of principles is not the same as his. The only polemical moment comes when he points out that contraries are mutually destructive, so that one of the two Platonist principles seeks its own destruction. But to assume that this is absurd begs the question against the Platonists, as I argued in the comments on 192a16–25. As for Argument 2: no doubt its aim is to uncover a logical confusion that might have misled some thinkers, including Platonists, to assume for every coming to be an eternal ὑποκείμενον—but the Platonists could simply reject the charge that they along with many others hold their

view from confusion rather than clear-eyed intuition. In this section I shall first look further into the issue of question-begging in I 9. Then I shall point out that even if I 9 taken by itself does beg questions against the Platonists, this is an unimportant defect in the context of Aristotle's project.

(24) In the *Physics* I survey of earlier philosophers Aristotle has several times said that one or another of their moves or positions manifests perplexity, confusion, or the condition of being stuck with having to say something one does not really want to say (185a17–19; 186a1; 191a23–4; 191b30–3; cf. 185a14, λύειν; 185b25, ἐθορυβοῦντο; 186a23, ἡ λύσις; 190b33, λύεται; 191b10–11 διὰ ταύτην τὴν ἄγνοιαν . . . προσηγνόησαν). It is noticeable that I 9 says no such things about the Platonists.[46] This suggests that their position is not for Aristotle a confused step in the right direction (the goal being a coherent natural philosophy). When Aristotle attributes confusion he generally means that the confused party is blocked from advancing along the path to the truth. When he dispels difficulties that hampered earlier thinkers, he is explaining how they failed to progress along the right road. More generally, I suggest that when he speaks of facing or solving an ἀπορία he has in mind a difficulty the solving of which helps him or would have helped others towards the right result (see, e.g., *Physics* IV, 211a10–11; IV, 217b29–30; 218a30; *De Anima* I, 403b20–4; *On Length of Life* 464b20–2; *Metaphysics* B, 995a24–b4; *Nicomachean Ethics* VII, 1145b2–4; 21; 1146a21–7; b6–8; 1154a22–5; *Eudemian Ethics* I, 1215a3–7; 20–1). Aristotle has little interest in solving problems internal to outlooks or projects in which he has no comradely involvement. It may be that he does not even count something as an ἀπορία (an ἀπορία *simpliciter*) if it only arises for an enterprise which he does not in some way share. To apply a familiar pattern: what is an ἀπορία *to* or *for* this or that person is not on that account an ἀπορία *simpliciter*. For the pattern see, e.g., *Physics* I, 184a18 (γνωριμώτερον, σαφέστερον); *NE* X, 1176a15–24 (ἡδύ); VII, 1152b29–31 (φαῦλον, αἱρετόν); 1153a26–30 (ἀγαθόν): if the judging subject is in some way defective, what is good etc. to or for her or him is not on that account good etc. *simpliciter*. (Having such a criterion for what counts as a real ἀπορία would not preclude Aristotle from showing, for the sake of *reductio*, that some misguided thinker's system generates an ἀπορία insoluble within the system.) The point of these remarks is to suggest that the absence of ἀπορία-vocabulary from the I 9 sketch of Platonism is a sign that for Aristotle the Platonist theory is no kind of step in the right direction—the right direction being along 'the path towards coming to be and perishing and change in general' (τῆς ὁδοῦ τῆς ἐπὶ τὴν γένεσιν καὶ φθορὰν καὶ ὅλως μεταβολήν, 191b32–3). The Platonists are not, in his view, making imperfect progress in an enterprise he shares: rather, their kind of project and his are fundamentally different. In such

[46] Cf. Cherniss 1944, 84.

a situation it is virtually impossible for either side to lodge an objection that is not in some way question-begging.

(25) Let us accept that I 9 by itself fails to win an argumentative advantage over the Platonists. But this is hardly a defect in the project of *Physics* I. It is obvious before one gets to I 9 that one of the book's most pressing concerns is to make intelligible the coming to be of organisms. This topic occurs at 187b16 (in connection with growth), at 190b3–5, and at 191b18–20.[47] In this particular arena the Platonist dyad makes a poor show, as we shall see. The question-begging of I 9 might be a weakness if it were still unclear to the reader of *Physics* I just what sort of work a theory of principles is expected to do, i.e. what the desired principles are supposed to be principles *of*. If we do not know what sort of work a theory of principles is expected to do, then for all we know the Platonistic dyad could turn out to be the right theory for the job. But we know, by the time we get to I 9, that a central part of the job is to explain the coming to be of organisms. And the Platonist dyad is not a lot of use for that purpose, as is about to be shown. Hence even if I 9 by itself is dialectically ineffective against the Platonists, this does not really matter.

(26) It seems that the Platonists, led by Plato, used the father-mother-offspring idea first and foremost as a model of certain timeless, metaphysical, dependence-relations. No doubt the most attractive thing about it was its promise of explaining how there are families of intelligibles (such as the numbers), pluralities that seem to embody a single rule or belong to a shared intelligible space. The explanation relies on the notion of inheritance: the plurality is the offspring of a father-form which propagates its unitary nature *via* a second principle, a matrix that both preserves and downgrades the unity by pluralizing it. This sort of explanation might seem above all to lend itself to the actual source of the metaphor, namely literal biological reproduction. If it does, then this type of theory not only achieves, as it were, a successful homecoming, but also turns out to be impressively comprehensive, extending from the domain of eternal and purely intelligible structures right down to the corporeal formations of transient living things.

(27) But the extension hits an obstacle. This is the fact that natural organic substances typically come into being as potentially on a level with what generated them. In general, the new creature is capable of becoming a parent itself. Literal biological generation is an arc beginning and ending with mature generative

[47] This case is relevant despite the difficulty of the example and possible need for emendation at 191b20–1. The coming to be of, e.g., water from air, and vice versa, is mentioned only once (188a16–17), and then only in the context of pointing out that Anaxagoras gets this wrong. Thus one has the impression that *Physics* I, so far as it is about substantial natural coming to be, is about the coming to be of organisms. Aristotle also speaks of the coming to be of artefacts such as statues at 190a25; b5–9; 14–17; 24; 191a7–12, but presumably here as in *Physics* II this is a lens for viewing the principles of organic substances.

members of the same species. The generated ends up completely sharing the parent's nature; its kind-name is the same as the parent's, not a derivative term. The Platonist dyad in its original context was proposed to explain the dependence of things such as numbers or number-forms on the One together with the Great and the Small, and the four physical elements on paternal forms together with the Receptacle. In such cases there can be no hope of the substrate's becoming fully invested with the formal principle that acts as both generator and exemplary *terminus ad quem*. The intelligible 'offspring' cannot come to be as ontologically strong as their parents. The indeterminate partner cannot mother and support a new version of the One capable of generating new numbers etc., or a new intelligible form of fire that generates its own progeny of sensible fire. For there can be only one One, and only one intelligible form of fire. This uniqueness implies an irrevocable merging of substrate with privation. While the substrate persists supporting the offspring, privation never goes away completely, because privation here is nothing but the offspring's essential failure ever to be a father itself—essential, since to be a father it would have to be *the* father, its own father. So the offspring never frees itself from privation any more than from its substrate. Hence there is reason to treat privation and substrate as a single principle, one of only two. But in the natural and sense-perceptible world, offspring wind up differing from their father not through inbuilt metaphysical inferiority, but through simple numerical difference. The offspring's privative inferiority gets to be shrugged off completely—or as completely as necessary for them to become generators as powerful and independent as their parents. Add the fact that the adult offspring's material constituents pre-existed (or have naturally grown out of matter that pre-existed) this individual's birth, and it is clear that there is a single continuous material line reaching back to before birth and persisting on in the new adult, a material line first lacking the form and later in possession of it (or possessed by) it. The coincidence of substrate and privation is merely temporary; hence they are different principles. The facts underlying this conclusion are undeniable even though ordinary discourse finds it hard to designate the *terminus a quo* of natural substantial genesis otherwise than privatively in relation to the form. This seemed a difficulty for the triadic scheme (see paragraph 8), but the difficulty recedes in light of fuller analysis.

(28) In order to extend their theory to cover the coming to be of natural organic substances the Platonists must sideline natural parents as causes of their offspring. The real generator, on the formal side, is the eternal, ontologically separate, Platonist form of, say, Human Being. The offspring are all the concrete human beings that ever will have existed. Despite the seeming parental and ancestral relations between various members of this class, they all, in a sense, belong to the same generation: they are all offspring of the separate form and are all equally closely related to it or distant from it. All have their being as

materiate images or shadows of the form, constituted by their essential reference to it as what they can only be like but not fully be (cf. *Timaeus* 52c2–5 on the εἰκών). The main objection to this view is that it purports to explain the coming to be of human individuals when we already have, from ordinary experience, a quite different adequate explanation of this:[48] each individual comes from a concrete biological parent which it resembles not as a two-dimensional image resembles the solid original, but as fully sharing the parental nature including the capacity to reproduce creatures as fully human as itself. There may be things that Platonist Forms—and even more abstract intelligibles such as the One—are needed in order to explain—things that demand an explanation and for which we don't already have a perfectly good one—but the coming to be of natural organic substances is not one of those things. If someone insisted on applying the dyadic Platonist account to this area, it would not be because doing so makes *this area* newly intelligible, but because of a pre-existing faith in the wholesale validity of the dyadic approach.

4. The Dyadism of Aristotelian Eternal Motion

(29) I turn now to the fact that in one domain of Aristotelian science there can be no question of explanation in terms of the privation-form-ὑποκείμενον triad. This is the domain of everlasting or eternal change-phenomena. The approach needed here is not in conflict with Aristotle's triadism in *Physics* I. Triadism applies to temporal comings to be, or to things (substances, qualitative conditions, etc.) that have being only through having come to be in a temporal sense, whereas everlasting change-phenomena never came to be in a temporal sense but nonetheless, according to Aristotle, demand to be explained. The prime examples are the eternally continuous rotations of the celestial spheres.[49] These movements do not come about from having not been, nor do they represent the absence of some more complete condition in which they will terminate.[50] Since every celestial rotation is necessarily absolutely continuous it would seem that explaining it means explaining it as somehow a whole. Yet even here Aristotle gives triadic analysis the best possible run for its money. He

[48] Aristotle draws a parallel with skilled processes such as medical healing and teaching, where he treats it as uncontroversial that a concrete agent (as opposed to a Platonic form) is the cause (*On Generation and Corruption* II 9, 335b20–4).

[49] It might be suggested that the endless series of sublunary γενέσεις and φθοραί is another example, but arguably this is not a single infinite phenomenon but a plurality consisting in an eternally repeated set of finite changes each answerable to the triadic model.

[50] Thus we have the problem, discussed since antiquity, that eternal motion does not conform to the *Physics* III 1 definition of κίνησις/μεταβολή as actuality of the non-*F*'s potential to be *F*. Aristotle seems not to notice any difficulty this creates for his argument to the incorporeal unmoved mover, which rests on applying to the eternal case the principle that every κίνησις depends on a κινοῦν.

wants to treat the celestial rotations as in some sense involving individually transient stretches of movement each with its own *a quo* and *ad quem*; thus the spheres, considered as ὑποκείμενα of these notional stretches, are possessed of 'topical matter', i.e. the potential at any given moment to be, part by part, somewhere *else* along the circuit (*Metaphysics* Λ, 1069b25–6; H, 1042b5–6; Θ, 1050b20–2).[51] But there is no suggestion that the overall motions of the spheres are to be explained as resulting from an infinitely repeated stringing together of successive distinct stretches of finite, transient, motion each with a prior explanation of its own. These sections of motion are only notionally distinct. The *explanandum* in the case of each sphere seems to be its motion considered as numerically one and eternal.

(30) Platonist explanation of how eternal things 'come to be' from other eternal things is dyadic because the ὑποκείμενον in these cases never attains the form, hence never disentangles itself from privation. This feature has an interesting parallel in the eternal motion of Aristotle's celestial spheres. Celestial motion too can be seen as expressing an interminable privation. The substrate in each case (the sphere) never breaks free of the lack that metaphysically fuels its movement. Thus we have the lineaments of the dyadic model. The incorporeal prime mover of *Physics* VIII 10 and *Metaphysics* Λ 7 is one principle, the substrate-privation is the other, and the phenomenon of which they are principles is the eternal motion itself. This phenomenon could be said in Platonist fashion to be timelessly generated (it is caused but not in such a way as to replace its own absence), although Aristotle would refuse the language of generation here.

(31) It is instructive to look at the differences between this situation and that of the literal coming to be of sublunary organisms as discussed in Section 3, paragraph 27, if only because it sheds light on the well-known puzzle concerning the causality of Aristotle's prime mover. Is this mover the efficient cause of celestial motion, is it the final/exemplary cause, or is it both? And if both, why duplicate the types of causality? In thinking about these different types of causality it is natural, I think, to start with sublunary cases. Here the parent as efficient and form-bearing cause intervenes on the matter, bringing about in it a process by which the not-being of the new substance is replaced by its being.[52] The parent

[51] See Charles 2000, esp. sections 5 and 6; also Fazzo 2013 for a probing discussion relating 1069b26 to the ἀπορία at *Metaphysics* B, 1000a6–7 ('Whether the principles of perishable and those of imperishable things are the same or different') and to Aristotle's concerns at 1075a25–34.

[52] The physical or temporal (as distinct from merely logical or mathematical) before-and-after of this is very clear in the summary of the four causes at *Physics* II, 198a21ff. At 33–5 Aristotle says: 'When it comes to *becoming* (περὶ γενεσέως), this above all is how people inquire about the causes: they ask what happened after what, and what was the first maker, and what the thing had done to it, and similarly what the sequel was at each stage.' The emphatic position of περὶ γενεσέως is missed in Charlton's 1970 translation. The phrase is emphatic because at a27–31 Aristotle has just drawn a contrast between moved movers and the absolutely (cf. 198b2) unmoved movers that are not subjects of physical science, and has stated that there are three fields of investigation: immobiles, things that move but are imperishable, and things that perish.

is also exemplary of what the new substance should be. The new substance can grow up to become another such parent, an efficient cause and exemplar of offspring in its own right. But the fact of its attaining the same status as its parent does not make it any less derivative from its parent. It is derivative in that its genesis came about through the parent's form-bearing action on matter previously devoid of the form and unable by itself to take on that form. Thus, central to our sense of the offspring's derivativeness is its existence now as something that *was-not* or *has not-been*. But while it is necessarily *derivative from* a parent, it is not necessarily *dependent on* a parent. Its coming into being did depend on the parent, but its being, now, is independent of the parent's. So the parent can perish and in a way be replaced by its offspring. In the characteristically Platonist scenario, by contrast, where paternal form eternally generates eternal or everlasting offspring in an eternal matrix, the offspring's derivativeness in relation to the form is not grounded in its having been brought to be from not-being through a moment of parental intervention on unformed material. So there is no way to understand the offspring's derivativeness from the generative form except as perpetual dependence on it. The form cannot be superseded, and the offspring cannot emerge into existence unsupported by the father. So far as the offspring does exist, its existence is an endless becoming. It is the substrate's unachievable aspiration to be, itself, the necessarily unique paternal form. The form, for its part, causes the existence of offspring simply by constituting a definite object for the endlessly aspiring substrate to aspire to be.

(32) So we have here a sort of combination of efficient—'paternal'—causation with exemplary or final causation.[53] In *Physics* I 9 Aristotle finds fault with the Platonists for their inability or refusal to disentangle ὑποκείμενον from privation. But this non-distinction of theirs is intimately connected with another: between efficient and final causation. For the roles of efficient and final cause are most obviously different when the efficient cause, characterized by form F, is actual and actually operative, yet its end, also characterized by F, is an F that is not yet actualized but actualiz*able*. But this situation implies that, when the process is not interfered with, the substrate will leave privation behind and take on the form, becoming a new F and making the original one causally irrelevant to the new F's existence. Conversely, I suggest, the case where the substrate necessarily fails to divest itself of privation is a case where efficient and

[53] On a familiar understanding of 'final cause' or 'that for the sake of which' (in the sense of τὸ οὗ ἕνεκα τινός, *Metaphysics* Λ, 1072b1–3; *De Anima* II, 415b2–3; 20–2), that for the sake of which *X* obtains is something that normally comes to be through *X*'s obtaining. In this sense an exemplar is not a final cause. Here, however, I follow many interpreters in taking 'final cause' to include any object or objective *Y* that explains the obtaining of *X* in a way that makes essential reference to the good of realizing or conforming to *Y*.

final cause cannot be pulled apart. For there is no contrastive moment when the first is active and the second not yet implemented, nor when the second is implemented and the first causally obsolete in relation to this particular coming to be. In sum, we have a dual non-differentiation, of privation from subjecthood and of efficient from final cause: and this is a necessary feature of causation by the eternal of the eternal. Clean separation, in either case, is possible only for transitory substances and transitory changes.

(33) These linked non-differentiations are not confined to Platonism. We find the causal one in Aristotle's explanatory approach to eternal celestial motion. Aristotle notoriously appears to shift position on whether the *cause* of this motion—the incorporeal unmoved mover—is an efficient or final/exemplary cause. At *Physics* VIII 10 it appears to be merely efficient, and efficiency seems to be suggested at *Metaphysics* Λ 6, 1071b13. But in *Metaphysics* Λ 7, however, the mover is very clearly a final or exemplary cause, and (in contrast to *Physics* VIII) its goodness, desirability, and divinity are a major theme (1072a26–b4; b14–30). But also in *Metaphysics* Λ 7 Aristotle proves the mover's incorporeality (more precisely, its having no magnitude) by the argument he used in the *Physics* (1073a5–11; cf. 267b17–26). Thus it would seem that anyway in *Metaphysics* Λ 7 the unmoved mover is both an efficient and exemplary cause.[54] However, if the present exposition is on the right track, it should be no surprise that Aristotle wants to have it both ways.

(34) Consider again the *effect* or 'offspring' of the incorporeal unmoved mover of *Metaphysics* Λ 7. It consists in the everlasting motion of the sphere. But the sphere is only notionally distinguishable from its motions, because its motion-activity is its life or the necessary expression of its life. It must be alive, since the mover is the object to it of ἔρως, a passionate longing translated into or given expression by the sphere's rotation (1072b3; 1072a26–9).[55] And since life for the living being is its existence, the very being of the spheres—not merely their notionally distinguishable motions—is the effect of the incorporeal movers. If our analysis above is a guide, we should expect the dependence of the sphere-in-motion on its cause to appear as a kind of inferiority, and this is surely what we find in *Metaphysics* Λ 7: the sort of inferiority whereby the subject is what it is through its unsatisfiable longing for unity with the originator. With the spheres this desire manifests itself as perfect (uniform, circular, necessarily unbroken, inexhaustible) physical motion, something that copies as fully as

[54] For recent discussion, see Judson 1994, 164–7; Graham 1999, 177–80; Berti 2000, 186–8; 205–6; Frede 2000, 38–47; Laks 2000, 240–43; Irrera 2011, 144–57; and Menn (forthcoming), part III gamma I in the online version.

[55] How the sphere's mental activity relates to its motion is a question on which Aristotle gives no guidance. For a suggestion see Broadie 2016, paragraphs 20–2.

anything physical can the life of an incorporeal being whose entire substance is the activity of thinking. But this physically perfect motion or life of the spheres is metaphysically imperfect because it falls short of the activity of the incorporeal mover: and the spheres' 'failure' in this respect is essential to their very being. In other words, the subject of this motion cannot emerge from the privation that grounds the motion.[56] We have here the Platonist dyad of principles that Aristotle wants no truck with in the context of *Physics* I 9.

(35) It has often been remarked that if the unmoved mover of *Metaphysics* Λ 7 is only a final cause and exemplar it need not, as we say, 'really exist': all that is necessary for explaining celestial motion is that such a being figure as an ideal to the spheres, and for this it need be no more than an intentional object of their thinking and desire.[57] This line of thought strongly suggests that if *per contra* Aristotle holds that the unmoved mover 'really exists', he is looking upon it as also in some way an efficient cause. Otherwise its real existence is redundant. And it is repugnant to suppose that a being that we know of only as a theoretical postulate, and whose place in our theory is that of the absolutely primary principle, should have anything theoretically redundant about it. In general, an originator's efficient causality (or the need on the part of the *explanandum* to have an efficient cause) guarantees that the theory is one according to which the originator really exists (or has existed). And an embargo on theoretical redundancy implies the converse for the unmoved mover of *Metaphysics* Λ 7. In other words, if the mover is assumed as really existent, this must be because it figures as efficient cause.[58] Now Aristotle clearly does assume in *Metaphysics* Λ that the unmoved mover really exists as distinct from being merely the content of an idea in the minds of the spheres.[59] It follows that he is committed to treating that mover as an eternally efficient, as well as exemplary, origin of celestial motion; and this notwithstanding the difficulty we have of forming a conception of how its efficiency 'works', given that this cause is not only incorporeal

[56] I am assuming here the most common interpretation of *Metaphysics* Λ, whereby the prime unmoved mover is distinct from the soul of the outermost sphere; the soul is a moved, hence not ultimate, mover (cf. *De Anima*, 433b14–18). This is contrary to the view defended in Broadie 1993.

[57] On this picture the souls of the spheres would be the efficient cause of the motion.

[58] For obviously it is not a material or formal cause of the motion.

[59] It is hard to read chapter 10 otherwise than on this basis. But the clinching evidence is at 1071a17–24, highlighted by Owens 1950 (on the text and interpretation see also Code 2000, 174–5). This passage states that a primary principle (like an ordinary substance) is a *this*, and that universals (which by definition are not *thises*) are not primary principles. The intentional object of thought and desire is surely a universal. Thus it is clearly implied here that the ultimate unmoved mover 'really exists'. An intentional object is still a universal even if it is such that were it to be to be instantiated it would be instantiated uniquely. The general point still holds if there is a plurality of unmoved movers, one for each sphere. Each, if a merely intentional object for its sphere, would be a universal, even though this universal, if instantiated, would be instantiated uniquely.

but 'separated' (1073a4).[60] Aristotle's fuss-free willingness to think along these lines may indicate that Platonists have already made this an easy way to go and he finds no reason to resist their lead. Both they and he envisage the cause as simultaneously efficient ('paternal' for the Platonists) and final/exemplary—although it might be better to say that the Platonists do not sharply distinguish these modes of causality, and that Aristotle in dealing with eternal things is happy to accept this vagueness. Many scholars see in *Metaphysics* Λ a bid to replace Platonist Forms with Aristotle's own incorporeal substances, the intellectually active (but like Platonist Forms, unmoved) movers.[61] This implies something of a shared framework. For example, perhaps both sides operate from much the same notion of causality at the level of eternals, even though they are far apart on what sort of eternal entity needs to be explained, and what sort fills the ultimate explanatory role.

Conclusion

Aristotle's attack on Platonist dyadism in *Physics* I 9 makes sense only if directed at a theory or theoretical move that competes for the same ground as his own triadic approach. On this basis I have argued that he must be targeting a Platonist ambition to extend their signature theory of principles to include the coming to be of natural organic substances. So what is at stake here between Aristotle and the Platonists is not the structure of explanation in general, but the right way of explaining a circumscribed, even though for Aristotle hugely important, area of physics. In *Physics* I 9 he does not mention this circumscription, and as far as I know it has received little attention from commentators on the chapter.[62] Aristotle is also silent on the similarly unnoticed fact (although *Physics* I 9 would not have been an obvious place for him to mention it) that eternal celestial motion, on his own account of this, lends itself very neatly to the Platonist

[60] At *Generation and Corruption* 335b6–24 Aristotle berates the Platonists for failing to recognize efficient causes of sublunary coming to be in addition to the Platonic form-causes: if the latter alone were enough (given the presence of participatory matter, 20 and 24), the forms should be generating (hence the offspring should be appearing) perpetually and non-stop instead of at intervals. This strongly suggests that if the generating and its effect were indeed perpetual and continuous there would be no place for a distinct efficient cause. There are two ways of interpreting this: either there would be no need for efficient causality at all, or the eternal Platonic form-causes would be by themselves efficient causes. The point holds if for 'Platonic form-causes' we substitute 'Aristotelian immobile movers'.

[61] Cf. 1071b12–22; 31–1072a3; 1075b16–20; 27–30. See e.g. Burnyeat 2001, 132; Laks 2000, 225 on the 'Platonic horizon of Aristotle's approach' in *Metaphysics* Λ 7. In fact Aristotle's solution to the problem of the cause of eternal motion must eschew not only Platonist Forms but also the self-moving first mover of the *Phaedrus*.

[62] Ross and Charlton do not mention it in their commentaries on *Physics* I 9.

dyadic scheme. This convergence is interesting not least because, as I have suggested, it indicates a Platonist background for the fact that the unmoved mover of *Metaphysics* M figures as both final and efficient cause. More generally, what we have seen is that Aristotle's principles of motion and change are a triad and also a dyad, depending on whether the subject matter is celestial and everlasting or sublunary and transient.

Appendix
On Evidence Relating to Section 4

(i) Certain passages may seem to tell against claims in Section 4. Thus *Physics* II, 198a24–8, has sometimes been thought to imply that Aristotle denies that efficient and exemplary causality are merged in the unmoved mover(s) of *Metaphysics* Λ. The passage says:

> The last three [sc. of the four causes] often coincide (ἔρχεται εἰς ἕν); for the what and that for the sake of which are one, while the primary source of motion is the same in species (τῷ εἴδει) as these. For man generates man—and so too, in general, with all the things which cause movement by themselves being moved; and such as are not of this kind are no longer inside the province of natural science. (tr. Hardie and Gaye 1984)

The formal, final, and efficient causes coincide in many cases, Aristotle says, and he seems to suggest that a (or: the) case where they do *not* coincide is that of the unmoved movers which are a topic of first philosophy, not physics. This may seem to contradict the idea that the unmoved mover is both efficient and final/exemplary cause of celestial motion.[63] But this conclusion rests on a misunderstanding of 'coincide' at 24–5. Aristotle's point is that 'in many cases' (in fact many sublunary cases) the same specific term denotes the formal, final, and efficient causes. He illustrates this with 'man generates man'. The first occurrence of 'man' stands for the efficient cause, the second for the final cause, and the numerically distinct parent and offspring share the form. By contrast, Aristotle indicates, the separate unmoved mover of *Metaphysics* Λ is *not* conspecific with what it causes. (It makes no difference whether we think of what it causes as the moving sphere or motion of the sphere.) This leaves it entirely open whether the separate unmoved mover functions as both efficient and exemplary cause.[64]

[63] Ross gets round the apparent problem by taking Aristotle to be denying the coincidence of all three mentioned causes at the level of changeless things: 'the formal cause ἐν τοῖς ἀκινήτοις is *not* an efficient or a final cause' (Ross 1936, 526, his emphasis). Ross 1924, vol. 1, cxxxiv, says that it is beyond doubt that the prime mover of *Metaphysics* Λ is efficient as well as final cause.

[64] At *On Generation and Corruption* I, 324b14–15, Aristotle says that the final cause is only 'metaphorically' (i.e. by a slide of meaning) called 'efficient' (ποιητικόν). This might be thought to be a general rejection of the compatibility of final and efficient causality in the same being. But the meaning is only that we do not properly use ποιητικόν of the final cause as such.

(ii) Section 4, paragraphs 30 and 34, argued that when it comes to explaining everlasting things such as celestial motion Aristotle falls back on a schema of two principles, exemplar and undifferentiated subject-privation, which thus far mirrors the Platonist one. There is a possible objection based on a passage in *On Generation and Corruption* II 9. This chapter comes after his discussion of the sublunary elements and compounds and before his proof (chapters 10–11) that coming to be is necessarily everlasting and cyclical. Chapter 9 mainly consists of an attack on previous philosophers for failing to recognize the distinct efficient causes of coming to be. The Platonists thought that form alone was enough to account for it, and the materialists thought the same about matter (335b4–336a12). Aristotle begins his attack by announcing:

> We must state the *number* and the *nature* of the principles of all coming to be alike, for a grasp of any universal facilitates the understanding of its specific forms. (335a24–8)

In the next sentence he specifies:

> (P) The principles [sc. of coming to be] are equal in number to, and identical in kind with, those in the sphere of the eternal and primary things. (335a28–9)[65]

Aristotle then goes on to discuss the inadequacy of matter and form (the latter is also said to be the final cause, 335b6–7), separately or together, to explain coming to be: there must be a third principle, the efficient cause (335a29–32; b7–336a12; cf. *Metaphysics* M, 1080a2–4). This leads into chapter 10, where the main theme is the sun's annual circling as efficient cause of perpetual coming to be near the centre of the cosmos. The argumentative upshot will be: given P, and given that the circling sun is an eternal efficient cause (336a14–18), efficient causes must be at work in the coming to be of transient things.

But it is not clear how exactly P contributes to the reasoning. P is rather grandly comprehensive, but Aristotle does not use it comprehensively to argue systematically from the number and kinds of principles operative at the eternal level to the same number and kinds at the level of becoming—and in fact this would have been a strange thing to do, since it is much more obvious that distinct principles of at any rate form and matter obtain for generable-perishables than for eternal things.

Still, P may seem to present a problem for the argument of Section 4. Taken at face value P validates deducing, one for one, the principles of coming to be from the principles of eternal things. So: given that form, ὑποκείμενον, and privation are principles of coming to be, there should be a matching triad of principles on the eternal level. These would be eternal form, eternal subject, and

[65] Translation follows Joachim (1984).

eternal privation distinct from the subject. But being eternal they would have to co-exist, so form would be compresent with its own privation! If Aristotle adhered to P he could only avoid this absurdity by renouncing the triadic scheme of *Physics* I and all that went into establishing it. Thus P meant seriously would be highly disruptive. But P is a much stronger claim than Aristotle needs in the context of *On Generation and Corruption* II 9. In the first place, 'principles' at 335a26 probably refers to the narrower category of 'causes', since the only principles mentioned are causes (and are called such at a33; b5; 336a2–3). If so, privation is not under discussion. Second, the ensuing proof in *On Generation and Corruption* II 10, so far as it is about causation at the eternal level, says nothing about eternal formal and material causes. Its concern is to pinpoint an eternal *efficient* cause of sublunary coming to be and perishing—with the immediate aim of showing that since this cause is eternal, the series of effects is too.[66] (Aristotle also gives a sort of final-causal explanation: coming to be is uninterrupted because in this way it approximates as closely as possible to eternal being, 336b27–34). The argument neither mentions nor needs any individually eternal ὑποκείμενον or formal cause that would raise the number of eternal principles to more than two.[67] In sum, Aristotle introduces P (along with the premise that there *is* an eternal efficient cause, sc. of perpetual coming to be and passing away) solely in order to extract the anti-Platonist and anti-materialist point that the coming to be of transient things depends on a distinct efficient cause as well as on matter and form.[68] There is nothing here to suggest that he ever envisaged for eternal things the triad of principles brought to light in *Physics* I.

[66] The eternal alternation of γένεσις and φθορά is not the logically immediate effect of the sun's eternal motion; rather, non-eternal alternations are produced one at a time by an eternally repeating cause.

[67] With regard to ὑποκείμενα, Aristotle has argued in *On Generation and Corruption* II, ch. 37 that the basic sublunary materials constantly come to be and perish into and out of each other. There is no statement to the effect that an eternal non-empirical *materia prima* underlies these transformations. For recent discussion of whether Aristotle recognizes *material prima* in this sense, see Algra 2004, Broadie 2004, Brunschwig 2004, 39–42, Charles 2004.

[68] *On Generation and Corruption* II 9–10 can also be seen as responding to a Platonist/materialist view that the on-goingness (let alone perpetuity) of the generations of sublunary things must be due to something eternal that is part of those things' own nature: hence either sublunary things are constituted from eternal materials or they are participants in eternal Platonist forms. Aristotle's solution in terms of an eternally recurring *efficient* cause (the circling sun) places the source of sublunary on-goingness right outside the things themselves, in a non-mysterious visible physical being and one whose matter and form in no way resemble theirs.

BIBLIOGRAPHY

Alexandru, S. (2014). *Aristotle's Metaphysics Lambda*. Leiden.

Algra, K. (1995). *Concepts of Space in Greek Thought*. Leiden.

Algra, K. (2004). '*On Generation and Corruption* I 3: Substantial Change and the Problem of Not-Being'. In *Aristotle: On Generation and Corruption Book I*, eds F. de Haas and J. Mansfeld, 91–122. Oxford.

Anagnostopoulos, A. (2013). 'Aristotle's Parmenidean Dilemma'. *Archiv für Geschichte der Philosophie* 95(3): 245–74.

Angioni, L. (2001). 'Explanation and Definition in *Physics* I 1'. *Apeiron* 34(4): 307–20.

Aubenque, P. (1962), (1966[2]). *Le problème de l'être chez Aristote: Essai sur la problématique aristotélicienne*. Paris.

Averroes (1550). *Aristotelis Stagiritae de physico auditu libri octo, cum Averrois Cordubensis variis in eosdem commentariis*. Venice.

Barnes, J. (ed.) (1984). *The Complete Works of Aristotle: The Revised Oxford Translation*. 2 vols. Princeton.

Barnes, J. (ed.) (1993). *Aristotle: Posterior Analytics*. With translation and commentary. 2nd ed. Oxford.

Barnes, J. (ed.) (2003). *Porphyry: Introduction*. Oxford.

Beere, J. (2009). *Doing and Being: An Interpretation of Aristotle's Metaphysics Theta*. Oxford.

Berti, E. (1973). 'La "riduzione dei contrari" in Aristotele'. In *Zetesis: Bijdragen op het Gebied van de Klassieke Filologie, Filosofie, Byzantinistiek, Patrologie en Theologie door Collega's en Vrienden aangeboden aan Prof. Dr. Emile de Strijcker naar aanleiding van zijn vijfenzenstigste Verjaardaag*, 122–46. Antwerp and Utrecht; reprinted in Berti (1975), 209–23.

Berti, E. (1975). *Studi aristotelici*. Methodos 7. L'Aquila.

Berti, E. (1977). *Aristotele: dalla dialettica alla filosofia prima*. Padua.

Berti, E. (1990). 'Être et non-être chez Aristote: contraires ou contradictoires?' *Revue de théologie et de philosophie* 40: 365–473; reprinted in Berti (2005), 309–17.

Berti, E. (2000). '*Metaphysics* Λ 6'. In *Aristotle's Metaphysics Lambda*, eds M. Frede and D. Charles, 181–206. Oxford.

Berti, E. (2005). *Nuovi studi aristotelici*. Vol. II: *Fisica, antropologia e metafisica*. Brescia.

Betegh, G. (2012). 'The Next Principle: Aristotle, *Metaphysics* A 3–4'. In *Aristotle's Metaphysics Alpha: Symposium Aristotelicum*, ed. C. Steel, 105–40. Oxford.

Bodéüs, R. (2002). *Aristote: Catégories*. Texte établi et traduit. Paris.

Boeri, M. D. (1993). *Aristoteles, Física. Livros I-II.* Traducción, introd. y comentario. Buenos Aires.

Bogen, J. (1992). 'Change and Contrariety in Aristotle'. *Phronesis* 37(1): 1–21.

Bolton, R. (1991). 'Aristotle's Method in Natural Science: *Physics* I'. In *Aristotle's Physics: A Collection of Essays*, ed. L. Judson, 1–29. Oxford.

Bonitz, H. (1870), (1961²). *Index Aristotelicus*. Berlin.

Bonitz, H. (1969 [1862]). *Aristotelische Studien*. Vol. I. Vienna; repr. 1969, Hildesheim.

Bostock, D. (1982). 'Aristotle on the Principles of Change in *Physics* I'. In *Language and Logos*, eds M. Schofield and M. Nussbaum, 179–96. Cambridge; repr. in Bostock (2006), 3–18.

Bostock, D. (2006). *Space, Time, Matter, and Form: Essays on Aristotle's Physics*. Oxford.

Broadie, S. (1993). 'Que fait le premier moteur d'Aristote?' *Revue philosophique* 183(2): 375–411.

Broadie, S. (2004). '*On Generation and Corruption* I 4: Distinguishing Alteration'. In *Aristotle: On Generation and Corruption Book I*, eds F. de Haas and J. Mansfeld, 123–50. Oxford.

Broadie, S. (2012). *Nature and Divinity in Plato's Timaeus*. Cambridge.

Broadie, S. (2016). 'Corporeal Gods'. In *ΣΩΜΑ: Körperkonzepte und körperliche Existenz in der antiken Philosophie und Literatur*, eds T. Buchheim, D. Meissner, and N. Wachsmann, 159–82. Hamburg.

Brunschwig, J. (2004). '*On Generation and Corruption* I: A False Start?' In *Aristotle: On Generation and Corruption Book I*, eds F. de Haas and J. Mansfeld, 25–63. Oxford.

Burnyeat, M. (2001). *A Map of Metaphysics Zeta*. Pittsburgh.

Burnyeat, M. (2003). '*Apology* 30b2–4: Socrates, Money, and the Grammar of γίγνεσθαι'. *Journal of Hellenic Studies* 123: 1–25.

Burnyeat, M. (2012). '*Episteme*'. In *Episteme, etc.: Essays in Honour of Jonathan Barnes*, eds B. Morison and K. Ierodiakonou, 3–29. Oxford.

Carteron, H. (1926). *Aristote: Physique*. Paris.

Castelli, L. M. (2018). '*Physics* I 3'. In *Aristotle's Physics Book I: A Systematic Exploration*, ed. D. Quarantotto, 82–105. Cambridge.

Caveing, M. (1997). *La Figure et le nombre. Recherches sur les premières mathématiques des Grecs*. Lille.

Centrone, B. (ed.) (2005). *Il Libro Iota (X) della Metafisica di Aristotele*. Sankt Augustin.

Cerami, C. (2009). 'Thomas d'Aquin lecteur critique du grand commentaire d'Averroès à *Phys*. I 1'. *Arabic Sciences and Philosophy* 19(2): 189–223.

Cerami, C. (2012). 'Mélange, *Minima Naturalia* et croissance animale dans le Commentaire Moyen d'Avarroès à *De Generatione et Corruptione* I 5'. In *La nature et le vide dans la physique medieval. Ètudes dédiées à Edward Grant*, eds J. Biard and S. Romnevaux, 137–64. Turnhout.

Cerami, C. (2018). 'Aristotle *Physics* I 4'. In *Aristotle's Physics Book I: A Systematic Exploration*, ed. D. Quarantotto, 106–29. Cambridge.

Charles, D. (2000). '*Metaphysics* Λ 2'. In *Aristotle's Metaphysics Lambda*, eds M. Frede and D. Charles, 81–110. Oxford.

Charles, D. (2004). 'Simple Genesis and Prime Matter'. In *Aristotle: On Generation and Corruption Book I*, eds F. de Haas and J. Mansfeld, 151–70. Oxford.

Charlton, W. (1984). *Physics Books I and II*. Translated with introduction, commentary, note on recent work, and revised bibliography. Oxford.

Cherniss, H. (1935). *Aristotle's Criticism of Presocratic Philosophy*. Baltimore.

Cherniss, H. (1944). *Aristotle's Criticism of Plato and the Academy*. Baltimore.

Chiaradonna, R. (2005). 'I contrari e i termini intermedi: *Metafisica* Iota 7'. In *Il Libro Iota (X) della* Metafisica *di Aristotele*, ed. B. Centrone, 157–69. Sankt Augustin.

Clark, E. V. (1973). 'What's in a Word? On the Child's Acquisition of Semantics in his First Language'. In *Cognitive Development and the Acquisition of Language*, ed. T. E. Moore, 65–110. New York.

Clarke, T. (2015). 'Aristotle and the Ancient Puzzle about Coming to Be'. *Oxford Studies in Ancient Philosophy* 49: 129–50.

Clarke, T. (2018). '*Physics* I 2'. In *Aristotle's Physics Book I: A Systematic Exploration*, ed. D. Quarantotto, 61–80. Cambridge.

Code, A. (1976). 'The Persistence of Aristotelian Matter'. *Philosophical Studies* 29: 357–67.

Code, A. (2000). '*Metaphysics* Λ 5'. In *Aristotle's Metaphysics Lambda*, eds M. Frede and D. Charles, 161–80. Oxford.

Code, A. (2018). '*Physics* I 6'. In *Aristotle's Physics Book I: A Systematic Exploration*, ed. D. Quarantotto, 154–77. Cambridge.

Cooper, J. (2004). 'Two Notes on Aristotle on Mixture'. In *Knowledge, Nature, and the Good. Essays on Ancient Philosophy*, 148–73. Princeton.

Cooper, J. (2012). 'Conclusion—and Retrospect (*Metaphysics* A 10)'. In *Aristotle's Metaphysics Alpha. Symposium Aristotelicum*, ed. C. Steel, 335–64. Oxford.

Cornford, F. (1935). *Plato's Cosmology*. London.

Couloubaritsis, L. (1980). *L'avènement de la science physique: Essai sur la Physique d'Aristote*. Bruxelles.

Couloubaritsis, L. (1983). 'L'être et l'un chez Aristote II'. *Revue de philosophie ancienne* 1(2): 143–95.

Couloubaritsis, L. (1997). *La Physique d'Aristote*. 2e édition révisée et augmentée de *L'avènement de la science physique*. Bruxelles.

Crubellier, M. (2000). '*Metaphysics* Λ 4'. In *Aristotle's Metaphysics Lambda*, eds M. Frede and D. Charles, 137–60. Oxford.

Da Rios, R. (1954). *Aristoxeni Elementa Harmonica*. Rome.

De Haas, F. and Mansfeld, J. (eds) (2004). *Aristotle: On Generation and Corruption Book I: Symposium Aristotelicum*. Oxford.

Delcomminette, S. (2006). *Le Philèbe de Platon: Introduction à la agathologie platonicienne*. Leiden.

Delcomminette, S. (2015). 'Aristote et le *Phédon*'. In *Ancient Readings of Plato's Phaedo*, eds S. Delcomminette, P. d'Hoine and M.-A. Gavray, 17–36. Leiden.

Denniston, J. D. (1954). *The Greek Particles*. 2nd ed. Oxford.

Diels, H. (1882). 'Zur Textgeschichte der aristotelischen *Physik*'. In *Abhandlungen der Königlichen Akademie der Wissenschaften zu Berlin, Phil.-hist. Kl.*, 1–42. Berlin.

Düring, I. (1961). *Aristotle's Protrepticus: An Attempt at Reconstruction*. Göteborg.

Düring, I. (1966). *Aristoteles: Darstellung und Interpretation seines Denkens.* Heidelberg.

Fait, P. (2013). 'The "False Validating Premiss" in Aristotle's Doctrine of Fallacies: An Interpretation of *Sophistical Refutations* 8'. In *Logical Analysis and History of Philosophy* 15. Special Issue: *Fallacious Arguments in Ancient Philosophy*, guest eds Ch. Rapp and P. S. Hasper, 238–66.

Falcon, A. (2018). '*Physics* I 1'. In *Aristotle's Physics Book I: A Systematic Exploration*, ed. D. Quarantotto, 41–59. Cambridge.

Fazzo, S. (2013). 'Heavenly Matter in Aristotle, *Metaphysics* Lambda 2'. *Phronesis* 58: 160–75.

Frede, M. (2000). 'Introduction'. In *Aristotle's Metaphysics Lambda*, eds M. Frede and D. Charles, 53–80. Oxford.

Frede, M. and Charles, D. (eds) (2000). *Aristotle's Metaphysics Lambda: Symposium Aristotelicum*. Oxford.

Freeland, C. (1991). 'Accidental Causes and Real Explanations'. In *Aristotle's Physics: A Collection of Essays*, ed. L. Judson, 49–72. Oxford.

Giardina, G. (2002). *I fondamenti della fisica. Analisi critica di Aristotele, Phys. I.* Catania.

Gill, M. L. and Lennox, J. (eds) (1994). *Self-Motion: From Aristotle to Newton.* Princeton.

Gourinat, J. B. (2004). 'L'intellect divin d'Aristote est-il cause efficiente?' *Bollettino filosofico* 10: 54–81.

Graham, D. W. (1987). *Aristotle's Two Systems*. Oxford.

Graham, D. W. (1999). *Aristotle, Physics Book VIII.* Translated with a commentary. Oxford.

Graham, D. W. (2003). 'A Testimony of Anaximenes in Plato'. *Classical Quarterly* 53(2): 327–37.

Graham, D. W. (2006). *Explaining the Cosmos: The Ionian Tradition of Scientific Philosophy*. Princeton.

Grube, G. M. A. (1997). '*Phaedo*'. In *Plato's Complete Works*, ed. J. M., Cooper 49–100. Indianapolis.

Happ, H. (1971). *Hyle. Studien zum aristotelischen Materie-Begriff*. Berlin.

Hardie, R. and Gaye, R. (1984). '*Physics*'. In *The Complete Works of Aristotle: The Revised Oxford Translation*, ed. J. Barnes, vol. I, 315–446. Princeton.

Heath, T. L. (1921). *A History of Greek Mathematics*. Vol. I: *From Thales to Euclid*. Oxford.

Heiberg, J. L. (1883–8). *Euclidis Elementa*. Leipzig.

Helmig, C. (2012). *Forms and Concepts*. Berlin.

Horstschäfer, T. M. (1998). *'Über Prinzipien': Eine Untersuchung zur methodischen und inhaltlichen Geschlossenheit des ersten Buches der Physik des Aristoteles*. Berlin and New York.

Ierodiakonou, K. (2005). 'Empedocles on Colour and Colour Vision'. *Oxford Studies in Ancient Philosophy* 29: 1–37.

Irrera, E. (2011). *Il bello come causalità metafisica in Aristotele*. Milano.

Irwin, T. (1996 [1988]). *Aristotle's First Principles*. Oxford (Italian trans. A. Giordani, Milano 1996).

Jaeger, W. (1957). *Aristotelis Metaphysica*. Recognovit Brevique Adnotatione Critica Instruxit. Oxford.

Joachim, H. H. (1984). '*On Generation and Corruption*'. In *The Complete Works of Aristotle: The Revised Oxford Translation*, ed. J. Barnes, vol. I, 512–54. Princeton.

Johansen, T. (2004). *Plato's Natural Philosophy: A Study of the Timaeus-Critias*. Cambridge.

Johansen, T. (2010). 'Should Aristotle Have Recognized Final Causes in Plato's *Timaeus*?' In *One Book, The Whole Universe: Plato's Timaeus Today*, eds R. Mohr and B. Sattler, 179–200. Las Vegas.

Judson, L. (ed.) (1991). *Aristotle's Physics: A Collection of Essays*. Oxford.

Judson, L. (1994). 'Heavenly Motion and the Unmoved Mover'. In *Self-Motion: From Aristotle to Newton*, eds M. L. Gill and J. Lennox, 155–71. Princeton.

Kelsey, S. (2006). 'Aristotle's *Physics* I 8'. *Phronesis* 51(4): 330–61.

Kelsey, S. (2008). 'The Place of I 7 in the Argument of *Physics* I'. *Phronesis* 53(2): 180–208.

Kelsey, S. (2010). 'Hylomorphism in Aristotle's *Physics*'. *Ancient Philosophy* 30(1): 107–24.

Kirk, G. S., Raven, J. E., and Schofield, M. (1983). *The Presocratic Philosophers*. 2nd ed. Cambridge.

Kirwan, C. (1993). *Aristotle, Metaphysics Books Γ, Δ, and E*. Translated with notes. 2nd ed. Oxford.

Krizan, M. (2013). 'Elemental Structure and the Transformation of the Elements in *On Generation and Corruption* II 4'. *Oxford Studies in Ancient Philosophy* 45: 195–224.

Laas, E. (1863). *Aristotelische Textes-Studien*. Berlin.

Lagnerini, A. (2015). 'Discussion Note on the Causality of the Immovable Mover'. *Rhizomata* 3(2): 200–13.

Laks, A. (1993). 'Mind's Crisis: On Anaxagoras' Nous'. In *Ancient Minds: Proceedings of the 1992 Spindel Conference*, ed. J. Ellis. *The Southern Journal of Philosophy* 31 (Supplement): 19–38.

Laks, A. (2000). '*Metaphysics* Λ 7'. In *Aristotle's Metaphysics Lambda*, eds M. Frede and D. Charles, 207–43. Oxford.

Laks, A. (2002). 'Les fonctions de l'intellect'. *Methodos* 2. <http://methodos.revues.org/224 accessed 6 Oct. 2017>.

Le Blond, J. M. (1938). *Eulogos et l'argument de convenance chez Aristote*. Paris.

Lefebvre, D. (2008). 'L'usage de la série des contraires en *Métaphysique* Γ 2, 1004b27–1005a6'. In *Aristote: Métaphysique Gamma. Édition, traduction, études*, eds M. Hecquet-Devienne and A. Stevens, 287–321. Louvain-la-Neuve.

Lennox, J. G. (2001). *Aristotle On the Parts of Animals*. Translated with introduction and commentary. Oxford.

Lennox, J.G. (2011). 'Aristotle on Norms of Inquiry', *HOPOS: The Journal of the International Society for the History of Philosophy of Science* 1: 23–46.

Lesher, J. H. (1995). 'Mind's Knowledge and Powers of Control in Anaxagoras DK B12'. *Phronesis* 40(2): 125–42.

Lettinck, P. (1994). *Aristotle's Physics and its Reception in the Arabic World*. Leiden.

Leunissen, M. (2010). *Explanation and Teleology in Aristotle's Science of Nature*. Cambridge.

Lloyd, G. (1987). 'The Alleged Fallacy of Hippocrates of Chios'. *Apeiron* 20(2): 103–28.

Lloyd, G. and Sivin, N. (2003). *The Way and the Word: Science and Medicine in Early China and Greece*. Yale.

Loux, M. J. (1992). 'Aristotle and Parmenides: An Interpretation of *Physics* A 8'. *Proceedings of the Boston Area Colloquium in Ancient Philosophy* 8: 281–319.

Makin, S. (2006). *Aristotle, Metaphysics Book Θ*. Translation with introduction and commentary. Oxford.

Mansfeld, J. (1986). 'Aristotle, Plato and the Preplatonic Doxography and Chronography'. In *Storiografia e dossografia nella filosofia antica*, ed. G. Cambiano, 1–59. Torino; reprinted in Mansfeld (1990), 22–83.

Mansfeld, J. (1990). *Studies in the Historiography of Greek Philosophy*, Assen.

Mansion, A. (1946). *Introduction à la physique aristotélicienne*, 2nd ed. Louvain.

Mansion, S. (1961). 'Le rôle de l'exposé et de la critique des philosophies antérieurs chez Aristote'. In *Aristote et les problèmes de méthode: Communications présentées au Symposium Aristotelicum tenu à Louvain*, 35–56. Louvain.

Mansion, S. (1979). ' "Plus connu en soi", "Plus connu pour nous": Une distinction épistémologique importante chez Aristote'. *Pensamiento* 35: 161–70.

Mendell, H. No date. *Vignettes of Ancient Mathematics*. Translations of ancient Greek mathematical texts, available at: <http://web.calstatela.edu/faculty/hmendel/Ancient%20Mathematics/VignettesAncientMath.html>

Menn, S. (1998). 'Collecting the Letters'. *Phronesis* 43(4): 291–305.

Menn, S. (2001). '*Metaphysics* Z 10–16 and the Argument-Structure of *Metaphysics* Z'. *Oxford Studies in Ancient Philosophy* 21: 83–134.

Menn, S. (2010). 'Simplicius on the *Theaetetus* (*In Physica* 17,38–18,23 Diels)'. *Phronesis* 55(3): 255–70.

Menn, S. (2011). 'On Myles Burnyeat's *Map of Metaphysics Zeta*'. *Ancient Philosophy* 31(1): 161–202.

Menn, S. (2012). '*Metaphysics* A7–A8 989a18: Critique of Earlier Philosophers on the Good and the Causes'. In *Aristotle's Metaphysics Alpha: Symposium Aristotelicum*, ed. C. Steel, 201–24. Oxford.

Menn, S. (forthcoming). *The Aim and the Argument of Aristotle's Metaphysics*. Oxford. Current drafts available at: <https://www.philosophie.hu-berlin.de/de/lehrbereiche/antike/mitarbeiter/menn/contents>

Mohr, R. and Sattler, B. (eds) (2010). *One Book, the Whole Universe: Plato's Timaeus Today*. Las Vegas.

Moravcsik, J. (1991). 'What Makes Reality Intelligible? Reflections on Aristotle's Theory of *Aitia*'. In *Aristotle's Physics: A Collection of Essays*, ed. L. Judson 31–48. Oxford.

Mumford, S. (2012). 'Contemporary Efficient Causation: Aristotelian Themes'. In *Efficient Causation: A History*, ed. T. M. Schmalz, 317–39. Oxford.

Murdoch, J. E. (2001). 'The Medieval and Renaissance Tradition of *Minima Naturalia*'. In *Late Medieval and Early Modern Corpuscular Matter Theories*, eds C. Lüthy, J. E. Murdoch, and W. R. Newman, 91–132. Leiden.

Natali, C. (1997). 'Causa motrice e causa finale nel libro Lambda della Metafisica di Aristotele'. *Méthexis* 10: 105–23.

Netz, R. (2004). 'Eudemus of Rhodes, Hippocrates of Chios and the Earliest Form of a Greek Mathematical Text'. *Centaurus* 46(4): 243–86.

O'Connor, S. (2015). 'The Subjects of Natural Generation in Aristotle's *Physics* I 7'. *Apeiron* 48(1): 45–75.

Ogle, W. (ed.) (1882). *Aristotle: On the Parts of Animals*. London.

Osborne, C. (2006). *Philoponus: On Aristotle Physics 1.1–3*. Translated with notes. London.

Owens, J. (1950). 'The Reality of the Aristotelian Separate Movers'. *Review of Metaphysics* 3(3): 319–37.

Pellegrin, P. (2000), (2002^2). *Aristote: Physique*. Traduction et présentation. Paris.

Philoponus (1887–8). *In Aristotelis Physica Commentaria*. Commentaria in Aristotelem Graeca (CAG), vols 16–17, ed. H. Vitelli. Berlin.

Philoponus (2009). *On Aristotle Physics 1.4–9*. Translated by C. Osborne. London.

Piaget, J. (1928). *Judgment and Reasoning in the Child*. New York.

Prantl, C. (1854). *Aristoteles' Acht Bücher Physik*. Leipzig.

Quarantotto, D. (2016). 'Aristotle's Way Away from Parmenides' Way. A Case of Scientific Controversy and Ancient Humour'. *Elenchos* 37: 209–28.

Quarantotto, D. (2018). 'The Role, Structure and Status of Aristotle's *Physics* I'. In *Aristotle's Physics Book I: A Systematic Exploration*, ed. D. Quarantotto, 1–40. Cambridge.

Quevedo, A. (1998). *La privación según Aristoteles*. Santa Fe de Bogotá.

Reeve, C. D. C. (2016). *Aristotle's Metaphysics*. Translated with introduction and notes. Indianapolis.

Ross, W. D. (ed.) (1924). *Aristotle's Metaphysics.* A revised text with introduction and commentary. 2 vols. Oxford; repr. with corrections 1953.

Ross, W. D. (ed.) (1936). *Aristotle Physics.* A revised text with introduction and commentary. Oxford.

Ross, W. D. (ed.) (1950). *Aristotelis Physica. Recognovit brevique adnotatione critica instruxit*. Oxford.

Ross, W. D. (ed.) (1955). *Aristotelis Fragmenta Selecta*. Oxford.

Ross, W. D. and Fobes, F. H. (eds) (1929). *Theophrastus, Metaphysics*. With a translation, introduction and commentary. Oxford.

Rossi, G. (2006). 'Desanudando argumentos: Las aplicaciones filosóficas de la dialéctica según las *Refutaciones Sofísticas*'. *Méthexis* 19: 79–109.

Rossi, G. (2014). 'Aristotle on Untying Arguments'. In *Zur modernen Deutung der aristotelischen Logik*, eds N. Öffenberger and A. Vigo, Bd. XI, 105–40. Hildesheim.

Rudio, F. (1907). *Der Bericht des Simplicius über die Quadraturen des Antiphon und Hippokrates*. Leipzig.

Sattler, B. (2012). 'A Likely Account of Necessity: Plato's Receptacle as a Physical and Metaphysical Foundation for Space'. *Journal of the History of Philosophy* 50(2): 159–95.

Sedley, D. (2012). 'Plato's Theory of Change at *Phaedo* 70–71'. In *Presocratics and Plato*, eds R. Patterson, V. Karasmanis, and A. Hermann, 147–63. Las Vegas.

Simplicius (1882). *In Aristotelis Physica*. Commentaria in Aristotelem Graeca (CAG), vol. 9, ed. H. Diels. Berlin.

Simplicius (2011). *Simplicius: On Aristotle Physics 1.3–4*. Translated by C. C. W. Taylor and P. M. Huby. London.

Smith, R. (1993). 'What Use Is Aristotle's *Organon*?' *Proceedings of the Boston Area Colloquium in Ancient Philosophy* 9: 261–85.

Smith, R. (2009a). '"None of the Arts that Gives Proofs about Some Nature is Interrogative": Questions and Aristotle's Concept of Science'. In *The Socratic*

Tradition: Questioning as Philosophy and as Method, ed. M. Sintonen, 25–49. London.

Smith, R. (2009b). 'Aristotle's Theory of Demonstration'. In *A Companion to Aristotle*, ed. G. Anagnostopoulos, 51–65. Chichester.

Solmsen, F. (1960). *Aristotle's System of the Physical World: A Comparison with his Predecessors*. Ithaca, NY.

Steel, C. (ed.) (2012). *Aristotle's Metaphysics Alpha: Symposium Aristotelicum*. Oxford.

Steel, C. (2012a). 'Plato as seen by Aristotle'. In *Aristotle's Metaphysics Alpha: Symposium Aristotelicum*, ed. C. Steel, 167–220. Oxford.

Suárez, F. (1597). *Opera Omnia*, vol. XXVI: *Disputationes Metaphysicae*. Paris.

Susemihl, F. (ed.) (1887). *Aristotelis Quae Feruntur Oeconomica*. Leipzig.

Tannery, P. (1876). 'Hippocrate de Chio et la quadrature des lunules'. In *Mémoires scientifiques*, eds J.-L. Heiberg and H.-G. Zeuthen, vol. 1, 46–52. Toulouse and Paris 1912.

Tannery, P. (1882). 'Sur les fragments d'Eudème de Rhodes relatifs à l'histoire des mathématiques'. In *Mémoires scientifiques*, eds J.-L. Heiberg and H.-G. Zeuthen, vol. 1, 168–77. Toulouse and Paris 1912.

Tannery, P. (1883). 'Le fragment d'Eudème sur la quadrature des lunules'. In *Mémoires scientifiques*, eds J.-L. Heiberg and H.-G. Zeuthen, vol. 1, 339–70. Toulouse and Paris 1912.

Tannery, P. (1902). 'Simplicius et la quadrature du cercle'. In *Mémoires scientifiques*, eds J.-L. Heiberg and H.-G. Zeuthen, vol. 3, 119–30. Toulouse and Paris 1915.

Taylor, A. E. (1928). *A Commentary on Plato's Timaeus*. Oxford.

Thomas Aquinas (1954). *In octo libros Physicorum Aristotelis expositio*, ed. P. M. Maggiòlo. Turin and Rome.

Tortstrik, A. (1867). '*Der Anfang der Physik* des Aristoteles'. *Jahrbücher für classische Philologie* 95: 236–44.

Van Melsen, A. G. (1960). *From Atomos to Atom*. New York.

Wagner, H. (1967). *Aristoteles: Physikvorlesung*. Berlin.

Waszink, J. H. (ed.) (1962). *Timaeus a Calcidio Translatus Commentarioque Instructus*. Plato Latinus, vol. 4. London.

Waterlow (= Broadie), S. (1982). *Nature, Change, and Agency in Aristotle's Physics*. Oxford.

Weidemann, H. (1986). 'Aristoteles und das Problem des kausalen Determinismus (*Met.* E 3)'. *Phronesis* 31(1): 27–50.

West, M. L. (1969). 'An Atomist Illustration in Aristotle'. *Philologus* 113(1–2): 150–1.

Wieland, W. (1960–1). 'Das Problem des Prinzipienforschung und die aristotelische *Physik*'. *Kant-Studien* 52: 206–19. Translated as 'Aristotle's *Physics* and the

Problem of Inquiry into Principles'. In *Articles on Aristotle*, eds J. Barnes, M. Schofield, and R. Sorabji, vol. 1, 127–40. London 1975.

Wieland, W. (1993 [1970]). *Die aristotelische Physik*. 2nd ed. Göttingen (Italian trans. C. Gentili, Bologna 1993).

Wilson, M. (2000). *Aristotle's Theory of the Unity of Science.* Toronto.

Zeller, E. and Mondolfo, R. (1969). *La filosofia dei Greci nel suo sviluppo storico.* Parte I: *Presocratici*, vol. V: *Empedocle, atomisti, Anassagora*. A cura di Antonio Capizzi. Florence.

Zeyl, D. (2000). *Plato, Timaeus*. Indianapolis.

Zeyl, D. (2010). 'Visualizing Platonic Space'. In *One Book, the Whole Universe: Plato's Timaeus Today*, eds R. Mohr and B. Sattler, 117–30. Las Vegas.

INDEX OF NAMES

1 Ancient, Medieval, and Renaissance Authors

2 Modern Authors

INDEX OF PASSAGES